P9-EEQ-961

THE STOLEN WHITE ELEPHANT *and Other Detective Stories*

THE OXFORD MARK TWAIN

Shelley Fisher Fishkin, Editor

The Celebrated Jumping Frog of Calaveras County, and Other Sketches
Introduction: Roy Blount Jr.
Afterword: Richard Bucci

The Innocents Abroad
Introduction: Mordecai Richler
Afterword: David E. E. Sloane

Roughing It
Introduction: George Plimpton
Afterword: Henry B. Wonham

The Gilded Age
Introduction: Ward Just
Afterword: Gregg Camfield

Sketches, New and Old
Introduction: Lee Smith
Afterword: Sherwood Cummings

The Adventures of Tom Sawyer
Introduction: E. L. Doctorow
Afterword: Albert E. Stone

A Tramp Abroad
Introduction: Russell Banks
Afterword: James S. Leonard

The Prince and the Pauper

Introduction: Judith Martin

Afterword: Everett Emerson

Life on the Mississippi

Introduction: Willie Morris

Afterword: Lawrence Howe

Adventures of Huckleberry Finn

Introduction: Toni Morrison

Afterword: Victor A. Doyno

A Connecticut Yankee in King Arthur's Court

Introduction: Kurt Vonnegut, Jr.

Afterword: Louis J. Budd

Merry Tales

Introduction: Anne Bernays

Afterword: Forrest G. Robinson

The American Claimant

Introduction: Bobbie Ann Mason

Afterword: Peter Messent

The £1,000,000 Bank-Note and Other New Stories

Introduction: Malcolm Bradbury

Afterword: James D. Wilson

Tom Sawyer Abroad

Introduction: Nat Hentoff

Afterword: M. Thomas Inge

The Tragedy of Pudd'nhead Wilson and the Comedy Those Extraordinary Twins

Introduction: Sherley Anne Williams

Afterword: David Lionel Smith

Personal Recollections of Joan of Arc
Introduction: Justin Kaplan
Afterword: Susan K. Harris

The Stolen White Elephant and Other Detective Stories
Introduction: Walter Mosley
Afterword: Lillian S. Robinson

How to Tell a Story and Other Essays
Introduction: David Bradley
Afterword: Pascal Covici, Jr.

Following the Equator and Anti-imperialist Essays
Introduction: Gore Vidal
Afterword: Fred Kaplan

The Man That Corrupted Hadleyburg and Other Stories and Essays
Introduction: Cynthia Ozick
Afterword: Jeffrey Rubin-Dorsky

The Diaries of Adam and Eve
Introduction: Ursula K. Le Guin
Afterword: Laura E. Skandera-Trombley

What Is Man?
Introduction: Charles Johnson
Afterword: Linda Wagner-Martin

The $30,000 Bequest and Other Stories
Introduction: Frederick Busch
Afterword: Judith Yaross Lee

Christian Science
Introduction: Garry Wills
Afterword: Hamlin Hill

Chapters from My Autobiography
Introduction: Arthur Miller
Afterword: Michael J. Kiskis

The Stolen White Elephant and Other Detective Stories

Mark Twain

FOREWORD

SHELLEY FISHER FISHKIN

INTRODUCTION

WALTER MOSLEY

AFTERWORD

LILLIAN S. ROBINSON

New York Oxford

OXFORD UNIVERSITY PRESS

1996

OXFORD UNIVERSITY PRESS
Oxford New York
Athens, Auckland, Bangkok, Bogotá, Bombay
Buenos Aires, Calcutta, Cape Town, Dar es Salaam
Delhi, Florence, Hong Kong, Istanbul, Karachi
Kuala Lumpur, Madras, Madrid, Melbourne
Mexico City, Nairobi, Paris, Singapore
Taipei, Tokyo, Toronto
and associated companies in
Berlin, Ibadan

Text design by Richard Hendel
Composition: David Thorne

Published by
Oxford University Press, Inc.
198 Madison Avenue, New York,
New York 10016

Library of Congress
Cataloging-in-Publication Data

Twain, Mark, 1835–1910.
Detective stories / by Mark Twain; with an
introduction by Walter Mosley and an afterword
by Lillian S. Robinson.
p. cm. — (The Oxford Mark Twain)
Contents: The stolen white elephant — Tom Sawyer,
detective — A double-barrelled detective story.
1 Detective and mystery stories, American 2. Sawyer,
Tom (Fictitious character)—Fiction. I. Title.
II. Series: Twain, Mark, 1835-1910. Works. 1996.
PS1302 1996
813'.4—dc20
96-16582
CIP
ISBN 0-19-510153-7 (trade ed.)
ISBN 0-19-511417-5 (lib. ed.)
ISBN 0-19-509088-8 (trade ed. set)
ISBN 0-19-511345-4 (lib. ed. set)

9 8 7 6 5 4 3 2 1

Printed in the United States of America
on acid-free paper

FRONTISPIECE
This photograph, dated July 17, 1903, was taken
in Samuel L. Clemens' Quarry Farm study in Elmira,
the year after "A Double Barrelled Detective Story"
was published. (The Mark Twain House, Hartford,
Connecticut)

CONTENTS

EDITOR'S NOTE

The Oxford Mark Twain consists of twenty-nine volumes of facsimiles of the first American editions of Mark Twain's works, with an editor's foreword, new introductions, afterwords, notes on the texts, and essays on the illustrations in volumes with artwork. The facsimiles have been reproduced from the originals unaltered, except that blank pages in the front and back of the books have been omitted, and any seriously damaged or missing pages have been replaced by pages from other first editions (as indicated in the notes on the texts).

In the foreword, introduction, afterword, and essays on the illustrations, the titles of Mark Twain's works have been capitalized according to modern conventions, as have the names of characters (except where otherwise indicated). In the case of discrepancies between the title of a short story, essay, or sketch as it appears in the original table of contents and as it appears on its own title page, the title page has been followed. The parenthetical numbers in the introduction, afterwords, and illustration essays are page references to the facsimiles.

FOREWORD

Shelley Fisher Fishkin

Samuel Clemens entered the world and left it with Halley's Comet, little dreaming that generations hence Halley's Comet would be less famous than Mark Twain. He has been called the American Cervantes, our Homer, our Tolstoy, our Shakespeare, our Rabelais. Ernest Hemingway maintained that "all modern American literature comes from one book by Mark Twain called *Huckleberry Finn*." President Franklin Delano Roosevelt got the phrase "New Deal" from *A Connecticut Yankee in King Arthur's Court. The Gilded Age* gave an entire era its name. "The future historian of America," wrote George Bernard Shaw to Samuel Clemens, "will find your works as indispensable to him as a French historian finds the political tracts of Voltaire."[1]

There is a Mark Twain Bank in St. Louis, a Mark Twain Diner in Jackson Heights, New York, a Mark Twain Smoke Shop in Lakeland, Florida. There are Mark Twain Elementary Schools in Albuquerque, Dayton, Seattle, and Sioux Falls. Mark Twain's image peers at us from advertisements for Bass Ale (his drink of choice was Scotch), for a gas company in Tennessee, a hotel in the nation's capital, a cemetery in California.

Ubiquitous though his name and image may be, Mark Twain is in no danger of becoming a petrified icon. On the contrary: Mark Twain lives. *Huckleberry Finn* is "the most taught novel, most taught long work, and most taught piece of American literature" in American schools from junior high to the graduate level.[2] Hundreds of Twain impersonators appear in theaters, trade shows, and shopping centers in every region of the country.[3] Scholars publish hundreds of articles as well as books about Twain every year, and he is the subject of daily exchanges on the Internet. A journalist somewhere in the world finds a reason to quote Twain just about every day. Television series such as *Bonanza*, *Star Trek: The Next Generation*, and *Cheers* broadcast episodes that feature Mark Twain as a character. Hollywood screenwriters regularly produce movies inspired by his works, and writers of mysteries and science fiction continue to weave him into their plots.[4]

A century after the American Revolution sent shock waves throughout Europe, it took Mark Twain to explain to Europeans and to his countrymen alike what that revolution had wrought. He probed the significance of this new land and its new citizens, and identified what it was in the Old World that America abolished and rejected. The founding fathers had thought through the political dimensions of making a new society; Mark Twain took on the challenge of interpreting the social and cultural life of the United States for those outside its borders as well as for those who were living the changes he discerned.

Americans may have constructed a new society in the eighteenth century, but they articulated what they had done in voices that were largely interchangeable with those of Englishmen until well into the nineteenth century. Mark Twain became the voice of the new land, the leading translator of what and who the "American" was — and, to a large extent, is. Frances Trollope's *Domestic Manners of the Americans,* a best-seller in England, Hector St. John de Crèvecoeur's *Letters from an American Farmer,* and Tocqueville's *Democracy in America* all tried to explain America to Europeans. But Twain did more than that: he allowed European readers to *experience* this strange "new world." And he gave his countrymen the tools to do two things they had not quite had the confidence to do before. He helped them stand before the cultural icons of the Old World unembarrassed, unashamed of America's lack of palaces and shrines, proud of its brash practicality and bold inventiveness, unafraid to reject European models of "civilization" as tainted or corrupt. And he also helped them recognize their own insularity, boorishness, arrogance, or ignorance, and laugh at it — the first step toward transcending it and becoming more "civilized," in the best European sense of the word.

Twain often strikes us as more a creature of our time than of his. He appreciated the importance and the complexity of mass tourism and public relations, fields that would come into their own in the twentieth century but were only fledgling enterprises in the nineteenth. He explored the liberating potential of humor and the dynamics of friendship, parenting, and marriage. He narrowed the gap between "popular" and "high" culture, and he meditated on the enigmas of personal and national identity. Indeed, it would be difficult to find an issue on the horizon today that Twain did not touch on somewhere in his work. Heredity versus environment? Animal rights? The boundaries of gender? The place of black voices in the cultural heritage of the United States? Twain was there.

With startling prescience and characteristic grace and wit, he zeroed in on many of the key challenges — political, social, and technological — that would face his country and the world for the next hundred years: the challenge of race relations in a society founded on both chattel slavery and ideals of equality, and the intractable problem of racism in American life; the potential of new technologies to transform our lives in ways that can be both exhilarating and terrifying — as well as unpredictable; the problem of imperialism and the difficulties entailed in getting rid of it. But he never lost sight of the most basic challenge of all: each man or woman's struggle for integrity in the face of the seductions of power, status, and material things.

Mark Twain's unerring sense of the right word and not its second cousin taught people to pay attention when he spoke, in person or in print. He said things that were smart and things that were wise, and he said them incomparably well. He defined the rhythms of our prose and the contours of our moral map. He saw our best and our worst, our extravagant promise and our stunning failures, our comic foibles and our tragic flaws. Throughout the world he is viewed as the most distinctively American of American authors — and as one of the most universal. He is assigned in classrooms in Naples, Riyadh, Belfast, and Beijing, and has been a major influence on twentieth-century writers from Argentina to Nigeria to Japan. The Oxford Mark Twain celebrates the versatility and vitality of this remarkable writer.

The Oxford Mark Twain reproduces the first American editions of Mark Twain's books published during his lifetime.[5] By encountering Twain's works in their original format—typography, layout, order of contents, and illustrations—readers today can come a few steps closer to the literary artifacts that entranced and excited readers when the books first appeared. Twain approved of and to a greater or lesser degree supervised the publication of all of this material.[6] The Mark Twain House in Hartford, Connecticut, generously loaned us its originals.[7] When more than one copy of a first American edition was available, Robert H. Hirst, general editor of the Mark Twain Project, in cooperation with Marianne Curling, curator of the Mark Twain House (and Jeffrey Kaimowitz, head of Rare Books for the Watkinson Library of Trinity College, Hartford, where the Mark Twain House collection is kept), guided our decision about which one to use.[8] As a set, the volumes also contain more than eighty essays commissioned especially for The Oxford Mark Twain, in which distinguished contributors reassess Twain's achievement as a writer and his place in the cultural conversation that he did so much to shape.

Each volume of The Oxford Mark Twain is introduced by a leading American, Canadian, or British writer who responds to Twain—often in a very personal way—as a fellow writer. Novelists, journalists, humorists, columnists, fabulists, poets, playwrights—these writers tell us what Twain taught them and what in his work continues to speak to them. Reading Twain's books, both famous and obscure, they reflect on the genesis of his art and the characteristics of his style, the themes he illuminated, and the aesthetic strategies he pioneered. Individually and collectively their contributions testify to the place Mark Twain holds in the hearts of readers of all kinds and temperaments.

Scholars whose work has shaped our view of Twain in the academy today have written afterwords to each volume, with suggestions for further reading. Their essays give us a sense of what was going on in Twain's life when he wrote the book at hand, and of how that book fits into his career. They explore how each book reflects and refracts contemporary events, and they show Twain responding to literary and social currents of the day, variously accepting, amplifying, modifying, and challenging prevailing paradigms. Sometimes they argue that works previously dismissed as quirky or eccentric departures actually address themes at the heart of Twain's work from the start. And as they bring new perspectives to Twain's composition strategies in familiar texts, several scholars see experiments in form where others saw only formlessness, method where prior critics saw only madness. In addition to elucidating the work's historical and cultural context, the afterwords provide an overview of responses to each book from its first appearance to the present.

Most of Mark Twain's books involved more than Mark Twain's words: unique illustrations. The parodic visual send-ups of "high culture" that Twain himself drew for *A Tramp Abroad*, the sketch of financial manipulator Jay Gould as a greedy and sadistic "Slave Driver" in *A Connecticut Yankee in King Arthur's Court*, and the memorable drawings of Eve in *Eve's Diary* all helped Twain's books to be sold, read, discussed, and preserved. In their essays for each volume that contains artwork, Beverly R. David and Ray Sapirstein highlight the significance of the sketches, engravings, and photographs in the first American editions of Mark Twain's works, and tell us what is known about the public response to them.

The Oxford Mark Twain invites us to read some relatively neglected works by Twain in the company of some of the most engaging literary figures of our time. Roy Blount Jr., for example, riffs in a deliciously Twain-like manner on "An Item Which the Editor Himself Could Not Understand," which may well rank as one of the least-known pieces Twain ever published. Bobbie Ann Mason celebrates the "mad energy" of Twain's most obscure comic novel, *The American Claimant*, in which the humor "hurtles beyond tall tale into simon-pure absurdity."[9] Garry Wills finds that *Christian Science* "gets us very close to the heart of American culture." Lee Smith reads "Political Economy" as a sharp and funny essay on language. Walter Mosley sees "The Stolen White Elephant," a story "reduced to a series of ridiculous telegrams related by an untrustworthy narrator caught up in an adventure that is as impossible as it is ludicrous," as a stunningly compact and economical satire of a world we still recognize as our own. Anne Bernays returns to "The Private History of a Campaign That Failed" and finds "an antiwar manifesto that is also confession, dramatic monologue, a plea for understanding and absolution, and a romp that gradually turns into atrocity even as we watch." After revisiting Captain Stormfield's heaven, Frederik Pohl finds that there "is no imaginable place more pleasant to spend eternity." Indeed, Pohl writes, "one would almost be willing to die to enter it."

While less familiar works receive fresh attention in The Oxford Mark Twain, new light is cast on the best-known works as well. Judith Martin ("Miss Manners") points out that it is by reading a court etiquette book that Twain's pauper learns how to behave as a proper prince. As important as etiquette may be in the palace, Martin notes, it is even more important in the slums.

> That etiquette is a sorer point with the ruffians in the street than with the proud dignitaries of the prince's court may surprise some readers. As in our own streets, etiquette is always a more volatile subject among those who cannot count on being treated with respect than among those who have the power to command deference.

And taking a fresh look at *Adventures of Huckleberry Finn,* Toni Morrison writes,

> much of the novel's genius lies in its quiescence, the silences that pervade it and give it a porous quality that is by turns brooding and soothing. It lies in . . . the subdued images in which the repetition of a simple word, such as "lonesome," tolls like an evening bell; the moments when nothing is said, when scenes and incidents swell the heart unbearably precisely because unarticulated, and force an act of imagination almost against the will.

Engaging Mark Twain as one writer to another, several contributors to The Oxford Mark Twain offer new insights into the processes by which his books came to be. Russell Banks, for example, reads *A Tramp Abroad* as "an important revision of Twain's incomplete first draft of *Huckleberry Finn*, a second draft, if you will, which in turn made possible the third and final draft." Erica Jong suggests that *1601,* a freewheeling parody of Elizabethan manners and mores, written during the same summer Twain began *Huckleberry Finn*, served as "a warm-up for his creative process" and "primed the pump for other sorts of freedom of expression." And Justin Kaplan suggests that "one of the transcendent figures standing behind and shaping" *Joan of Arc* was Ulysses S. Grant, whose memoirs Twain had recently published, and who, like Joan, had risen unpredictably "from humble and obscure origins" to become a "military genius" endowed with "the gift of command, a natural eloquence, and an equally natural reserve."

As a number of contributors note, Twain was a man ahead of his times. *The Gilded Age* was the first "Washington novel," Ward Just tells us, because "Twain was the first to see the possibilities that had eluded so many others." Commenting on *The Tragedy of Pudd'nhead Wilson,* Sherley Anne Williams observes that "Twain's argument about the power of environment in shaping character runs directly counter to prevailing sentiment where the negro was concerned." Twain's fictional technology, wildly fanciful by the standards of his day, predicts developments we take for granted in ours. DNA cloning, fax machines, and photocopiers are all prefigured, Bobbie Ann Mason tells us, in *The American Claimant.* Cynthia Ozick points out that the "telelectrophonoscope" we meet in "From the 'London Times' of 1904" is suspiciously like what we know as "television." And Malcolm Bradbury suggests that in the "phrenophones" of "Mental Telegraphy" "the Internet was born."

Twain turns out to have been remarkably prescient about political affairs as well. Kurt Vonnegut sees in *A Connecticut Yankee* a chilling foreshadowing (or perhaps a projection from the Civil War) of "all the high-tech atrocities which followed, and which follow still." Cynthia Ozick suggests that "The Man That Corrupted Hadleyburg," along with some of the other pieces collected under that title — many of them written when Twain lived in a Vienna ruled by Karl Lueger, a demagogue Adolf Hitler would later idolize — shoot up moral flares that shed an eerie light on the insidious corruption, prejudice, and hatred that reached bitter fruition under the Third Reich. And Twain's portrait in this book of "the dissolving Austria-Hungary of the 1890s," in Ozick's view, presages not only the Sarajevo that would erupt in 1914 but also "the disintegrated components of the former Yugoslavia" and "the *fin-de-siècle* Sarajevo of our own moment."

Despite their admiration for Twain's ambitious reach and scope, contributors to The Oxford Mark Twain also recognize his limitations. Mordecai Richler, for example, thinks that "the early pages of *Innocents Abroad* suffer from being a tad broad, proffering more burlesque than inspired satire," perhaps because Twain was "trying too hard for knee-slappers." Charles Johnson notes that the Young Man in Twain's philosophical dialogue about free will and determinism (*What Is Man?*) "caves in far too soon," failing to challenge what through late-twentieth-century eyes looks like "pseudoscience" and suspect essentialism in the Old Man's arguments.

Some contributors revisit their first encounters with Twain's works, recalling what surprised or intrigued them. When David Bradley came across "Fenimore Cooper's Literary Offences" in his college library, he "did not at first realize that Twain was being his usual ironic self with all this business about the 'nineteen rules governing literary art in the domain of romantic fiction,' but by the time I figured out there was no such list outside Twain's own head, I had decided that the rules made *sense*. . . . It seemed to me they were a pretty good blueprint for writing — Negro writing included." Sherley Anne Williams remembers that part of what attracted her to *Pudd'nhead Wilson* when she first read it thirty years ago was "that Twain, writing at the end of the nineteenth century, could imagine negroes as characters, albeit white ones, who actually thought for and of themselves, whose actions were the product of their thinking rather than the spontaneous ephemera of physical instincts that stereotype assigned to blacks." Frederik Pohl recalls his first reading of *Huckleberry Finn* as "a watershed event" in his life, the first book he read as a child in which "bad people" ceased to exercise a monopoly on doing "bad things." In *Huckleberry Finn* "some seriously bad things — things like the possession and mistreatment of black slaves, like stealing and lying, even like killing other people in duels — were quite often done by people who not only thought of themselves as exemplarily moral but, by any other standards I knew how to apply, actually *were* admirable citizens." The world that

Tom and Huck lived in, Pohl writes, "was filled with complexities and contradictions," and resembled "the world I appeared to be living in myself."

Other contributors explore their more recent encounters with Twain, explaining why they have revised their initial responses to his work. For Toni Morrison, parts of *Huckleberry Finn* that she "once took to be deliberate evasions, stumbles even, or a writer's impatience with his or her material," now strike her "as otherwise: as entrances, crevices, gaps, seductive invitations flashing the possibility of meaning. Unarticulated eddies that encourage diving into the novel's undertow — the real place where writer captures reader." One such "eddy" is the imprisonment of Jim on the Phelps farm. Instead of dismissing this portion of the book as authorial bungling, as she once did, Morrison now reads it as Twain's commentary on the 1880s, a period that "saw the collapse of civil rights for blacks," a time when "the nation, as well as Tom Sawyer, was deferring Jim's freedom in agonizing play." Morrison believes that Americans in the 1880s were attempting "to bury the combustible issues Twain raised in his novel," and that those who try to kick Huck Finn out of school in the 1990s are doing the same: "The cyclical attempts to remove the novel from classrooms extend Jim's captivity on into each generation of readers."

Although imitation-Hemingway and imitation-Faulkner writing contests draw hundreds of entries annually, no one has ever tried to mount a faux-Twain competition. Why? Perhaps because Mark Twain's voice is too much a part of who we are and how we speak even today. Roy Blount Jr. suggests that it is impossible, "at least for an American writer, to parody Mark Twain. It would be like doing an impression of your father or mother: he or she is already there in your voice."

Twain's style is examined and celebrated in The Oxford Mark Twain by fellow writers who themselves have struggled with the nuances of words, the structure of sentences, the subtleties of point of view, and the trickiness of opening lines. Bobbie Ann Mason observes, for example, that "Twain loved the sound of words and he knew how to string them by sound, like different shades of one color: 'The earl's barbaric eye,' 'the Usurping Earl,' 'a double-dyed humbug.'" Twain "relied on the punch of plain words" to show writers how to move beyond the "wordy romantic rubbish" so prevalent in nineteenth-century fiction, Mason says; he "was one of the first writers in America to deflower literary language." Lee Smith believes that "American writers have benefited as much from the way Mark Twain opened up the possibilities of first-person narration as we have from his use of vernacular language." (She feels that "the ghost of Mark Twain was hovering someplace in the background" when she decided to write her novel *Oral History* from the standpoint of multiple first-person narrators.) Frederick Busch maintains that "A Dog's Tale" "boasts one of the great opening sentences" of all time: "My father was a St. Bernard, my mother was a collie, but I am a Presbyterian." And Ursula Le Guin marvels at the ingenuity of the following sentence that she encounters in *Extracts from Adam's Diary*.

> . . . This made her sorry for the creatures which live in there, which she calls fish, for she continues to fasten names on to things that don't need them and don't come when they are called by them, which is a matter of no consequence to her, as she is such a numskull anyway; so she got a lot of them out and brought them in last night and put them in my bed to keep warm, but I have noticed them now and then all day, and I don't see that they are any happier there than they were before, only quieter.[10]

Le Guin responds,

> Now, that is a pure Mark-Twain-tour-de-force sentence, covering an immense amount of territory in an effortless, aimless ramble that seems to be heading nowhere in particular and ends up with breathtaking accuracy at the gold mine. Any sensible child would find that funny, perhaps not following all its divagations but delighted by the swing of it, by the word "numskull," by the idea of putting fish in the bed; and as that child grew older and reread it, its reward would only grow; and if that grown-up child had to write an essay on the piece and therefore earnestly studied and pored over this sentence, she would end up in unmitigated admiration of its vocabulary, syntax, pacing, sense, and rhythm, above all the beautiful

> timing of the last two words; and she would, and she does, still find it funny.

The fish surface again in a passage that Gore Vidal calls to our attention, from *Following the Equator*: "'The Whites always mean well when they take human fish out of the ocean and try to make them dry and warm and happy and comfortable in a chicken coop,' which is how, through civilization, they did away with many of the original inhabitants. Lack of empathy is a principal theme in Twain's meditations on race and empire."

Indeed, empathy — and its lack — is a principal theme in virtually all of Twain's work, as contributors frequently note. Nat Hentoff quotes the following thoughts from Huck in *Tom Sawyer Abroad*:

> I see a bird setting on a dead limb of a high tree, singing with its head tilted back and its mouth open, and before I thought I fired, and his song stopped and he fell straight down from the limb, all limp like a rag, and I run and picked him up and he was dead, and his body was warm in my hand, and his head rolled about this way and that, like his neck was broke, and there was a little white skin over his eyes, and one little drop of blood on the side of his head; and laws! I could n't see nothing more for the tears; and I hain't never murdered no creature since that war n't doing me no harm, and I ain't going to.[11]

"The Humane Society," Hentoff writes, "has yet to say anything as powerful — and lasting."

Readers of The Oxford Mark Twain will have the pleasure of revisiting Twain's Mississippi landmarks alongside Willie Morris, whose own lower Mississippi Valley boyhood gives him a special sense of connection to Twain. Morris knows firsthand the mosquitoes described in *Life on the Mississippi* — so colossal that "two of them could whip a dog" and "four of them could hold a man down"; in Morris's own hometown they were so large during the flood season that "local wags said they wore wristwatches." Morris's Yazoo City and Twain's Hannibal shared a "rough-hewn democracy . . . complicated by all the visible textures of caste and class, . . . harmless boyhood fun and mischief right along with . . . rank hypocrisies, churchgoing sanctimonies, racial hatred, entrenched and unrepentant greed."

For the West of Mark Twain's *Roughing It*, readers will have George Plimpton as their guide. "What a group these newspapermen were!" Plimpton writes about Twain and his friends Dan De Quille and Joe Goodman in Virginia City, Nevada. "Their roisterous carryings-on bring to mind the kind of frat-house enthusiasm one associates with college humor magazines like the *Harvard Lampoon*." Malcolm Bradbury examines Twain as "a living example of what made the American so different from the European." And Hal Holbrook, who has interpreted Mark Twain on stage for some forty years, describes how Twain "played" during the civil rights movement, during the Vietnam War, during the Gulf War, and in Prague on the eve of the demise of Communism.

Why do we continue to read Mark Twain? What draws us to him? His wit? His compassion? His humor? His bravura? His humility? His understanding of who and what we are in those parts of our being that we rarely open to view? Our sense that he knows we can do better than we do? Our sense that he knows we can't? E. L. Doctorow tells us that children are attracted to *Tom Sawyer* because in this book "the young reader confirms his own hope that no matter how troubled his relations with his elders may be, beneath all their disapproval is their underlying love for him, constant and steadfast." Readers in general, Arthur Miller writes, value Twain's "insights into America's always uncertain moral life and its shifting but everlasting hypocrisies"; we appreciate the fact that he "is not using his alienation from the public illusions of his hour in order to reject the country implicitly as though he could live without it, but manifestly in order to correct it." Perhaps we keep reading Mark Twain because, in Miller's words, he "wrote much more like a father than a son. He doesn't seem to be sitting in class taunting the teacher but standing at the head of it challenging his students to acknowledge their own humanity, that is, their immemorial attraction to the untrue."

Mark Twain entered the public eye at a time when many of his countrymen considered "American culture" an oxymoron; he died four years before a world conflagration that would lead many to question whether the contradiction in terms was not "European civilization" instead. In between he worked in journalism, printing, steamboating, mining, lecturing, publishing, and editing, in virtually every region of the country. He tried his hand at humorous sketches, social satire, historical novels, children's books, poetry, drama, science fiction, mysteries, romance, philosophy, travelogue, memoir, polemic, and several genres no one had ever seen before or has ever seen since. He invented a self-pasting scrapbook, a history game, a vest strap, and a gizmo for keeping bed sheets tucked in; he invested in machines and processes designed to revolutionize typesetting and engraving, and in a food supplement called "Plasmon." Along the way he cheerfully impersonated himself and prior versions of himself for doting publics on five continents while playing out a charming rags-to-riches story followed by a devastating riches-to-rags story followed by yet another great American comeback. He had a long-running real-life engagement in a sumptuous comedy of manners, and then in a real-life tragedy not of his own design: during the last fourteen years of his life almost everyone he ever loved was taken from him by disease and death.

Mark Twain has indelibly shaped our views of who and what the United States is as a nation and of who and what we might become. He understood the nostalgia for a "simpler" past that increased as that past receded — and he saw through the nostalgia to a past that was just as complex as the present. He recognized better than we did ourselves our potential for greatness and our potential for disaster. His fictions brilliantly illuminated the world in which he lived, changing it — and us — in the process. He knew that our feet often danced to tunes that had somehow remained beyond our hearing; with perfect pitch he played them back to us.

My mother read *Tom Sawyer* to me as a bedtime story when I was eleven. I thought Huck and Tom could be a lot of fun, but I dismissed Becky Thatcher as a bore. When I was twelve I invested a nickel at a local garage sale in a book that contained short pieces by Mark Twain. That was where I met Twain's Eve. Now, *that's* more like it, I decided, pleased to meet a female character I could identify *with* instead of against. Eve had spunk. Even if she got a lot wrong, you had to give her credit for trying. "The Man That Corrupted

Hadleyburg" left me giddy with satisfaction: none of my adolescent reveries of getting even with my enemies were half as neat as the plot of the man who got back at that town. "How I Edited an Agricultural Paper" set me off in uncontrollable giggles.

People sometimes told me that I looked like Huck Finn. "It's the freckles," they'd explain—not explaining anything at all. I didn't read *Huckleberry Finn* until junior year in high school in my English class. It was the fall of 1965. I was living in a small town in Connecticut. I expected a sequel to *Tom Sawyer*. So when the teacher handed out the books and announced our assignment, my jaw dropped: "Write a paper on how Mark Twain used irony to attack racism in *Huckleberry Finn*."

The year before, the bodies of three young men who had gone to Mississippi to help blacks register to vote—James Chaney, Andrew Goodman, and Michael Schwerner—had been found in a shallow grave; a group of white segregationists (the county sheriff among them) had been arrested in connection with the murders. America's inner cities were simmering with pent-up rage that began to explode in the summer of 1965, when riots in Watts left thirty-four people dead. None of this made any sense to me. I was confused, angry, certain that there was something missing from the news stories I read each day: the why. Then I met Pap Finn. And the Phelpses.

Pap Finn, Huck tells us, "had been drunk over in town" and "was just all mud." He erupts into a drunken tirade about "a free nigger . . . from Ohio—a mulatter, most as white as a white man," with "the whitest shirt on you ever see, too, and the shiniest hat; and there ain't a man in town that's got as fine clothes as what he had."

> . . . they said he was a p'fessor in a college, and could talk all kinds of languages, and knowed everything. And that ain't the wust. They said he could *vote*, when he was at home. Well, that let me out. Thinks I, what is the country a-coming to? It was 'lection day, and I was just about to go and vote, myself, if I warn't too drunk to get there; but when they told me there was a State in this country where they'd let that nigger vote, I drawed out. I says I'll never vote agin. Them's the very words I said. . . . And to see the

> cool way of that nigger — why, he wouldn't a give me the road if I hadn't shoved him out o' the way.[12]

Later on in the novel, when the runaway slave Jim gives up his freedom to nurse a wounded Tom Sawyer, a white doctor testifies to the stunning altruism of his actions. The Phelpses and their neighbors, all fine, upstanding, well-meaning, churchgoing folk,

> agreed that Jim had acted very well, and was deserving to have some notice took of it, and reward. So every one of them promised, right out and hearty, that they wouldn't curse him no more.
>
> Then they come out and locked him up. I hoped they was going to say he could have one or two of the chains took off, because they was rotten heavy, or could have meat and greens with his bread and water, but they didn't think of it.[13]

Why did the behavior of these people tell me more about why Watts burned than anything I had read in the daily paper? And why did a drunk Pap Finn railing against a black college professor from Ohio whose vote was as good as his own tell me more about white anxiety over black political power than anything I had seen on the evening news?

Mark Twain knew that there was nothing, absolutely *nothing*, a black man could do — including selflessly sacrificing his freedom, the only thing of value he had — that would make white society see beyond the color of his skin. And Mark Twain knew that depicting racists with chilling accuracy would expose the viciousness of their world view like nothing else could. It was an insight echoed some eighty years after Mark Twain penned Pap Finn's rantings about the black professor, when Malcolm X famously asked, "Do you know what white racists call black Ph.D.'s?" and answered, "'*Nigger!*'"[14]

Mark Twain taught me things I needed to know. He taught me to understand the raw racism that lay behind what I saw on the evening news. He taught me that the most well-meaning people can be hurtful and myopic. He taught me to recognize the supreme irony of a country founded in freedom that continued to deny freedom to so many of its citizens. Every time I hear of another effort to kick Huck Finn out of school somewhere, I recall everything that Mark Twain taught *this* high school junior, and I find myself jumping into the fray.[15] I remember the black high school student who called CNN during the phone-in portion of a 1985 debate between Dr. John Wallace, a black educator spearheading efforts to ban the book, and myself. She accused Dr. Wallace of insulting her and all black high school students by suggesting they weren't smart enough to understand Mark Twain's irony. And I recall the black cameraman on the *CBS Morning News* who came up to me after he finished shooting another debate between Dr. Wallace and myself. He said he had never read the book by Mark Twain that we had been arguing about—but now he really wanted to. One thing that puzzled him, though, was why a white woman was defending it and a black man was attacking it, because as far as he could see from what we'd been saying, the book made whites look pretty bad.

As I came to understand *Huckleberry Finn* and *Pudd'nhead Wilson* as commentaries on the era now known as the nadir of American race relations, those books pointed me toward the world recorded in nineteenth-century black newspapers and periodicals and in fiction by Mark Twain's black contemporaries. My investigation of the role black voices and traditions played in shaping Mark Twain's art helped make me aware of their role in shaping all of American culture.[16] My research underlined for me the importance of changing the stories we tell about who we are to reflect the realities of what we've been.[17]

Ever since our encounter in high school English, Mark Twain has shown me the potential of American literature and American history to illuminate each other. Rarely have I found a contradiction or complexity we grapple with as a nation that Mark Twain had not puzzled over as well. He insisted on taking America seriously. And he insisted on *not* taking America seriously: "I think that there is but a single specialty with us, only one thing that can be called by the wide name 'American,'" he once wrote. "That is the national devotion to ice-water."[18]

Mark Twain threw back at us our dreams and our denial of those dreams, our greed, our goodness, our ambition, and our laziness, all rattling around together in that vast echo chamber of our talk — that sharp, spunky American talk that Mark Twain figured out how to write down without robbing it of its energy and immediacy. Talk shaped by voices that the official arbiters of "culture" deemed of no importance — voices of children, voices of slaves, voices of servants, voices of ordinary people. Mark Twain listened. And he made us listen. To the stories he told us, and to the truths they conveyed. He still has a lot to say that we need to hear.

Mark Twain lives — in our libraries, classrooms, homes, theaters, movie houses, streets, and most of all in our speech. His optimism energizes us, his despair sobers us, and his willingness to keep wrestling with the hilarious and horrendous complexities of it all keeps us coming back for more. As the twenty-first century approaches, may he continue to goad us, chasten us, delight us, berate us, and cause us to erupt in unrestrained laughter in unexpected places.

NOTES

1. Ernest Hemingway, *Green Hills of Africa* (New York: Charles Scribner's Sons, 1935), 22. George Bernard Shaw to Samuel L. Clemens, July 3, 1907, quoted in Albert Bigelow Paine, *Mark Twain: A Biography* (New York: Harper and Brothers, 1912), 3:1398.

2. Allen Carey-Webb, "Racism and *Huckleberry Finn*: Censorship, Dialogue and Change," *English Journal* 82, no. 7 (November 1993):22.

3. See Louis J. Budd, "Impersonators," in J. R. LeMaster and James D. Wilson, eds., *The Mark Twain Encyclopedia* (New York: Garland Publishing Company, 1993), 389–91.

4. See Shelley Fisher Fishkin, "Ripples and Reverberations," part 3 of *Lighting Out for the Territory: Reflections on Mark Twain and American Culture* (New York: Oxford University Press, 1996).

5. There are two exceptions. Twain published chapters from his autobiography in the *North American Review* in 1906 and 1907, but this material was not published in book form in Twain's lifetime; our volume reproduces the material as it appeared in the *North American Review*. The other exception is our final volume, *Mark Twain's Speeches*, which appeared two months after Twain's death in 1910.

An unauthorized handful of copies of *1601* was privately printed by an Alexander Gunn of Cleveland at the instigation of Twain's friend John Hay in 1880. The first American edition authorized by Mark Twain, however, was printed at the United States Military Academy at West Point in 1882; that is the edition reproduced here.

It should further be noted that four volumes — *The Stolen White Elephant and Other Detective Stories*, *Following the Equator and Anti-imperialist Essays*, *The Diaries of Adam and Eve*, and *1601, and Is Shakespeare Dead?* — bind together material originally published separately. In each case the first American edition of the material is the version that has been reproduced, always in its entirety. Because Twain constantly recycled and repackaged previously published works in his collections of short pieces, a certain amount of duplication is unavoidable. We have selected volumes with an eye toward keeping this duplication to a minimum.

Even the twenty-nine-volume Oxford Mark Twain has had to leave much out. No edition of Twain can ever claim to be "complete," for the man was too prolix, and the file drawers of both ephemera and as yet unpublished texts are deep.

6. With the possible exception of *Mark Twain's Speeches*. Some scholars suspect Twain knew about this book and may have helped shape it, although no hard evidence to that effect has yet surfaced. Twain's involvement in the production process varied greatly from book to book. For a fuller sense of authorial intention, scholars will continue to rely on the superb definitive editions of Twain's works produced by the Mark Twain Project at the University of California at Berkeley as they become available. Dense with annotation documenting textual emendation and related issues, these editions add immeasurably to our understanding of Mark Twain and the genesis of his works.

7. Except for a few titles that were not in its collection. The American Antiquarian Society in Worcester, Massachusetts, provided the first edition of *King Leopold's Soliloquy*; the Elmer Holmes Bobst Library of New York University furnished the 1906–7 volumes of the *North American Review* in which *Chapters from My Autobiography* first appeared; the Harry Ransom Humanities Research Center at the University of Texas at Austin made their copy of the West Point edition of *1601* available; and the Mark Twain Project provided the first edition of *Extract from Captain Stormfield's Visit to Heaven.*

8. The specific copy photographed for Oxford's facsimile edition is indicated in a note on the text at the end of each volume.

9. All quotations from contemporary writers in this essay are taken from their introductions to the volumes of The Oxford Mark Twain, and the quotations from Mark Twain's works are taken from the texts reproduced in The Oxford Mark Twain.

10. *The Diaries of Adam and Eve*, The Oxford Mark Twain [hereafter OMT] (New York: Oxford University Press, 1996), p. 33.

11. *Tom Sawyer Abroad*, OMT, p. 74.

12. *Adventures of Huckleberry Finn*, OMT, p. 49–50.

13. Ibid., p. 358.

14. Malcolm X, *The Autobiography of Malcolm X*, with the assistance of Alex Haley (New York: Grove Press, 1965), p. 284.

15. I do not mean to minimize the challenge of teaching this difficult novel, a challenge for which all teachers may not feel themselves prepared. Elsewhere I have developed some concrete strategies for approaching the book in the classroom, including teaching it in the context of the history of American race relations and alongside books by black writers. See Shelley Fisher Fishkin, "Teaching *Huckleberry Finn*," in James S. Leonard, ed., *Making Mark Twain Work in the Classroom* (Durham: Duke University Press, forthcoming). See also Shelley Fisher Fishkin, *Was Huck Black? Mark Twain and African-American Voices* (New York: Oxford University Press, 1993), pp. 106–8, and a curriculum kit in preparation at the Mark Twain House in Hartford, containing teaching suggestions from myself, David Bradley, Jocelyn Chadwick-Joshua, James Miller, and David E. E. Sloane.

16. See Fishkin, *Was Huck Black?* See also Fishkin, "Interrogating 'Whiteness,' Complicating 'Blackness': Remapping American Culture," in Henry Wonham, ed., *Criticism and the Color Line: Desegregating American Literary Studies* (New Brunswick: Rutgers UP, 1996, pp. 251–90 and in shortened form in *American Quarterly* 47, no. 3 (September 1995):428–66.

17. I explore the roots of my interest in Mark Twain and race at greater length in an essay entitled "Changing the Story," in Jeffrey Rubin-Dorsky and Shelley Fisher Fishkin, eds., *People of the Book: Thirty Scholars Reflect on Their Jewish Identity* (Madison: U of Wisconsin Press, 1996), pp. 47–63.

18. "What Paul Bourget Thinks of Us," *How to Tell a Story and Other Essays*, OMT, p. 197.

INTRODUCTION
Reflections on the Detective Stories of Mark Twain and an Idea or Two More About His Fiction

Walter Mosley

Every novel, short story, or poem should be a discovery, happened upon by the author while in a fit of tenacity, inspiration, and the love of at least one reader. This work does not define but insinuates a world view, a state of mind, or a people, or family — a history too rich to be described in literal detail.

Writing is the magical possibility that exists in the cracks of our reality. It creates from smudges of ink a world that exposes and at the same time transcends everyday experience.

Good writing is sloppy and uneven, rutted and without predictable meter; it is imperfect, broken. Good writing is never what it seems to be; even if it appears beautiful, its underbelly is crude, petty, ambitious, and pedestrian.

Good writing is a streetwalker, as common as rain. Any pleasure it brings is most likely vulgar or coarse — with a twist, a saving grace maybe; a hope that being human isn't as bad as we fear.

Good writing is, of course, genius. Not the kind of genius gauged by SAT scores and IQ tests. Not the marketing genius of record moguls creating musical stars who can't carry a tune. Not the kind of writing that says, *If you can understand what I'm saying, then you are smarter than your neighbors.* Good writing is the genius of language. Mark Twain once said that writing is easy — all the words are in the dictionary. What he meant, I think, is that language in all its richness and all its history belongs to us all, us'n.

The writer chooses the right words to tell her tale. The reader might not have chosen those words but he knows them. These peers — the writer and the reader — come together in a moment of truth that transports and elates.

There's no telling how many truths a good tale can spin — at least as many as there are readers.

Huckleberry Finn is an example of good writing, a book that shoots from the hip, aimed at sympathetic hearts with love and at antagonistic ones with murderous intent. It's a book about flawed men with passions and the desire to achieve justice. It's a book about racism, stupidity, and the ability to share ideas until the truth is obvious, even if it is unattainable.

Huckleberry Finn is a book about the world we live in today, even though it is fiction and even though it was published one hundred and eleven years ago. It's also a book of journeys. First there's the trek of man and boy making their way through a strange world that is their home and their school, a place that's treacherous, where the company is unruly, and worse, unfamiliar. The map is a river, but the destination is dependent upon the ability of the writer and the reader, and of Huck and Jim, to *catch on* — to make sense out of what they see and what they are told is true.

For Jim it's a desperate ride similar to the one that so many slaves took on the Underground Railroad: a train that was no train traveling on tracks that were never laid. It was the only road to freedom. The precedent for taking this road was the idea, or ideal, of freedom, but the road traveled was different for every man and woman who took it.

The ticket was always stolen because the rules said that a Negro could not be free.

2

Mark Twain broke the rules because they reinforced bigotry, racism, and the banal comfort of thoughtlessness (stupidity). Twain's religious folk want to hog-tie the wonders of the world, while his scientists attempt to become God. Law and government and bureaucracy are potential dictators for Mr. Clemens, while childhood selfishness may be the closest we ever come to the truth.

Mark Twain broke the rules, and in doing so created a new and powerful American voice. Twain is irreverent, satirical, and critical of our world and our assumptions. Like any great humorist, he makes us laugh at ourselves.

In works like *A Double Barrelled Detective Story* and "The Stolen White Elephant" Twain breaks with realistic fiction. The second story is reduced to a series of ridiculous telegrams related by an untrustworthy narrator caught up in an adventure that is as impossible as it is ludicrous. In short, Twain has re-created our world in a few brief lines of prose. He uses the device of telecommunication to show us that science has not delivered us from our folly. He shows us that the law can be more corrupt than it is misguided and still hold us in thrall to its uniform.

A Double Barrelled Detective Story is more serious fare. It is a story about spousal abuse and revenge, a metaphysical story in which the unborn child of a brutalized pregnant woman takes on the olfactory sense of the dogs that have been set upon his mother. It's the story of a son searching for a father whom he has been taught to hate but comes to love. It's a story of mistaken identity and redemption. It's also a story in which Sherlock Holmes is an educated buffoon and the pawn of his malevolent nephew.

Twain will not let us take comfort in the seriousness and truth of his tales. He pulls us in and then laughs at us. He makes us indignant and then shows us how we fool ourselves with righteousness.

How does he do it? I don't know exactly. I understand a few of the tools that he employs. I see how he subverts convention by shading serious topics with ridicule. I see how he tells a story in a way that is in some sense a lie, but a lie that makes us laugh — and there is nothing more honest than laughter.

Mark Twain was proficient at a form of storytelling that I call "mob fiction." This mob fiction is hard to describe. It's as if somebody were guiding you though the terrain of the story he's telling. But it's not just you; there's a whole crowd of people pushing and shoving along as the narrator shows us various scenes and events. It's like a guided tour in a foreign land where you and I are simple tourists who don't speak the native tongue.

"This piece of rock here is five thousand years old," the narrator says. "It was skimmed by the pharaoh Nip-n-Tuck across the Nile at its greatest width and was brought to Peoria by the great adventurer Felix Orcat after he was struck low with fever at the Battle of Two Towns Running . . ."

We don't know what to think. He's brought us this far. The next line might be from the man next to you, whispering to his wife.

"I happen to know for a fact that that's not true," the brush salesman from Atlanta says. "Felix Orcat lost a limb in the Battle of Two Towns Running, but he never ran a fever in all his eighty-seven years."

The dialogues around you continue, and the experience of the story becomes a mob event even though it's just you and the words.

Words were very important to Mark Twain. "Use the right word, not its second cousin," he wrote in 1895, in "Fenimore Cooper's Literary Offences," in which he pilloried Cooper's *Deerslayer*.

> It has no invention; it has no order, system, sequence, or result; it has no lifelikeness, no thrill, no stir, no seeming of reality; its characters are confusedly drawn, and by their acts and words they prove that they are not the sort of people the author claims that they are; its humor is pathetic; its pathos is funny; its conversations are—oh! indescribable; its love-scenes odious; its English a crime against the language.
>
> Counting these out, what is left is Art. I think we must all admit that.

This essay, along with "Fenimore Cooper's Further Literary Offenses," is all a creative writer would ever have to read in order to understand the workings of his or her craft.

3

Mark Twain is the rough uncle that nobody likes to see at the family reunion. He smokes cigars, drinks rank whiskey, speaks his mind, and doesn't follow convention. The young people love him, while those who have something invested in pretension or propriety (or silence) would rather be attending his funeral.

Mark Twain's fiction is harsh, and no one, including this writer, is immune to his words.

There is racism in lines of Mark Twain. The racism that most people want to keep quiet. *Don't use words that might seem derogatory*, an ancient aunt might tell you today, or a hundred years ago today. But that same aunt wouldn't let a *colored* man into her home, her life.

Every day white people, of all classes and castes, use insulting terms for people like me. I know it. You know it. If you're white you hear it in taxis, or at home around the dinner table, and if you're rich, in boardrooms or at the club. If you're black you're passed over by taxis. If you're black, white women grab their purses when you come too near. If you're black, you probably haven't seen the inside of those boardrooms; and if you have you had to abandon the dream of yourself to get inside.

Mark Twain knew these truths before my father's father was born. He spoke the language of poor whites, and blacks. He said in print what many white men, then and now, say secretly, out of earshot.

You might, if you only read *Tom Sawyer, Detective*, think that Mark Twain was an unrepentant and callous racist. I doubt that he was. *Huckleberry Finn* was the heart of the man. Huckleberry wanted to understand the nature of justice. He started out believing, as we know Twain did, that dark men were naturally the property of whites. He was raised in a nation that held this belief. But Huckleberry learned from his association with Jim that the truth was far from that.

To follow Mark Twain's path and his language is difficult and often distasteful. I cringed when I read certain passages of these stories. I wouldn't accept such language from most writers, contemporary or otherwise.

Is there racism in Twain's works? Definitely. Does that racism reflect, to any degree, racism in tea man? I would say probably so. But I would go on to say that Mark Twain was making a strong attempt to understand his shortcomings and the shortcomings of his nation.

He was trying to make things better, not worse.

I wouldn't suggest these stories to a hardcore mystery reader. But I would suggest them to anyone who wants a view of the mind and flaws of a bygone people — a people whose imperfections we can still see in our daily lives.

As in most of his work, Mark Twain is a mixed bag as far as the detective genre is concerned. He runs roughshod over the conventions of the genre, when he is not spoofing them outrageously (but not always successfully). His mysteries are like the rest of his fiction: Mr. Human Nature with a bare butt and he thinks he's dressed to the nines. And for that they're worth our attention, even if they're not Mark Twain at his vernacular rule-breaking best.

THE
STOLEN
WHITE ELEPHANT

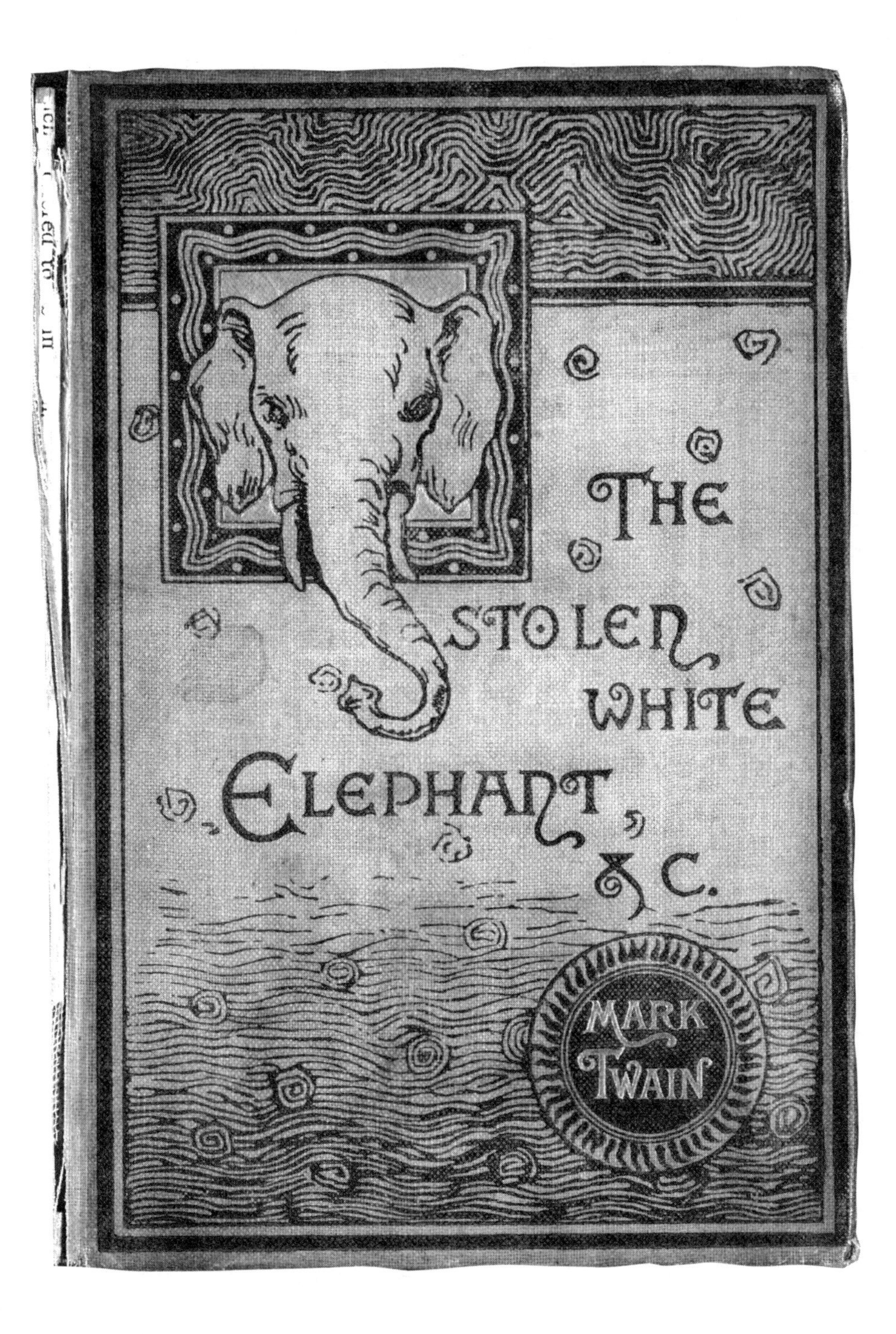
THE
STOLEN
WHITE
ELEPHANT,
&C.
MARK
TWAIN

THE STOLEN WHITE ELEPHANT

THE

STOLEN WHITE ELEPHANT

ETC.

BY MARK TWAIN

BOSTON
JAMES R. OSGOOD AND COMPANY
1882

UNIVERSITY PRESS:
JOHN WILSON AND SON, CAMBRIDGE.

CONTENTS.

THE STOLEN WHITE ELEPHANT.[1]

I.

THE following curious history was related to me by a chance railway acquaintance. He was a gentleman more than seventy years of age, and his thoroughly good and gentle face and earnest and sincere manner imprinted the unmistakable stamp of truth upon every statement which fell from his lips. He said : —

You know in what reverence the royal white elephant of Siam is held by the people of that country. You know it is sacred to kings, only kings may possess it, and that it is indeed in a measure even superior to kings, since it receives not merely honor but worship. Very well; five years ago, when the troubles concerning the frontier line arose between Great Britain and Siam, it was presently manifest that Siam had been in the wrong. Therefore every reparation was quickly

[1] Left out of "A Tramp Abroad," because it was feared that some of the particulars had been exaggerated, and that others were not true. Before these suspicions had been proven groundless, the book had gone to press. — M. T.

made, and the British representative stated that he was satisfied and the past should be forgotten. This greatly relieved the King of Siam, and partly as a token of gratitude, but partly also, perhaps, to wipe out any little remaining vestige of unpleasantness which England might feel toward him, he wished to send the Queen a present, — the sole sure way of propitiating an enemy, according to Oriental ideas. This present ought not only to be a royal one, but transcendently royal. Wherefore, what offering could be so meet as that of a white elephant? My position in the Indian civil service was such that I was deemed peculiarly worthy of the honor of conveying the present to her Majesty. A ship was fitted out for me and my servants and the officers and attendants of the elephant, and in due time I arrived in New York harbor and placed my royal charge in admirable quarters in Jersey City. It was necessary to remain awhile in order to recruit the animal's health before resuming the voyage.

All went well during a fortnight, — then my calamities began. The white elephant was stolen! I was called up at dead of night and informed of this fearful misfortune. For some moments I was beside myself with terror and anxiety; I was helpless. Then I grew calmer and collected my faculties. I soon saw my course, — for indeed there was but the one course for an intelligent man to pursue. Late as it was, I flew to New York and got a policeman to conduct me to the headquarters of the detective force. Fortunately I arrived in time, though the chief of the force, the celebrated Inspector Blunt, was just on the point of leaving for his home. He was a man of middle size and compact frame, and when he was thinking deeply he had a way of knitting his brows and tapping his forehead reflectively with his finger, which impressed you at once with the conviction that you stood in the presence of a person of no common order. The very sight of him gave me confidence and made me hopeful. I stated my errand. It did not flurry him in the least; it had no more visible effect upon his iron self-possession than if I had told him somebody had stolen my dog. He motioned me to a seat, and said calmly, —

"Allow me to think a moment, please."

So saying, he sat down at his office table and leaned his head upon his hand. Several clerks were at work at the other end of the room; the scratching of their pens was all the sound I heard during the next six or seven minutes. Meantime the inspector sat there, buried in thought. Finally he raised his head, and there was that in the firm lines of his face which showed me that his brain had done its work and his plan was made. Said he, — and his voice was low and impressive, —

"This is no ordinary case. Every step must be warily taken; each step must be made sure before the next is ventured. And secrecy must be observed, — secrecy profound and absolute. Speak to no one about the matter, not even the reporters. I will take care of *them;* I will see that they get only what it may suit my ends to let them know." He touched a bell;

a youth appeared. "Alaric, tell the reporters to remain for the present." The boy retired. "Now let us proceed to business, — and systematically. Nothing can be accomplished in this trade of mine without strict and minute method."

He took a pen and some paper. "Now — name of the elephant?"

"Hassan Ben Ali Ben Selim Abdallah Mohammed Moisé Alhammal Jamsetjejeebhoy Dhuleep Sultan Ebu Bhudpoor."

"Very well. Given name?"

"Jumbo."

"Very well. Place of birth?"

"The capital city of Siam."

"Parents living?"

"No, — dead."

"Had they any other issue besides this one?"

"None. He was an only child."

"Very well. These matters are sufficient under that head. Now please describe the elephant, and leave out no particular, however insignificant, — that is, insignificant from *your* point of view. To men in my profession there *are* no insignificant particulars; they do not exist."

I described, — he wrote. When I was done, he said, —

"Now listen. If I have made any mistakes, correct me."

He read as follows: —

"Height, 19 feet; length from apex of forehead to insertion of tail, 26 feet; length of trunk, 16 feet;

length of tail, 6 feet; total length, including trunk and tail, 48 feet; length of tusks, $9\frac{1}{2}$ feet; ears in keeping with these dimensions; footprint resembles the mark left when one up-ends a barrel in the snow; color of the elephant, a dull white; has a hole the size of a plate in each ear for the insertion of jewelry, and possesses the habit in a remarkable degree of squirting water upon spectators and of maltreating with his trunk not only such persons as he is acquainted with, but even entire strangers; limps slightly with his right hind leg, and has a small scar in his left armpit caused by a former boil; had on, when stolen, a castle containing seats for fifteen persons, and a gold-cloth saddle-blanket the size of an ordinary carpet."

There were no mistakes. The inspector touched the bell, handed the description to Alaric, and said, —

"Have fifty thousand copies of this printed at once and mailed to every detective office and pawnbroker's shop on the continent." Alaric retired. "There, — so far, so good. Next, I must have a photograph of the property."

I gave him one. He examined it critically, and said, —

"It must do, since we can do no better; but he has his trunk curled up and tucked into his mouth. That is unfortunate, and is calculated to mislead, for of course he does not usually have it in that position." He touched his bell.

"Alaric, have fifty thousand copies of this photograph made, the first thing in the morning, and mail them with the descriptive circulars."

Alaric retired to execute his orders. The inspector said, —

"It will be necessary to offer a reward, of course. Now as to the amount?"

"What sum would you suggest?"

"To *begin* with, I should say, — well, twenty-five thousand dollars. It is an intricate and difficult business; there are a thousand avenues of escape and opportunities of concealment. These thieves have friends and pals everywhere — "

"Bless me, do you know who they are?"

The wary face, practised in concealing the thoughts and feelings within, gave me no token, nor yet the replying words, so quietly uttered: —

"Never mind about that. I may, and I may not. We generally gather a pretty shrewd inkling of who our man is by the manner of his work and the size of the game he goes after. We are not dealing with a pickpocket or a hall thief, now, make up your mind to that. This property was not 'lifted' by a novice. But, as I was saying, considering the amount of travel which will have to be done, and the diligence with which the thieves will cover up their traces as they move along, twenty-five thousand may be too small a sum to offer, yet I think it worth while to start with that."

So we determined upon that figure, as a beginning. Then this man, whom nothing escaped which could by any possibility be made to serve as a clew, said: —

"There are cases in detective history to show that criminals have been detected through peculiarities in their appetites. Now, what does this elephant eat, and how much?"

"Well, as to *what* he eats, — he will eat *anything.* He will eat a man, he will eat a Bible, — he will eat anything *between* a man and a Bible."

"Good, — very good indeed, but too general. Details are necessary, — details are the only valuable things in our trade. Very well, — as to men. At one meal, — or, if you prefer, during one day, — how many men will he eat, if fresh?"

"He would not care whether they were fresh or not; at a single meal he would eat five ordinary men."

"Very good; five men; we will put that down. What nationalities would he prefer?"

"He is indifferent about nationalities. He prefers acquaintances, but is not prejudiced against strangers."

"Very good. Now, as to Bibles. How many Bibles would he eat at a meal?"

"He would eat an entire edition."

"It is hardly succinct enough. Do you mean the ordinary octavo, or the family illustrated?"

"I think he would be indifferent to illustrations; that is, I think he would not value illustrations above simple letter-press."

"No, you do not get my idea. I refer to bulk. The ordinary octavo Bible weighs about two pounds and a half, while the great quarto with the illustrations weighs ten or twelve. How many Doré Bibles would he eat at a meal?"

"If you knew this elephant, you could not ask. He would take what they had."

"Well, put it in dollars and cents, then. We must get at it somehow. The Doré costs a hundred dollars a copy, Russia leather, bevelled."

"He would require about fifty thousand dollars' worth, — say an edition of five hundred copies."

"Now that is more exact. I will put that down. Very well; he likes men and Bibles; so far, so good. What else will he eat? I want particulars."

"He will leave Bibles to eat bricks, he will leave bricks to eat bottles, he will leave bottles to eat clothing, he will leave clothing to eat cats, he will leave cats to eat oysters, he will leave oysters to eat ham, he will leave ham to eat sugar, he will leave sugar to eat pie, he will leave pie to eat potatoes, he will leave potatoes to eat bran, he will leave bran to eat hay, he will leave hay to eat oats, he will leave oats to eat rice, for he was mainly raised on it. There is nothing whatever that he will not eat but European butter, and he would eat that if he could taste it."

"Very good. General quantity at a meal, — say about — "

"Well, anywhere from a quarter to half a ton."

"And he drinks — "

"Everything that is fluid. Milk, water, whiskey, molasses, castor oil, camphene, carbolic acid, — it is no use to go into particulars; whatever fluid occurs to you set it down. He will drink anything that is fluid, except European coffee."

"Very good. As to quantity?"

"Put it down five to fifteen barrels, — his thirst varies; his other appetites do not."

"These things are unusual. They ought to furnish quite good clews toward tracing him."

He touched the bell.

"Alaric, summon Captain Burns."

Burns appeared. Inspector Blunt unfolded the whole matter to him, detail by detail. Then he said in the clear, decisive tones of a man whose plans are clearly defined in his head, and who is accustomed to command, —

"Captain Burns, detail Detectives Jones, Davis, Halsey, Bates, and Hackett to shadow the elephant."

"Yes, sir."

"Detail Detectives Moses, Dakin, Murphy, Rogers, Tupper, Higgins, and Bartholomew to shadow the thieves."

"Yes, sir."

"Place a strong guard — a guard of thirty picked men, with a relief of thirty — over the place from whence the elephant was stolen, to keep strict watch there night and day, and allow none to approach — except reporters — without written authority from me."

"Yes, sir."

"Place detectives in plain clothes in the railway, steamship, and ferry depots, and upon all roadways leading out of Jersey City, with orders to search all suspicious persons."

"Yes, sir."

"Furnish all these men with photograph and accompanying description of the elephant, and instruct them to search all trains and outgoing ferry-boats and other vessels."

"Yes, sir."

"If the elephant should be found, let him be seized, and the information forwarded to me by telegraph."

"Yes, sir."

"Let me be informed at once if any clews should be found, — footprints of the animal, or anything of that kind."

"Yes, sir."

"Get an order commanding the harbor police to patrol the frontages vigilantly."

"Yes, sir."

"Despatch detectives in plain clothes over all the railways, north as far as Canada, west as far as Ohio, south as far as Washington."

"Yes, sir."

"Place experts in all the telegraph offices to listen to all messages; and let them require that all cipher despatches be interpreted to them."

"Yes, sir."

"Let all these things be done with the utmost secrecy, — mind, the most impenetrable secrecy."

"Yes, sir."

"Report to me promptly at the usual hour."

"Yes, sir."

"Go!"

"Yes, sir."

He was gone.

Inspector Blunt was silent and thoughtful a moment, while the fire in his eye cooled down and faded out. Then he turned to me and said in a placid voice, —

"I am not given to boasting, it is not my habit; but — we shall find the elephant."

I shook him warmly by the hand and thanked him; and I *felt* my thanks, too. The more I had seen of the man the more I liked him, and the more I admired him and marvelled over the mysterious wonders of his profession. Then we parted for the night, and I went home with a far happier heart than I had carried with me to his office.

II.

NEXT morning it was all in the newspapers, in the minutest detail. It even had additions, — consisting of Detective This, Detective That, and Detective The Other's "Theory" as to how the robbery was done, who the robbers were, and whither they had flown with their booty. There were eleven of these theories, and they covered all the possibilities; and this single fact shows what independent thinkers detectives are. No two theories were alike, or even much resembled each other, save in one striking particular, and in that one all the eleven theories were absolutely agreed. That was, that although the rear of my building was torn out and the only door remained locked, the elephant had not been removed through the rent, but by some other (undiscovered) outlet. All agreed that the robbers had made that rent only to mislead the detectives. That never would have occurred to me or to any other layman, perhaps, but it had not deceived the detectives for a moment.

Thus, what I had supposed was the only thing that had no mystery about it was in fact the very thing I had gone furthest astray in. The eleven theories all named the supposed robbers, but no two named the same robbers; the total number of suspected persons was thirty-seven. The various newspaper accounts all closed with the most important opinion of all, — that of Chief Inspector Blunt. A portion of this statement read as follows: —

"The chief knows who the two principals are, namely, 'Brick' Duffy and 'Red' McFadden. Ten days before the robbery was achieved he was already aware that it was to be attempted, and had quietly proceeded to shadow these two noted villains; but unfortunately on the night in question their track was lost, and before it could be found again the bird was flown, — that is, the elephant.

"Duffy and McFadden are the boldest scoundrels in the profession; the chief has reasons for believing that they are the men who stole the stove out of the detective headquarters on a bitter night last winter, — in consequence of which the chief and every detective present were in the hands of the physicians before morning, some with frozen feet, others with frozen fingers, ears, and other members."

When I read the first half of that I was more astonished than ever at the wonderful sagacity of this strange man. He not only saw everything in the present with a clear eye, but even the future could not be hidden from him. I was soon at his office, and said I could not help wishing he had had those men arrested, and so prevented the trouble and loss; but his reply was simple and unanswerable: —

"It is not our province to prevent crime, but to punish it. We cannot punish it until it is committed."

I remarked that the secrecy with which we had begun had been marred by the newspapers; not only all our facts but all our plans and purposes had been revealed; even all the suspected persons had been named; these would doubtless disguise themselves now, or go into hiding.

"Let them. They will find that when I am ready for them my hand will descend upon them, in their secret places, as unerringly as the hand of fate. As to the newspapers, we *must* keep in with them. Fame, reputation, constant public mention,—these are the detective's bread and butter. He must publish his facts, else he will be supposed to have none; he must publish his theory, for nothing is so strange or striking as a detective's theory, or brings him so much wondering respect; we must publish our plans, for these the journals insist upon having, and we could not deny them without offending. We must constantly show the public what we are doing, or they will believe we are doing nothing. It is much pleasanter to have a newspaper say, 'Inspector Blunt's ingenious and extraordinary theory is as follows,' than to have it say some harsh thing, or, worse still, some sarcastic one."

"I see the force of what you say. But I noticed that in one part of your remarks in the papers this morning you refused to reveal your opinion upon a certain minor point."

"Yes, we always do that; it has a good effect. Besides, I had not formed any opinion on that point, any way."

I deposited a considerable sum of money with the inspector, to meet current expenses, and sat down to wait for news. We were expecting the telegrams to begin to arrive at any moment now. Meantime I re-read the newspapers and also our descriptive circular, and observed that our $25,000 reward seemed to be offered only to detectives. I said I thought it ought to be offered to anybody who would catch the elephant. The inspector said: —

"It is the detectives who will find the elephant, hence the reward will go to the right place. If other people found the animal, it would only be by watching the detectives and taking advantage of clews and indications stolen from them, and that would entitle the detectives to the reward, after all. The proper office of a reward is to stimulate the men who deliver up their time and their trained sagacities to this sort of work, and not to confer benefits upon chance citizens who stumble upon a capture without having earned the benefits by their own merits and labors."

This was reasonable enough, certainly. Now the telegraphic machine in the corner began to click, and the following despatch was the result: —

FLOWER STATION, N. Y., 7.30 A. M.

Have got a clew. Found a succession of deep tracks across a farm near here. Followed them two miles east without result; think elephant went west. Shall now shadow him in that direction.

DARLEY, *Detective.*

"Darley 's one of the best men on the force," said the inspector. "We shall hear from him again before long."

Telegram No. 2 came: —

BARKER'S, N. J., 7.40 A. M.

Just arrived. Glass factory broken open here during night, and eight hundred bottles taken. Only water in large quantity near here is five miles distant. Shall strike for there. Elephant will be thirsty. Bottles were empty.

BAKER, *Detective.*

"That promises well, too," said the inspector. "I told you the creature's appetites would not be bad clews."

Telegram No. 3: —

TAYLORVILLE, L. I., 8.15 A. M.

A haystack near here disappeared during night. Probably eaten. Have got a clew, and am off.

HUBBARD, *Detective.*

"How he does move around!" said the inspector. "I knew we had a difficult job on hand, but we shall catch him yet."

FLOWER STATION, N. Y., 9 A. M.

Shadowed the tracks three miles westward. Large, deep, and ragged. Have just met a farmer who says they are not elephant tracks. Says they are holes where he dug up saplings for shade-trees when ground was frozen last winter. Give me orders how to proceed.

DARLEY, *Detective.*

"Aha! a confederate of the thieves! The thing grows warm," said the inspector.

He dictated the following telegram to Darley : —

Arrest the man and force him to name his pals. Continue to follow the tracks, — to the Pacific, if necessary.

Chief BLUNT.

Next telegram : —

CONEY POINT, PA., 8.45 A. M.

Gas office broken open here during night and three months' unpaid gas bills taken. Have got a clew and am away.

MURPHY, *Detective.*

" Heavens ! " said the inspector ; " would he eat gas bills ? "

" Through ignorance, — yes ; but they cannot support life. At least, unassisted."

Now came this exciting telegram : —

IRONVILLE, N. Y., 9.30 A. M.

Just arrived. This village in consternation. Elephant passed through here at five this morning. Some say he went east, some say west, some north, some south, — but all say they did not wait to notice particularly. He killed a horse ; have secured a piece of it for a clew. Killed it with his trunk ; from style of blow, think he struck it left-handed. From position in which horse lies, think elephant travelled northward along line of Berkley railway. Has four and a half hours' start, but I move on his track at once.

HAWES, *Detective.*

I uttered exclamations of joy. The inspector was as self-contained as a graven image. He calmly touched his bell.

" Alaric, send Captain Burns here."

Burns appeared.

"How many men are ready for instant orders?"

"Ninety-six, sir."

"Send them north at once. Let them concentrate along the line of the Berkley road north of Ironville."

"Yes, sir."

"Let them conduct their movements with the utmost secrecy. As fast as others are at liberty, hold them for orders."

"Yes, sir."

"Go!"

"Yes, sir."

Presently came another telegram: —

SAGE CORNERS, N. Y., 10.30.

Just arrived. Elephant passed through here at 8.15. All escaped from the town but a policeman. Apparently elephant did not strike at policeman, but at the lamp-post, Got both. I have secured a portion of the policeman as clew.

STUMM, *Detective.*

"So the elephant has turned westward," said the inspector. "However, he will not escape, for my men are scattered all over that region."

The next telegram said: —

GLOVER'S, 11.15.

Just arrived. Village deserted, except sick and aged. Elephant passed through three quarters of an hour ago. The anti-temperance mass meeting was in session; he put his trunk in at a window and washed it out with water from cistern. Some swallowed it — since dead; several drowned. Detectives Cross and O'Shaughnessy were passing through town, but going south, — so missed elephant. Whole region for many miles around in terror, — people flying from their homes. Wherever they turn they meet elephant, and many are killed.

BRANT, *Detective.*

I could have shed tears, this havoc so distressed me. But the inspector only said, —

"You see, — we are closing in on him. He feels our presence; he has turned eastward again."

Yet further troublous news was in store for us. The telegraph brought this: —

HOGANPORT, 12.19.

Just arrived. Elephant passed through half an hour ago, creating wildest fright and excitement. Elephant raged around streets; two plumbers going by, killed one — other escaped. Regret general.

O'FLAHERTY, *Detective.*

"Now he is right in the midst of my men," said the inspector. "Nothing can save him."

A succession of telegrams came from detectives who were scattered through New Jersey and Pennsylvania, and who were following clews consisting of ravaged barns, factories, and Sunday school libraries, with high hopes, — hopes amounting to certainties, indeed. The inspector said, —

"I wish I could communicate with them and order them north, but that is impossible. A detective only visits a telegraph office to send his report; then he is off again, and you don't know where to put your hand on him."

Now came this despatch: —

BRIDGEPORT, CT., 12.15.

Barnum offers rate of $4,000 a year for exclusive privilege of using elephant as travelling advertising medium from now till detectives find him. Wants to paste circus-posters on him. Desires immediate answer.

BOGGS, *Detective.*

"That is perfectly absurd!" I exclaimed.

"Of course it is," said the inspector. "Evidently Mr. Barnum, who thinks he is so sharp, does not know me, — but I know him."

Then he dictated this answer to the despatch: —

Mr. Barnum's offer declined. Make it $7,000 or nothing.

Chief BLUNT.

"There. We shall not have to wait long for an answer. Mr. Barnum is not at home; he is in the telegraph office, — it is his way when he has business on hand. Inside of three — "

DONE. — P. T. BARNUM.

So interrupted the clicking telegraphic instrument. Before I could make a comment upon this extraordinary episode, the following despatch carried my thoughts into another and very distressing channel: —

BOLIVIA, N. Y., 12.50.

Elephant arrived here from the south and passed through toward the forest at 11.50, dispersing a funeral on the way, and diminishing the mourners by two. Citizens fired some small cannon-balls into him, and then fled. Detective Burke and I arrived ten minutes later, from the north, but mistook some excavations for footprints, and so lost a good deal of time; but at last we struck the right trail and followed it to the woods. We then got down on our hands and knees and continued to keep a sharp eye on the track, and so shadowed it into the brush. Burke was in advance. Unfortunately the animal had stopped to rest; therefore, Burke having his head down, intent upon the track, butted up against the elephant's hind legs before he was aware of his vicinity. Burke instantly rose to his feet, seized the tail, and exclaimed joyfully, "I claim the re——" but got no further, for a single blow of the huge trunk laid the brave fellow's fragments low in death. I fled rearward, and the elephant turned and shadowed me to the edge of the wood, making tremendous speed, and I should inevitably have been lost, but that the remains of the funeral providentially intervened again and diverted his attention. I have just learned that nothing of that funeral is now left; but this is no loss, for there is an abundance of material for another. Meantime, the elephant has disappeared again.

MULROONEY, *Detective.*

We heard no news except from the diligent and confident detectives scattered about New Jersey, Pennsylvania, Delaware, and Virginia, — who were all following fresh and encouraging clews, — until shortly after 2 P.M., when this telegram came: —

BAXTER CENTRE, 2.15.

Elephant been here, plastered over with circus-bills, and broke up a revival, striking down and damaging many who were on the point of entering upon a better life. Citizens penned him up, and established a guard. When Detective Brown and I arrived, some time after, we entered enclosure and proceeded to identify elephant by photograph and description. All marks tallied exactly except one, which we could not see, — the boil-scar under armpit. To make sure, Brown crept under to look, and was immediately brained,

—that is, head crushed and destroyed, though nothing issued from debris. All fled; so did elephant, striking right and left with much effect. Has escaped, but left bold blood-track from cannon-wounds. Rediscovery certain. He broke southward, through a dense forest.

BRENT, *Detective.*

That was the last telegram. At nightfall a fog shut down which was so dense that objects but three feet away could not be discerned. This lasted all night. The ferry-boats and even the omnibuses had to stop running.

III.

NEXT morning the papers were as full of detective theories as before; they had all our tragic facts in detail also, and a great many more which they had received from their telegraphic correspondents. Column after column was occupied, a third of its way down, with glaring head-lines, which it made my heart sick to read. Their general tone was like this:—

"THE WHITE ELEPHANT AT LARGE! HE MOVES UPON HIS FATAL MARCH! WHOLE VILLAGES DESERTED BY THEIR FRIGHT-STRICKEN OCCUPANTS! PALE TERROR GOES BEFORE HIM, DEATH AND DEVASTATION FOLLOW AFTER! AFTER THESE, THE DETECTIVES. BARNS DESTROYED, FACTORIES GUTTED, HARVESTS DEVOURED, PUBLIC ASSEMBLAGES DISPERSED, ACCOMPANIED BY SCENES OF CARNAGE IMPOSSIBLE TO DESCRIBE! THEORIES OF THIRTY-FOUR OF THE MOST DISTINGUISHED DETECTIVES ON THE FORCE! THEORY OF CHIEF BLUNT!"

"There!" said Inspector Blunt, almost betrayed into excitement, "this is magnificent! This is the greatest windfall that any detective organization ever had. The fame of it will travel to the ends of the earth, and endure to the end of time, and my name with it."

But there was no joy for me. I felt as if I had committed all those red crimes, and that the elephant was only my irresponsible agent. And how the list had grown! In one place he had "interfered with an election and killed five repeaters." He had followed this act with the destruction of two poor fellows, named O'Donohue and McFlannigan, who had "found a refuge in the home of the oppressed of all lands only the day before, and were in the act of exercising for the first time the noble right of American citizens at the polls, when stricken down by the relentless hand of the Scourge of Siam." In another, he had "found a crazy sensation-preacher preparing his next season's heroic attacks on the dance, the theatre, and other things which can't strike back, and had stepped on him." And in still another place he had "killed a lightning-rod agent." And so the list went on, growing redder and redder, and more and more heart-breaking. Sixty persons had been killed, and two hundred and forty wounded. All the accounts bore just testimony to the activity and devotion of the detectives, and all closed with the remark that "three hundred thousand citizens and four detectives saw the dread creature, and two of the latter he destroyed."

I dreaded to hear the telegraphic instrument begin

to click again. By and by the messages began to pour in, but I was happily disappointed in their nature. It was soon apparent that all trace of the elephant was lost. The fog had enabled him to search out a good hiding-place unobserved. Telegrams from the most absurdly distant points reported that a dim vast mass had been glimpsed there through the fog at such and such an hour, and was "undoubtedly the elephant." This dim vast mass had been glimpsed in New Haven, in New Jersey, in Pennsylvania, in interior New York, in Brooklyn, and even in the city of New York itself! But in all cases the dim vast mass had vanished quickly and left no trace. Every detective of the large force scattered over this huge extent of country sent his hourly report, and each and every one of them had a clew, and was shadowing something, and was hot upon the heels of it.

But the day passed without other result.

The next day the same.

The next just the same.

The newspaper reports began to grow monotonous with facts that amounted to nothing, clews which led to nothing, and theories which had nearly exhausted the elements which surprise and delight and dazzle.

By advice of the inspector I doubled the reward.

Four more dull days followed. Then came a bitter blow to the poor, hard-working detectives, — the journalists declined to print their theories, and coldly said, "Give us a rest."

Two weeks after the elephant's disappearance I raised the reward to $75,000 by the inspector's advice.

It was a great sum, but I felt that I would rather sacrifice my whole private fortune than lose my credit with my government. Now that the detectives were in adversity, the newspapers turned upon them, and began to fling the most stinging sarcasms at them. This gave the minstrels an idea, and they dressed themselves as detectives and hunted the elephant on the stage in the most extravagant way. The caricaturists made pictures of detectives scanning the country with spy-glasses, while the elephant, at their backs, stole apples out of their pockets. And they made all sorts of ridiculous pictures of the detective badge, — you have seen that badge printed in gold on the back of detective novels, no doubt, — it is a wide-staring eye, with the legend, "WE NEVER SLEEP." When detectives called for a drink, the would-be facetious bar-keeper resurrected an obsolete form of expression and said, "Will you have an eye-opener?" All the air was thick with sarcasms.

But there was one man who moved calm, untouched, unaffected, through it all. It was that heart of oak, the Chief Inspector. His brave eye never drooped, his serene confidence never wavered. He always said, —

"Let them rail on; he laughs best who laughs last."

My admiration for the man grew into a species of worship. I was at his side always. His office had become an unpleasant place to me, and now became daily more and more so. Yet if he could endure it I meant to do so also; at least, as long as I could. So I came regularly, and stayed, — the only outsider who seemed to be capable of it. Everybody wondered how I could ; and often it seemed to me that I must desert, but at such times I looked into that calm and apparently unconscious face, and held my ground.

About three weeks after the elephant's disappearance I was about to say, one morning, that I should *have* to strike my colors and retire, when the great detective arrested the thought by proposing one more superb and masterly move.

This was to compromise with the robbers. The fertility of this man's invention exceeded anything I have ever seen, and I have had a wide intercourse with the world's finest minds. He said he was confident he could compromise for $100,000 and recover the elephant. I said I believed I could scrape the amount together, but what would become of the poor detectives who had worked so faithfully? He said, —

"In compromises they always get half."

This removed my only objection. So the inspector wrote two notes, in this form : —

DEAR MADAM, — Your husband can make a large sum of money (and be entirely protected from the law) by making an immediate appointment with me.

Chief BLUNT.

He sent one of these by his confidential messenger to the "reputed wife" of Brick Duffy, and the other to the reputed wife of Red McFadden.

Within the hour these offensive answers came : —

YE OWLD FOOL : brick McDuffys bin ded 2 yere.

BRIDGET MAHONEY.

CHIEF BAT, — Red McFadden is hung and in heving 18 month. Any Ass but a detective knose that.

MARY O'HOOLIGAN.

"I had long suspected these facts," said the inspector; "this testimony proves the unerring accuracy of my instinct."

The moment one resource failed him he was ready with another. He immediately wrote an advertisement for the morning papers, and I kept a copy of it: —

A. — xwblv. 242 N. Tjnd — fz328wmlg. Ozpo,—; 2 m! ogw. Mum.

He said that if the thief was alive this would bring him to the usual rendezvous. He further explained that the usual rendezvous was a place where all business affairs between detectives and criminals were conducted. This meeting would take place at twelve the next night.

We could do nothing till then, and I lost no time in getting out of the office, and was grateful indeed for the privilege.

At 11 the next night I brought $100,000 in bank-notes and put them into the chief's hands, and shortly afterward he took his leave, with the brave old undimmed confidence in his eye. An almost intolerable hour dragged to a close; then I heard his welcome tread, and rose gasping and tottered to meet him. How his fine eyes flamed with triumph! He said, —

"We've compromised! The jokers will sing a different tune to-morrow! Follow me!"

He took a lighted candle and strode down into the vast vaulted basement where sixty detectives always slept, and where a score were now playing cards to while the time. I followed close after him. He walked swiftly down to the dim remote end of the place, and just as I succumbed to the pangs of suffocation and was swooning away he stumbled and fell over the outlying members of a mighty object, and I heard him exclaim as he went down, —

"Our noble profession is vindicated. Here is your elephant!"

I was carried to the office above and restored with carbolic acid. The whole detective force swarmed in, and such another season of triumphant rejoicing ensued as I had never witnessed before. The reporters were called, baskets of champagne were opened, toasts were drunk, the handshakings and congratulations were continuous and enthusiastic. Naturally the chief was the hero of the hour, and his happiness was so complete and had been so patiently and worthily and bravely won that it made me happy to see it, though I stood there a homeless beggar, my priceless charge dead, and my position in my country's service lost to me through what would always seem my fatally careless execution of a great trust. Many an eloquent eye testified its deep admiration for the chief, and many a detective's voice murmured, "Look at him,— just the king of the profession,—only give him a clew, it 's all he wants, and there ain't anything hid that he can't find." The dividing of the $50,000 made great pleasure; when it was finished the chief made a little speech while he put his share in his pocket, in which he said, "Enjoy it, boys, for you've earned it; and more than that you've earned for the detective profession undying fame."

A telegram arrived, which read: —

MONROE, MICH., 10 P. M.

First time I've struck a telegraph office in over three weeks. Have followed those footprints, horseback, through the woods, a thousand miles to here, and they get stronger and bigger and fresher every day. Don't worry — inside of another week I'll have the elephant. This is dead sure.

DARLEY, *Detective.*

The chief ordered three cheers for "Darley, one of the finest minds on the force," and then commanded that he be telegraphed to come home and receive his share of the reward.

So ended that marvellous episode of the stolen elephant. The newspapers were pleasant with praises once more, the next day, with one contemptible exception. This sheet said, "Great is the detective! He may be a little slow in finding a little thing like a mislaid elephant, — he may hunt him all day and sleep with his rotting carcass all night for three weeks, but he will find him at last — if he can get the man who mislaid him to show him the place!"

Poor Hassan was lost to me forever. The cannon-shots had wounded him fatally, he had crept to that unfriendly place in the fog, and there, surrounded by his enemies and in constant danger of detection, he had wasted away with hunger and suffering till death gave him peace.

The compromise cost me $100,000; my detective expenses were $42,000 more; I never applied for a place again under my government; I am a ruined man and a wanderer in the earth, — but my admiration for that man, whom I believe to be the greatest detective the world has ever produced, remains undimmed to this day, and will so remain unto the end.

SOME RAMBLING NOTES OF AN IDLE EXCURSION.

I.

ALL the journeyings I had ever done had been purely in the way of business. The pleasant May weather suggested a novelty, namely, a trip for pure recreation, the bread-and-butter element left out. The Reverend said he would go, too: a good man, one of the best of men, although a clergyman. By eleven at night we were in New Haven and on board the New York boat. We bought our tickets, and then went wandering around, here and there, in the solid comfort of being free and idle, and of putting distance between ourselves and the mails and telegraphs.

After a while I went to my state-room and undressed, but the night was too enticing for bed. We were moving down the bay now, and it was pleasant to stand at the window and take the cool night-breeze and watch the gliding lights on shore. Presently, two elderly men sat down under that window and began a conversation. Their talk was properly no business of mine, yet I was feeling friendly toward the world and willing to be entertained. I soon gathered that they were brothers, that they were from a small Connecticut village, and that the matter in hand concerned the cemetery. Said one, —

"Now, John, we talked it all over amongst ourselves, and this is what we 've done. You see, everybody was a-movin' from the old buryin' ground, and our folks was most about left to theirselves, as you may say. They was crowded, too, as you know; lot wa' n't big enough in the first place; and last year, when Seth's wife died, we could n't hardly tuck her in. She sort o' overlaid Deacon Shorb's lot, and he soured on her, so to speak, and on the rest of us, too. So we talked it over, and I was for a lay-out in the new simitery on the hill. They wa' n't unwilling, if it was cheap. Well, the two best and biggest plots was No. 8 and No. 9, — both of a size; nice comfortable room for twenty-six, — twenty-six full-growns, that is; but you reckon in children and other shorts, and strike an everage, and I should say you might lay in thirty, or may be thirty-two or three, pretty genteel, — no crowdin' to signify."

"That 's a plenty, William. Which one did you buy?"

"Well, I 'm a-comin' to that, John. You see, No. 8 was thirteen dollars, No. 9 fourteen —"

"I see. So 's 't you took No. 8."

"You wait. I took No. 9. And I 'll tell you for why. In the first place, Deacon Shorb wanted it. Well, after the way he 'd gone on about Seth's wife overlappin' his prem'ses, I 'd 'a' beat him out of that No. 9 if I 'd 'a' had to stand two dollars extra, let alone one. That 's the way I felt about it. Says

I, what 's a dollar, any way? Life 's on'y a pilgrimage, says I; we ain't here for good, and we can't take it with us, says I. So I just dumped it down, knowin' the Lord don't suffer a good deed to go for nothin', and cal'latin' to take it out o' somebody in the course o' trade. Then there was another reason, John. No. 9 's a long way the handiest lot in the simitery, and the likeliest for situation. It lays right on top of a knoll in the dead centre of the buryin' ground; and you can see Millport from there, and Tracy's, and Hopper Mount, and a raft o' farms, and so on. There ain't no better outlook from a buryin' plot in the State. Si Higgins says so, and I reckon he ought to know. Well, and that ain't all. 'Course Shorb had to take No. 8; wa' n't no help for 't. Now, No. 8 jines on to No. 9, but it 's on the slope of the hill, and every time it rains it 'll soak right down on to the Shorbs. Si Higgins says 't when the deacon's time comes, he better take out fire and marine insurance both on his remains."

Here there was the sound of a low, placid, duplicate chuckle of appreciation and satisfaction.

"Now, John, here 's a little rough draught of the ground, that I 've made on a piece of paper. Up here in the left-hand corner we 've bunched the departed; took them from the old grave-yard and stowed them one along side o' t' other, on a first-come-first-served plan, no partialities, with Gran'ther Jones for a starter, on'y because it happened so, and windin' up indiscriminate with Seth's twins. A little crowded towards the end of the lay-out, may be, but we reckoned 't wa' n't best to scatter the twins. Well, next comes the livin'. Here, where it 's marked A, we 're goin' to put Mariar and her family, when they 're called; B, that 's for Brother Hosea and his'n; C, Calvin and tribe. What 's left is these two lots here, — just the gem of the whole patch for general style and outlook; they 're for me and my folks, and you and yourn. Which of them would you ruther be buried in?"

"I swan you 've took me mighty unexpected, William! It sort of started the shivers. Fact is, I was thinkin' so busy about makin' things comfortable for the others, I had n't thought about being buried myself."

"Life 's on'y a fleetin' show, John, as the sayin' is. We 've all got to go, sooner or later. To go with a clean record 's the main thing. Fact is, it 's the on'y thing worth strivin' for, John."

"Yes, that 's so, William, that 's so; there ain't no getting around it. Which of these lots would you recommend?"

"Well, it depends, John. Are you particular about outlook?"

"I don't say I am, William; I don't say I ain't. Reely, I don't know. But mainly, I reckon, I 'd set store by a south exposure."

"That 's easy fixed, John. They 're both south exposure. They take the sun, and the Shorbs get the shade."

"How about sile, William?"

"D 's a sandy sile, E 's mostly loom."

"You may gimme E, then, William; a sandy sile caves in, more or less, and costs for repairs."

"All right; set your name down here, John, under E. Now, if you don't mind payin' me your share of the fourteen dollars, John, while we're on the business, everything's fixed."

After some higgling and sharp bargaining the money was paid, and John bade his brother good-night and took his leave. There was silence for some moments; then a soft chuckle welled up from the lonely William, and he muttered: "I declare for 't, if I have n't made a mistake! It's D that's mostly loom, not E. And John's booked for a sandy sile, after all."

There was another soft chuckle, and William departed to his rest, also.

The next day, in New York, was a hot one. Still we managed to get more or less entertainment out of it. Toward the middle of the afternoon we arrived on board the stanch steamship Bermuda, with bag and baggage, and hunted for a shady place. It was blazing summer weather, until we were half way down the harbor. Then I buttoned my coat closely; half an hour later I put on a spring overcoat and buttoned that. As we passed the light-ship I added an ulster and tied a handkerchief around the collar to hold it snug to my neck. So rapidly had the summer gone and winter come again!

By nightfall we were far out at sea, with no land in sight. No telegrams could come here, no letters, no news. This was an uplifting thought. It was still more uplifting to reflect that the millions of harassed people on shore behind us were suffering just as usual.

The next day brought us into the midst of the Atlantic solitudes, — out of smoke-colored soundings into fathomless deep blue; no ships visible anywhere over the wide ocean; no company but Mother Cary's chickens wheeling, darting, skimming the waves in the sun. There were some sea-faring men among the passengers, and conversation drifted into matters concerning ships and sailors. One said that "true as the needle to the pole" was a bad figure, since the needle seldom pointed to the pole. He said a ship's compass was not faithful to any particular point, but was the most fickle and treacherous of the servants of man. It was forever changing. It changed every day in the year; consequently the amount of the daily variation had to be ciphered out and allowance made for it, else the mariner would go utterly astray. Another said there was a vast fortune waiting for the genius who should invent a compass that would not be affected by the local influences of an iron ship. He said there was only one creature more fickle than a wooden ship's compass, and that was the compass of an iron ship. Then came reference to the well-known fact that an experienced mariner can look at the compass of a new iron vessel, thousands of miles from her birthplace, and tell which way her head was pointing when she was in process of building.

Now an ancient whale-ship master fell to talking about the sort of crews they used to have in his early days. Said he, —

"Sometimes we'd have a batch of college students. Queer lot. Ignorant? Why, they didn't know the cat-heads from the main brace. But if you took them for fools you 'd get bit, sure. They 'd learn more in a month than another man would in a year. We had one, once, in the Mary Ann, that came aboard with gold spectacles on. And besides, he was rigged out from main truck to keelson in the nobbiest clothes that ever saw a fo'castle. He had a chest full, too: cloaks, and broadcloth coats, and velvet vests: everything swell, you know; and did n't the salt water fix them out for him? I guess not! Well, going to sea, the mate told him to go aloft and help shake out the fore-to'gallants'l. Up he shins to the foretop, with his spectacles on, and in a minute down he comes again, looking insulted. Says the mate, 'What did you come down for?' Says the chap, 'P'r'aps you did n't notice that there ain't any ladders above there.' You see we had n't any shrouds above the foretop. The men bursted out in a laugh such as I guess you never heard the like of. Next night, which was dark and rainy, the mate ordered this chap to go aloft about something, and I 'm dummed if he did n't start up with an umbrella and a lantern! But no matter; he made a mighty good sailor before the voyage was done, and we had to hunt up something else to laugh at. Years afterwards, when I had forgot all about him, I comes into Boston, mate of a ship, and was loafing around town with the second mate, and it so happened that we stepped into the Revere House, thinking may be we would chance the salt-horse in that big dining-room for a flyer, as the boys say. Some fellows were talking just at our elbow, and one says, 'Yonder 's the new governor of Massachusetts, — at that table over there, with the ladies.' We took a good look, my mate and I, for we had n't either of us ever seen a governor before. I looked and looked at that face, and then all of a sudden it popped on me! But I did n't give any sign. Says I, 'Mate, I 've a notion to go over and shake hands with him.' Says he, 'I think I see you doing it, Tom.' Says I, 'Mate, I 'm a-going to do it.' Says he, 'Oh, yes, I guess so! May be you don't want to bet you will, Tom?' Says I, 'I don't mind going a V on it, mate.' Says he, 'Put it up.' 'Up she goes,' says I, planking the cash. This surprised him. But he covered it, and says, pretty sarcastic, 'Had n't you better take your grub with the governor and the ladies, Tom?' Says I, 'Upon second thoughts, I will.' Says he, 'Well, Tom, you *are* a dum fool.' Says I, 'May be I am, may be I ain't; but the main question is, do you want to risk two and a half that I won't do it?' 'Make it a V,' says he. 'Done,' says I. I started, him a-giggling and slapping his hand on his thigh, he felt so good. I went over there and leaned my knuckles on the table a minute and looked the governor in the face, and says I, 'Mister Gardner, don't you know me?' He stared, and I stared, and he stared. Then all of a sudden he sings out, 'Tom Bowling, by the holy poker! Ladies, it 's old Tom Bowling, that you 've heard me talk about, — shipmate of mine in the Mary Ann.' He rose up and shook hands with me ever so hearty — I sort of glanced around and took a realizing sense of my mate's saucer eyes, — and then says the governor, 'Plant yourself,

Tom, plant yourself; you can't cat your anchor again till you've had a feed with me and the ladies!' I planted myself alongside the governor, and canted my eye around towards my mate. Well, sir, his dead-lights were bugged out like tompions; and his mouth stood that wide open that you could have laid a ham in it without him noticing it."

There was great applause at the conclusion of the old captain's story; then, after a moment's silence, a grave, pale young man said, —

"Had you ever met the governor before?"

The old captain looked steadily at this inquirer a while, and then got up and walked aft without making any reply. One passenger after another stole a furtive glance at the inquirer, but failed to make him out, and so gave him up. It took some little work to get the talk-machinery to running smoothly again after this derangement; but at length a conversation sprang up about that important and jealously guarded instrument, a ship's time-keeper, its exceeding delicate accuracy, and the wreck and destruction that have sometimes resulted from its varying a few seemingly trifling moments from the true time; then, in due course, my comrade, the Reverend, got off on a yarn, with a fair wind and everything drawing. It was a true story, too, — about Captain Rounceville's shipwreck, — true in every detail. It was to this effect: —

Captain Rounceville's vessel was lost in mid-Atlantic, and likewise his wife and his two little children. Captain Rounceville and seven seamen escaped with life, but with little else. A small, rudely constructed raft was to be their home for eight days. They had neither provisions nor water. They had scarcely any clothing; no one had a coat but the captain. This coat was changing hands all the time, for the weather was very cold. Whenever a man became exhausted with the cold, they put the coat on him and laid him down between two shipmates until the garment and their bodies had warmed life into him again. Among the sailors was a Portuguese who knew no English. He seemed to have no thought of his own calamity, but was concerned only about the captain's bitter loss of wife and children. By day, he would look his dumb compassion in the captain's face; and by night, in the darkness and the driving spray and rain, he would seek out the captain and try to comfort him with caressing pats on the shoulder. One day, when hunger and thirst were making their sure inroads upon the men's strength and spirits, a floating barrel was seen at a distance. It seemed a great find, for doubtless it contained food of some sort. A brave fellow swam to it, and after long and exhausting effort got it to the raft. It was eagerly opened. It was a barrel of magnesia! On the fifth day an onion was spied. A sailor swam off and got it. Although perishing with hunger, he brought it in its integrity and put it into the captain's hand. The history of the sea teaches that among starving, shipwrecked men selfishness is rare, and a wonder-compelling magnanimity the rule. The onion was equally divided into eight parts, and eaten with deep thanksgivings. On the eighth day a distant ship was sighted. Attempts were made to hoist an oar, with

Captain Rounceville's coat on it for a signal. There were many failures, for the men were but skeletons now, and strengthless. At last success was achieved, but the signal brought no help. The ship faded out of sight and left despair behind her. By and by another ship appeared, and passed so near that the castaways, every eye eloquent with gratitude, made ready to welcome the boat that would be sent to save them. But this ship also drove on, and left these men staring their unutterable surprise and dismay into each other's ashen faces. Late in the day, still another ship came up out of the distance, but the men noted with a pang that her course was one which would not bring her nearer. Their remnant of life was nearly spent; their lips and tongues were swollen, parched, cracked with eight days' thirst; their bodies starved; and here was their last chance gliding relentlessly from them; they would not be alive when the next sun rose. For a day or two past the men had lost their voices, but now Captain Rounceville whispered, "Let us pray." The Portuguese patted him on the shoulder in sign of deep approval. All knelt at the base of the oar that was waving the signal-coat aloft, and bowed their heads. The sea was tossing; the sun rested, a red, rayless disk, on the sea-line in the west. When the men presently raised their heads they would have roared a hallelujah if they had had a voice: the ship's sails lay wrinkled and flapping against her masts, she was going about! Here was rescue at last, and in the very last instant of time that was left for it. No, not rescue yet, — only the imminent prospect of it. The red disk sank under the sea and darkness blotted out the ship. By and by came a pleasant sound, — oars moving in a boat's rowlocks. Nearer it came, and nearer, — within thirty steps, but nothing visible. Then a deep voice: "Hol-*lo!*" The castaways could not answer; their swollen tongues refused voice. The boat skirted round and round the raft, started away — the agony of it! — returned, rested the oars, close at hand, listening, no doubt. The deep voice again: "Hol-*lo!* Where are ye, shipmates?" Captain Rounceville whispered to his men, saying: "Whisper your best, boys! now — all at once!" So they sent out an eightfold whisper in hoarse concert: "Here!" There was life in it if it succeeded; death if it failed. After that supreme moment Captain Rounceville was conscious of nothing until he came to himself on board the saving ship. Said the Reverend, concluding, —

"There was one little moment of time in which that raft could be visible from that ship, and only one. If that one little fleeting moment had passed unfruitful, those men's doom was sealed. As close as that does God shave events foreordained from the beginning of the world. When the sun reached the water's edge that day, the captain of that ship was sitting on deck reading his prayer-book. The book fell; he stooped to pick it up, and happened to glance at the sun. In that instant that far-off raft appeared for a second against the red disk, its needle-like oar and diminutive signal cut sharp and black against the bright surface, and in the next instant was thrust away into the dusk again. But that ship, that captain, and that pregnant instant had had their work appointed for them in the dawn of time and could not fail of the performance. The chronometer of God never errs!"

There was deep, thoughtful silence for some moments. Then the grave, pale young man said, —

"What is the chronometer of God?"

II.

At dinner, six o'clock, the same people assembled whom we had talked with on deck and seen at luncheon and breakfast this second day out, and at dinner the evening before. That is to say, three journeying ship-masters, a Boston merchant, and a returning Bermudian who had been absent from his Bermuda thirteen years; these sat on the starboard side. On the port side sat the Reverend in the seat of honor; the pale young man next to him; I next; next to me an aged Bermudian, returning to his sunny islands after an absence of twenty-seven years. Of course our captain was at the head of the table, the purser at the foot of it. A small company, but small companies are pleasantest.

No racks upon the table; the sky cloudless, the sun brilliant, the blue sea scarcely ruffled: then what had become of the four married couples, the three bachelors, and the active and obliging doctor from the rural districts of Pennsylvania? — for all these were on deck when we sailed down New York harbor. This is the explanation. I quote from my note-book: —

Thursday, 3.30 P. M. Under way, passing the Battery. The large party, of four married couples, three bachelors, and a cheery, exhilarating doctor from the wilds of Pennsylvania, are evidently travelling together. All but the doctor grouped in camp-chairs on deck.

Passing principal fort. The doctor is one of those people who has an infallible preventive of sea-sickness; is flitting from friend to friend administering it and saying, "Don't you be afraid; I *know* this medicine; absolutely infallible; prepared under my own supervision." Takes a dose himself, intrepidly.

4.15 P. M. Two of those ladies have struck their colors, notwithstanding the "infallible." They have gone below. The other two begin to show distress.

5 P. M. Exit one husband and one bachelor. These still had their infallible in cargo when they started, but arrived at the companion-way without it.

5.10. Lady No. 3, two bachelors, and one married man have gone below with their own opinion of the infallible.

5.20. Passing Quarantine Hulk. The infallible has done the business for all the party except the Scotchman's wife and the author of that formidable remedy.

Nearing the Light-Ship. Exit the Scotchman's wife, head drooped on stewardess's shoulder.

Entering the open sea. Exit doctor!

The rout seems permanent; hence the smallness of the company at table since the voyage began. Our captain is a grave, handsome Hercules of thirty-five, with a brown hand of such majestic size that one cannot eat for admiring it and wondering if a single kid or calf could furnish material for gloving it.

Conversation not general; drones along between couples. One catches a sentence here and there. Like this, from Bermudian of thirteen years' absence: "It is the nature of women to ask trivial, irrelevant, and pursuing questions, — questions that pursue you from a beginning in nothing to a run-to-cover in nowhere." Reply of Bermudian of twenty-seven years' absence: "Yes; and to think they have logical, analytical minds and argumentative ability. You see 'em begin to whet up whenever they smell argument in the air." Plainly these be philosophers.

Twice since we left port our engines have stopped for a couple of minutes at a time. Now they stop again. Says the pale young man, meditatively, "There! — that engineer is sitting down to rest again."

Grave stare from the captain, whose mighty jaws cease to work, and whose harpooned potato stops in mid-air on its way to his open, paralyzed mouth. Presently says he in measured tones, "Is it your idea that the engineer of this ship propels her by a crank turned by his own hands?"

The pale young man studies over this a moment, then lifts up his guileless eyes, and says, "Don't he?"

Thus gently falls the death-blow to further conversation, and the dinner drags to its close in a reflective silence, disturbed by no sounds but the murmurous wash of the sea and the subdued clash of teeth.

After a smoke and a promenade on deck, where is no motion to discompose our steps, we think of a game of whist. We ask the brisk and capable stewardess from Ireland if there are any cards in the ship.

"Bless your soul, dear, indeed there is. Not a whole pack, true for ye, but not enough missing to signify."

However, I happened by accident to bethink me of a new pack in a morocco case, in my trunk, which I had placed there by mistake, thinking it to be a flask of something. So a party of us conquered the tedium of the evening with a few games and were ready for bed at six bells, mariner's time, the signal for putting out the lights.

There was much chat in the smoking-cabin on the upper deck after luncheon to-day, mostly whaler yarns from those old sea-captains. Captain Tom Bowling was garrulous. He had that garrulous attention to minor detail which is born of secluded farm life or life at sea on long voyages, where there is little to do and time no object. He would sail along till he was right in the most exciting part of a yarn, and then say, "Well, as I was saying, the rudder was fouled, ship driving before the gale, head-on, straight for the iceberg, all hands holding their breath, turned to stone, top-hamper giving way, sails blown to ribbons, first one stick going, then another, boom! smash! crash! duck your head and stand from under! when up comes Johnny Rogers, capstan bar in hand, eyes a-blazing, hair a-flying . . . no, 't wa' n't Johnny Rogers . . . lemme see . . . seems to me Johnny Rogers wa' n't along that voyage; he was along *one* voyage, I know that mighty well, but somehow it seems to me that he signed the articles for this voyage, but — but — whether he come along or not, or got left, or something happened — "

And so on and so on, till the excitement all cooled down and nobody cared whether the ship struck the iceberg or not.

In the course of his talk he rambled into a criticism upon New England degrees of merit in ship-building. Said he, "You get a vessel built away down Maine-way; Bath, for instance; what 's the result? First thing you do, you want to heave her down for repairs, — *that 's* the result! Well, sir, she hain't been hove down a week till you can heave a dog through her seams. You send that vessel to sea, and what 's the result? She wets her oakum the first trip! Leave it to any man if 't ain't so. Well, you let *our* folks build you a vessel — down New Bedford-way. What 's the result? Well, sir, you might take that ship and heave her down, and keep her hove down six months, and she 'll never shed a tear!"

Everybody, landsmen and all, recognized the descriptive neatness of that figure, and applauded, which greatly pleased the old man. A moment later, the meek eyes of the pale young fellow heretofore mentioned came up slowly, rested upon the old man's face a moment, and the meek mouth began to open.

"Shet your head!" shouted the old mariner.

It was a rather startling surprise to everybody, but it was effective in the matter of its purpose. So the conversation flowed on instead of perishing.

There was some talk about the perils of the sea, and a landsman delivered himself of the customary nonsense about the poor mariner wandering in far oceans, tempest-tossed, pursued by dangers, every storm-blast and thunderbolt in the home skies moving the friends by snug firesides to compassion for that poor mariner, and prayers for his succor. Captain Bowling put up with this for a while, and then burst out with a new view of the matter.

"Come, belay there! I have read this kind of rot all my life in poetry and tales and such like rubbage. Pity for the poor mariner! sympathy for the poor mariner! All right enough, but not in the way the poetry puts it. Pity for the mariner's wife! all right again, but not in the way the poetry puts it. Look-a-here! whose life 's the safest in the whole world? The poor mariner's. You look at the statistics, you 'll see. So don't you fool away any sympathy on the poor mariner's dangers and privations and sufferings. Leave that to the poetry muffs. Now you look at the other side a minute. Here is Captain Brace, forty years old, been at sea thirty. On his way now to take command of his ship and sail south from Bermuda. Next week he 'll be under way: easy times; comfortable quarters; passengers, sociable company; just enough to do to keep his mind healthy and not tire him; king over his ship, boss of everything and everybody; thirty years' safety to learn him that his profession ain't a dangerous one. Now you look back at his home. His wife 's a feeble woman; she 's a stranger in New York; shut up in blazing hot or freezing cold lodgings, according to the season; don't know anybody hardly; no company but her lonesomeness and her thoughts; husband gone six months at a time. She has borne eight children; five of them she has buried without her husband ever setting eyes on them. She watched them all the long nights till they died, — he comfortable on the sea; she followed them to the grave, she heard the clods fall that broke her heart, — he comfortable on the sea; she mourned at home, weeks and weeks, missing them every day and every hour, — he cheerful at sea, knowing nothing about it. Now look at it a minute, — turn it over in your mind and size it: five children born, she among strangers, and him not by to hearten her; buried, and him not by to comfort her; think of that! Sympathy for the poor mariner's perils is rot; give it to his wife's hard lines, where it belongs! Poetry makes out that all the wife worries about is the dangers her husband's running. She's got substantialer things to worry over, I tell you. Poetry's always pitying the poor mariner on account of his perils at sea; better a blamed sight pity him for the nights he can't sleep for thinking of how he had to leave his wife in her very birth pains, lonesome and friendless, in the thick of disease and trouble and death. If there's one thing that can make me madder than another, it's this sappy, damned maritime poetry!"

Captain Brace was a patient, gentle, seldom-speaking man, with a pathetic something in his bronzed face that had been a mystery up to this time, but stood interpreted now, since we had heard his story. He had voyaged eighteen times to the Mediterranean, seven times to India, once to the arctic pole in a discovery-ship, and "between times" had visited all the remote seas and ocean corners of the globe. But he said that twelve years ago, on account of his family, he "settled down," and ever since then had ceased to roam. And what do you suppose was this simple-hearted, life-long wanderer's idea of settling down and ceasing to roam? Why, the making of two five-month voyages a year between Surinam and Boston for sugar and molasses!

Among other talk, to-day, it came out that whale-ships carry no doctor. The captain adds the doctorship to his own duties. He not only gives medicines, but sets broken limbs after notions of his own, or saws them off and sears the stump when amputation seems best. The captain is provided with a medicine-chest, with the medicines numbered instead of named. A book of directions goes with this. It describes diseases and symptoms, and says, "Give a teaspoonful of No. 9 once an hour," or "Give ten grains of No. 12 every half hour," etc. One of our sea-captains came across a skipper in the North Pacific who was in a state of great surprise and perplexity. Said he, —

"There 's something rotten about this medicine-chest business. One of my men was sick, — nothing much the matter. I looked in the book: it said, give him a teaspoonful of No. 15. I went to the medicine-chest, and I see I was out of No. 15. I judged I 'd got to get up a combination somehow that would fill the bill; so I hove into the fellow half a teaspoonful of No. 8 and half a teaspoonful of No. 7, and I 'll be hanged if it did n't kill him in fifteen minutes! There 's something about this medicine-chest system that 's too many for me!"

There was a good deal of pleasant gossip about old Captain "Hurricane" Jones, of the Pacific Ocean, — peace to his ashes! Two or three of us present had known him; I, particularly well, for I had made four sea-voyages with him. He was a very remarkable man. He was born in a ship; he picked up what little education he had among his shipmates; he began life in the forecastle, and climbed grade by grade to the captaincy. More than fifty years of his sixty-five were spent at sea. He had sailed all oceans, seen all lands, and borrowed a tint from all climates. When a man has been fifty years at sea he necessarily knows nothing of men, nothing of the world but its surface, nothing of the world's thought, nothing of the world's learning but its A B C, and that blurred and distorted by the unfocused lenses of an untrained mind. Such a man is only a gray and bearded child. That is what old Hurricane Jones was, — simply an innocent, lovable old infant. When his spirit was in repose he was as sweet and gentle as a girl; when his wrath was up he was a hurricane that made his nickname seem tamely descriptive. He was formidable in a fight, for he was of powerful build and dauntless courage. He was frescoed from head to heel with pictures and mottoes tattooed in red and blue India ink. I was with him one voyage when he got his last vacant space tattooed; this vacant space was around his left ankle. During three days he stumped about the ship with his ankle bare and swollen, and this legend gleaming red and angry out from a clouding of India ink : "Virtue is its own R'd." (There was a lack of room.) He was deeply and sincerely pious, and swore like a fish-woman. He considered swearing blameless, because sailors would not understand an order unillumined by it. He was a profound Biblical scholar, — that is, he thought he was. He believed everything in the Bible, but he had his own methods of arriving at his beliefs. He was of the "advanced" school of thinkers, and applied natural laws to the interpretation of all miracles, somewhat on the plan of the people who make the six days of creation six geological epochs, and so forth. Without being aware of it, he was a rather severe satire on modern scientific religionists. Such a man as I have been describing is rabidly fond of disquisition and argument; one knows that without being told it.

One trip the captain had a clergyman on board, but did not know he was a clergyman, since the passenger list did not betray the fact. He took a great liking to this Rev. Mr. Peters, and talked with him a great deal: told him yarns, gave him toothsome scraps of personal history, and wove a glittering streak of profanity through his garrulous fabric that was refreshing to a spirit weary of the dull neutralities of undecorated speech. One day the captain said, "Peters, do you ever read the Bible?"

"Well — yes."

"I judge it ain't often, by the way you say it. Now, you tackle it in dead earnest once, and you'll find it 'll pay. Don't you get discouraged, but hang right on. First, you won't understand it; but by and by things will begin to clear up, and then you would n't lay it down to eat."

"Yes, I have heard that said."

"And it 's so, too. There ain't a book that begins with it. It lays over 'em all, Peters. There 's some pretty tough things in it, — there ain't any getting around that, — but you stick to them and think them out, and when once you get on the inside everything 's plain as day."

"The miracles, too, captain?"

"Yes, sir! the miracles, too. Every one of them. Now, there 's that business with the prophets of Baal; like enough that stumped you?"

"Well, I don't know but — "

"Own up, now; it stumped you. Well, I don't wonder. You had n't had any experience in ravelling such things out, and naturally it was too many for you. Would you like to have me explain that thing to you, and show you how to get at the meat of these matters?"

"Indeed, I would, captain, if you don't mind."

Then the captain proceeded as follows: "I 'll do it with pleasure. First, you see, I read and read, and thought and thought, till I got to understand what sort of people they were in the old Bible times, and then after that it was all clear and easy. Now, this was the way I put it up, concerning Isaac[1] and the prophets of Baal. There was some mighty sharp men

[1] This is the captain's own mistake.

amongst the public characters of that old ancient day, and Isaac was one of them. Isaac had his failings,— plenty of them, too; it ain't for me to apologize for Isaac; he played it on the prophets of Baal, and like enough he was justifiable, considering the odds that was against him. No, all I say is, 't wa' n't any miracle, and that I 'll show you so 's 't you can see it yourself.

"Well, times had been getting rougher and rougher for prophets, — that is, prophets of Isaac's denomination. There was four hundred and fifty prophets of Baal in the community, and only one Presbyterian; that is, if Isaac *was* a Presbyterian, which I reckon he was, but it don't say. Naturally, the prophets of Baal took all the trade. Isaac was pretty low-spirited, I reckon, but he was a good deal of a man, and no doubt he went a-prophesying around, letting on to be doing a land-office business, but 't wa' n't any use; he could n't run any opposition to amount to anything. By and by things got desperate with him; he sets his head to work and thinks it all out, and then what does he do? Why, he begins to throw out hints that the other parties are this and that and t' other,— nothing very definite, may be, but just kind of undermining their reputation in a quiet way. This made talk, of course, and finally got to the king. The king asked Isaac what he meant by his talk. Says Isaac, 'Oh, nothing particular; only, can they pray down fire from heaven on an altar? It ain't much, may be, your majesty, only can they *do* it? That 's the idea.' So the king was a good deal disturbed, and he went to the prophets of Baal, and they said, pretty airy, that if he had an altar ready, *they* were ready; and they intimated he better get it insured, too.

"So next morning all the children of Israel and their parents and the other people gathered themselves together. Well, here was that great crowd of prophets of Baal packed together on one side, and Isaac walking up and down all alone on the other, putting up his job. When time was called, Isaac let on to be comfortable and indifferent; told the other team to take the first innings. So they went at it, the whole four hundred and fifty, praying around the altar, very hopeful, and doing their level best. They prayed an hour, — two hours, — three hours, — and so on, plumb till noon. It wa' n't any use; they had n't took a trick. Of course they felt kind of ashamed before all those people, and well they might. Now, what would a magnanimous man do? Keep still, would n't he? Of course. What did Isaac do? He gravelled the prophets of Baal every way he could think of. Says he, 'You don't speak up loud enough; your god 's asleep, like enough, or may be he 's taking a walk; you want to holler, you know,'—or words to that effect; I don't recollect the exact language. Mind, I don't apologize for Isaac; he had his faults.

"Well, the prophets of Baal prayed along the best they knew how all the afternoon, and never raised a spark. At last, about sundown, they were all tuckered out, and they owned up and quit.

"What does Isaac do, now? He steps up and says to some friends of his, there, 'Pour four barrels of water on the altar!' Everybody was astonished; for the other side had prayed at it dry, you know, and got whitewashed. They poured it on. Says he, 'Heave on four more barrels.' Then he says, 'Heave on four more.' Twelve barrels, you see, altogether. The water ran all over the altar, and all down the sides, and filled up a trench around it that would hold a couple of hogsheads, — 'measures,' it says; I reckon it means about a hogshead. Some of the people were going to put on their things and go, for they allowed he was crazy. They did n't know Isaac. Isaac knelt down and began to pray: he strung along, and strung along, about the heathen in distant lands, and about the sister churches, and about the state and the country at large, and about those that 's in authority in the government, and all the usual programme, you know, till everybody had got tired and gone to thinking about something else, and then, all of a sudden, when nobody was noticing, he outs with a match and rakes it on the under side of his leg, and pff! up the whole thing blazes like a house afire! Twelve barrels of *water? Petroleum*, sir, PETROLEUM! that 's what it was!"

"Petroleum, captain?"

"Yes, sir; the country was full of it. Isaac knew all about that. You read the Bible. Don't you worry about the tough places. They ain't tough when you come to think them out and throw light on them. There ain't a thing in the Bible but what is true; all you want is to go prayerfully to work and cipher out how 't was done."

At eight o'clock on the third morning out from New

York, land was sighted. Away across the sunny waves one saw a faint dark stripe stretched along under the horizon, — or pretended to see it, for the credit of his eyesight. Even the Reverend said he saw it, a thing which was manifestly not so. But I never have seen any one who was morally strong enough to confess that he could not see land when others claimed that they could.

By and by the Bermuda Islands were easily visible. The principal one lay upon the water in the distance, a long, dull-colored body, scalloped with slight hills and valleys. We could not go straight at it, but had to travel all the way around it, sixteen miles from shore, because it is fenced with an invisible coral reef. At last we sighted buoys, bobbing here and there, and then we glided into a narrow channel among them, "raised the reef," and came upon shoaling blue water that soon further shoaled into pale green, with a surface scarcely rippled. Now came the resurrection hour: the berths gave up their dead. Who are these pale spectres in plug hats and silken flounces that file up the companion-way in melancholy procession and step upon the deck? These are they which took the infallible preventive of sea-sickness in New York harbor and then disappeared and were forgotten. Also there came two or three faces not seen before until this moment. One's impulse is to ask, "Where did *you* come aboard?"

We followed the narrow channel a long time, with land on both sides, — low hills that might have been green and grassy, but had a faded look instead. However, the land-locked water was lovely, at any rate, with its glittering belts of blue and green where moderate soundings were, and its broad splotches of rich brown where the rocks lay near the surface. Everybody was feeling so well that even the grave, pale young man (who, by a sort of kindly common consent, had come latterly to be referred to as "the Ass") received frequent and friendly notice, — which was right enough, for there was no harm in him.

At last we steamed between two island points whose rocky jaws allowed only just enough room for the vessel's body, and now before us loomed Hamilton on her clustered hillsides and summits, the whitest mass of terraced architecture that exists in the world, perhaps.

It was Sunday afternoon, and on the pier were gathered one or two hundred Bermudians, half of them black, half of them white, and all of them nobbily dressed, as the poet says.

Several boats came off to the ship, bringing citizens. One of these citizens was a faded, diminutive old gentleman, who approached our most ancient passenger with a childlike joy in his twinkling eyes, halted before him, folded his arms, and said, smiling with all his might and with all the simple delight that was in him, "You don't know me, John! Come, out with it, now; you know you don't!"

The ancient passenger scanned him perplexedly, scanned the napless, threadbare costume of venerable fashion that had done Sunday-service no man knows how many years, contemplated the marvellous stove-pipe hat of still more ancient and venerable pattern, with its poor pathetic old stiff brim canted up "gallusly" in the wrong places, and said, with a hesitation that indicated strong internal effort to "place" the gentle old apparition, "Why . . . let me see . . . plague on it . . . there 's *something* about you that . . . er . . . er . . . but I 've been gone from Bermuda for twenty-seven years, and . . . hum, hum . . . I don't seem to get at it, somehow, but there 's something about you that is just as familiar to me as — "

"Likely it might be his hat," murmured the Ass, with innocent, sympathetic interest.

III.

So the Reverend and I had at last arrived at Hamilton, the principal town in the Bermuda Islands. A wonderfully white town; white as snow itself. White as marble; white as flour. Yet looking like none of these, exactly. Never mind, we said; we shall hit upon a figure by and by that will describe this peculiar white.

It was a town that was compacted together upon the sides and tops of a cluster of small hills. Its outlying borders fringed off and thinned away among the cedar forests, and there was no woody distance of curving coast, or leafy islet sleeping upon the dimpled, painted sea, but was flecked with shining white points, — half-concealed houses peeping out of the foliage.

The architecture of the town was mainly Spanish, inherited from the colonists of two hundred and fifty years ago. Some ragged-topped cocoa-palms, glimpsed here and there, gave the land a tropical aspect.

There was an ample pier of heavy masonry; upon this, under shelter, were some thousands of barrels containing that product which has carried the fame of Bermuda to many lands, the potato. With here and there an onion. That last sentence is facetious; for they grow at least two onions in Bermuda to one potato. The onion is the pride and joy of Bermuda. It is her jewel, her gem of gems. In her conversation, her pulpit, her literature, it is her most frequent and eloquent figure. In Bermudian metaphor it stands for perfection, — perfection absolute.

The Bermudian weeping over the departed exhausts praise when he says, "He was an onion!" The Bermudian extolling the living hero bankrupts applause when he says, "He is an onion!" The Bermudian setting his son upon the stage of life to dare and do for himself climaxes all counsel, supplication, admonition, comprehends all ambition, when he says, "Be an onion!"

When parallel with the pier, and ten or fifteen steps outside it, we anchored. It was Sunday, bright and sunny. The groups upon the pier — men, youths, and boys — were whites and blacks in about equal proportion. All were well and neatly dressed, many of them nattily, a few of them very stylishly. One would have to travel far before he would find another town of twelve thousand inhabitants that could represent itself so respectably, in the matter of clothes, on a freight-pier, without premeditation or effort. The women and young girls, black and white, who occasionally passed by, were nicely clad, and many were elegantly and fashionably so. The men did not affect summer clothing much, but the girls and women did, and their white garments were good to look at, after so many months of familiarity with sombre colors.

Around one isolated potato barrel stood four young gentlemen, two black, two white, becomingly dressed, each with the head of a slender cane pressed against his teeth, and each with a foot propped up on the barrel. Another young gentleman came up, looked longingly at the barrel, but saw no rest for his foot there, and turned pensively away to seek another barrel. He wandered here and there, but without result. Nobody sat upon a barrel, as is the custom of the idle in other lands, yet all the isolated barrels were humanly occupied. Whosoever had a foot to spare put it on a barrel, if all the places on it were not already taken. The habits of all peoples are determined by their circumstances. The Bermudians lean upon barrels because of the scarcity of lamp-posts.

Many citizens came on board and spoke eagerly to the officers, — inquiring about the Turco-Russian war news, I supposed. However, by listening judiciously I found that this was not so. They said, "What is the price of onions?" or, "How 's onions?" Naturally enough this was their first interest; but they dropped into the war the moment it was satisfied.

We went ashore and found a novelty of a pleasant

nature : there were no hackmen, hacks, or omnibuses on the pier or about it anywhere, and nobody offered his services to us, or molested us in any way. I said it was like being in heaven. The Reverend rebukingly and rather pointedly advised me to make the most of it, then. We knew of a boarding-house, and what we needed now was somebody to pilot us to it. Presently a little barefooted colored boy came along, whose raggedness was conspicuously un-Bermudian. His rear was so marvellously bepatched with colored squares and triangles that one was half persuaded he had got it out of an atlas. When the sun struck him right, he was as good to follow as a lightning-bug. We hired him and dropped into his wake. He piloted us through one picturesque street after another, and in due course deposited us where we belonged. He charged nothing for his map, and but a trifle for his services; so the Reverend doubled it. The little chap received the money with a beaming applause in his eye which plainly said, "This man 's an onion!"

We had brought no letters of introduction; our names had been misspelt in the passenger list; nobody knew whether we were honest folk or otherwise. So we were expecting to have a good private time in case there was nothing in our general aspect to close boarding-house doors against us. We had no trouble. Bermuda has had but little experience of rascals, and is not suspicious. We got large, cool, well-lighted rooms on a second floor, overlooking a bloomy display of flowers and flowering shrubs, — calla and annunciation lilies, lantanas, heliotrope, jessamine, roses, pinks, double geraniums, oleanders, pomegranates, blue morning-glories of a great size, and many plants that were unknown to me.

We took a long afternoon walk, and soon found out that that exceedingly white town was built of blocks of white coral. Bermuda is a coral island, with a six-inch crust of soil on top of it, and every man has a quarry on his own premises. Everywhere you go you see square recesses cut into the hillsides, with perpendicular walls unmarred by crack or crevice, and perhaps you fancy that a house grew out of the ground there, and has been removed in a single piece from the mould. If you do, you err. But the material for a house has been quarried there. They cut right down through the coral, to any depth that is convenient, — ten to twenty feet, — and take it out in great square blocks. This cutting is done with a chisel that has a handle twelve or fifteen feet long, and is used as one uses a crowbar when he is drilling a hole, or a dasher when he is churning. Thus soft is this stone. Then with a common handsaw they saw the great blocks into handsome, huge bricks that are two feet long, a foot wide, and about six inches thick. These stand loosely piled during a month to harden; then the work of building begins. The house is built of these blocks; it is roofed with broad coral slabs an inch thick, whose edges lap upon each other, so that the roof looks like a succession of shallow steps or terraces; the chimneys are built of the coral blocks, and sawed into graceful and picturesque patterns; the ground-floor veranda is paved with coral blocks; also the walk to the gate;

the fence is built of coral blocks, — built in massive panels, with broad capstones and heavy gate-posts, and the whole trimmed into easy lines and comely shape with the saw. Then they put a hard coat of whitewash, as thick as your thumb nail, on the fence and all over the house, roof, chimneys, and all; the sun comes out and shines on this spectacle, and it is time for you to shut your unaccustomed eyes, lest they be put out. It is the whitest white you can conceive of, and the blindingest. A Bermuda house does not look like marble; it is a much intenser white than that; and besides, there is a dainty, indefinable something else about its look that is not marble-like. We put in a great deal of solid talk and reflection over this matter of trying to find a figure that would describe the unique white of a Bermuda house, and we contrived to hit upon it at last. It is exactly the white of the icing of a cake, and has the same unemphasized and scarcely perceptible polish. The white of marble is modest and retiring compared with it.

After the house is cased in its hard scale of whitewash, not a crack, or sign of a seam, or joining of the blocks, is detectable, from base-stone to chimney-top; the building looks as if it had been carved from a single block of stone, and the doors and windows sawed out afterwards. A white marble house has a cold, tomb-like, unsociable look, and takes the conversation out of a body and depresses him. Not so with a Bermuda house. There is something exhilarating, even hilarious, about its vivid whiteness when the sun plays upon it. If it be of picturesque shape and graceful contour,—and many of the Bermudian dwellings are,—it will so fascinate you that you will keep your eyes on it until they ache. One of those clean-cut, fanciful chimneys,—too pure and white for this world,—with one side glowing in the sun and the other touched with a soft shadow, is an object that will charm one's gaze by the hour. I know of no other country that has chimneys worthy to be gazed at and gloated over. One of those snowy houses, half-concealed and half-glimpsed through green foliage, is a pretty thing to see; and if it takes one by surprise and suddenly, as he turns a sharp corner of a country road, it will wring an exclamation from him, sure.

Wherever you go, in town or country, you find those snowy houses, and always with masses of bright-colored flowers about them, but with no vines climbing their walls; vines cannot take hold of the smooth, hard whitewash. Wherever you go, in the town or along the country roads, among little potato farms and patches or expensive country-seats, these stainless white dwellings, gleaming out from flowers and foliage, meet you at every turn. The least little bit of a cottage is as white and blemishless as the stateliest mansion. Nowhere is there dirt or stench, puddle or hog-wallow, neglect, disorder, or lack of trimness and neatness. The roads, the streets, the dwellings, the people, the clothes,—this neatness extends to everything that falls under the eye. It is the tidiest country in the world. And very much the tidiest, too.

Considering these things, the question came up, Where do the poor live? No answer was arrived at.

Therefore, we agreed to leave this conundrum for future statesmen to wrangle over.

What a bright and startling spectacle one of those blazing white country palaces, with its brown-tinted window caps and ledges, and green shutters, and its wealth of caressing flowers and foliage, would be in black London! And what a gleaming surprise it would be in nearly any American city one could mention, too!

Bermuda roads are made by cutting down a few inches into the solid white coral — or a good many feet, where a hill intrudes itself — and smoothing off the surface of the road-bed. It is a simple and easy process. The grain of the coral is coarse and porous; the road-bed has the look of being made of coarse white sugar. Its excessive cleanness and whiteness are a trouble in one way: the sun is reflected into your eyes with such energy as you walk along, that you want to sneeze all the time. Old Captain Tom Bowling found another difficulty. He joined us in our walk, but kept wandering unrestfully to the roadside. Finally he explained. Said he, "Well, I chew, you know, and the road 's so plaguy clean."

We walked several miles that afternoon in the bewildering glare of the sun, the white roads, and the white buildings. Our eyes got to paining us a good deal. By and by a soothing, blessed twilight spread its cool balm around. We looked up in pleased surprise and saw that it proceeded from an intensely black negro who was going by. We answered his military salute in the grateful gloom of his near presence, and then passed on into the pitiless white glare again.

The colored women whom we met usually bowed and spoke; so did the children. The colored men commonly gave the military salute. They borrow this fashion from the soldiers, no doubt; England has kept a garrison here for generations. The younger men's custom of carrying small canes is also borrowed from the soldiers, I suppose, who always carry a cane, in Bermuda as everywhere else in Britain's broad dominions.

The country roads curve and wind hither and thither in the delightfulest way, unfolding pretty surprises at every turn: billowy masses of oleander that seem to float out from behind distant projections like the pink cloud-banks of sunset; sudden plunges among cottages and gardens, life and activity, followed by as sudden plunges into the sombre twilight and stillness of the woods; flitting visions of white fortresses and beacon towers pictured against the sky on remote hill-tops; glimpses of shining green sea caught for a moment through opening headlands, then lost again; more woods and solitude; and by and by another turn lays bare, without warning, the full sweep of the inland ocean, enriched with its bars of soft color, and graced with its wandering sails.

Take any road you please, you may depend upon it you will not stay in it half a mile. Your road is everything that a road ought to be: it is bordered with trees, and with strange plants and flowers; it is shady and pleasant, or sunny and still pleasant; it carries you by the prettiest and peacefulest and most homelike of homes, and through stretches of forest that lie in a deep hush sometimes, and sometimes are alive with the music of birds; it curves always, which is a continual promise, whereas straight roads reveal everything at a glance and kill interest. Your road is all this, and yet you will not stay in it half a mile, for the reason that little seductive, mysterious roads are always branching out from it on either hand, and as these curve sharply also and hide what is beyond, you cannot resist the temptation to desert your own chosen road and explore them. You are usually paid for your trouble; consequently, your walk inland always turns out to be one of the most crooked, involved, purposeless, and interesting experiences a body can imagine. There is enough of variety. Sometimes you are in the level open, with marshes thick grown with flag-lances that are ten feet high on the one hand, and potato and onion orchards on the other; next, you are on a hilltop, with the ocean and the Islands spread around you; presently the road winds through a deep cut, shut in by perpendicular walls thirty or forty feet high, marked with the oddest and abruptest stratum lines, suggestive of sudden and eccentric old upheavals, and garnished with here and there a clinging adventurous flower, and here and there a dangling vine; and by and by your way is along the sea edge, and you may look down a fathom or two through the transparent water and watch the diamond-like flash and play of the light upon the rocks and sands on the bottom until you are tired of it, — if you are so constituted as to be able to get tired of it.

You may march the country roads in maiden meditation, fancy free, by field and farm, for no dog will plunge out at you from unsuspected gate, with breath-taking surprise of ferocious bark, notwithstanding it is a Christian land and a civilized. We saw upwards of a million cats in Bermuda, but the people are very abstemious in the matter of dogs. Two or three nights we prowled the country far and wide, and never once were accosted by a dog. It is a great privilege to visit such a land. The cats were no offence when properly distributed, but when piled they obstructed travel.

As we entered the edge of the town that Sunday afternoon, we stopped at a cottage to get a drink of water. The proprietor, a middle-aged man with a good face, asked us to sit down and rest. His dame brought chairs, and we grouped ourselves in the shade of the trees by the door. Mr. Smith — that was not his name, but it will answer — questioned us about ourselves and our country, and we answered him truthfully, as a general thing, and questioned him in return. It was all very simple and pleasant and sociable. Rural, too; for there was a pig and a small donkey and a hen anchored out, close at hand, by cords to their legs, on a spot that purported to be grassy. Presently, a woman passed along, and although she coldly said nothing she changed the drift of our talk. Said Smith: —

"She did n't look this way, you noticed? Well, she is our next neighbor on one side, and there 's another family that 's our next neighbors on the other side; but there 's a general coolness all around now, and we don't speak. Yet these three families, one generation and another, have lived here side by side and been as friendly as weavers for a hundred and fifty years, till about a year ago."

"Why, what calamity could have been powerful enough to break up so old a friendship?"

"Well, it was too bad, but it could n't be helped. It happened like this: About a year or more ago, the rats got to pestering my place a good deal, and I set up a steel-trap in the back yard. Both of these neighbors run considerable to cats, and so I warned them about the trap, because their cats were pretty sociable around here nights, and they might get into trouble without my intending it. Well, they shut up their cats for a while, but you know how it is with people; they got careless, and sure enough one night the trap took Mrs. Jones's principal tomcat into camp, and finished him up. In the morning Mrs. Jones comes here with the corpse in her arms, and cries and takes on the same as if it was a child. It was a cat by the name of Yelverton, — Hector G. Yelverton, — a troublesome old rip, with no more principle than an Injun, though you could n't make *her* believe it. I said all a man could to comfort her, but no, nothing would do but I must pay for him. Finally, I said I warn't investing in cats now as much as I was, and with that she walked off in a huff, carrying the remains with her. That closed our intercourse with the Joneses. Mrs. Jones joined another church and took her tribe with her. She said she would not hold fellowship with assassins. Well, by and by comes

Mrs. Brown's turn, — she that went by here a minute ago. She had a disgraceful old yellow cat that she thought as much of as if he was twins, and one night he tried that trap on his neck, and it fitted him so, and was so sort of satisfactory, that he laid down and curled up and stayed with it. Such was the end of Sir John Baldwin."

"Was that the name of the cat?"

"The same. There's cats around here with names that would surprise you. Maria" (to his wife), "what was that cat's name that eat a keg of ratsbane by mistake over at Hooper's, and started home and got struck by lightning and took the blind staggers and fell in the well and was most drowned before they could fish him out?"

"That was that colored Deacon Jackson's cat. I only remember the last end of its name, which was Hold-The-Fort-For-I-Am-Coming Jackson."

"Sho! that ain't the one. That's the one that eat up an entire box of Seidlitz powders, and then had n't any more judgment than to go and take a drink. He was considered to be a great loss, but I never could see it. Well, no matter about the names. Mrs. Brown wanted to be reasonable, but Mrs. Jones would n't let her. She put her up to going to law for damages. So to law she went, and had the face to claim seven shillings and sixpence. It made a great stir. All the neighbors went to court. Everybody took sides. It got hotter and hotter, and broke up all the friendships for three hundred yards around, — friendships that had lasted for generations and generations.

"Well, I proved by eleven witnesses that the cat was of a low character and very ornery, and warn't worth a cancelled postage-stamp, any way, taking the average of cats here; but I lost the case. What could I expect? The system is all wrong here, and is bound to make revolution and bloodshed some day. You see, they give the magistrate a poor little starvation salary, and then turn him loose on the public to gouge for fees and costs to live on. What is the natural result? Why he never looks into the justice of a case, — never once. All he looks at is which client has got the money. So this one piled the fees and costs and everything on to me. I could pay specie, don't you see? and he knew mighty well that if he put the verdict on to Mrs. Brown, where it belonged, he 'd have to take his swag in currency."

"Currency? Why, has Bermuda a currency?"

"Yes, — onions. And they were forty per cent discount, too, then, because the season had been over as much as three months. So I lost my case. I had to pay for that cat. But the general trouble the case made was the worst thing about it. Broke up so much good feeling. The neighbors don't speak to each other now. Mrs. Brown had named a child after me. But she changed its name right away. She is a Baptist. Well, in the course of baptizing it over again, it got drowned. I was hoping we might get to be friendly again some time or other, but of course this drowning the child knocked that all out of the question. It would have saved a world of heart-break and ill blood if she had named it dry."

I knew by the sigh that this was honest. All this trouble and all this destruction of confidence in the purity of the bench on account of a seven-shilling lawsuit about a cat! Somehow, it seemed to "size" the country.

At this point we observed that an English flag had just been placed at half-mast on a building a hundred yards away. I and my friends were busy in an instant trying to imagine whose death, among the island dignitaries, could command such a mark of respect as this. Then a shudder shook them and me at the same moment, and I knew that we had jumped to one and the same conclusion: "The governor has gone to England; it is for the British admiral!"

At this moment Mr. Smith noticed the flag. He said with emotion, —

"That 's on a boarding-house. I judge there 's a boarder dead."

A dozen other flags within view went to half-mast.

"It 's a boarder, sure," said Smith.

"But would they half-mast the flags here for a boarder, Mr. Smith?"

"Why, certainly they would, if he was *dead*."

That seemed to size the country again.

IV.

The early twilight of a Sunday evening in Hamilton, Bermuda, is an alluring time. There is just enough of whispering breeze, fragrance of flowers, and sense of repose to raise one's thoughts heavenward; and just enough amateur piano music to keep him reminded of the other place. There are many venerable pianos in Hamilton, and they all play at twilight. Age enlarges and enriches the powers of some musical instruments, — notably those of the violin, — but it seems to set a piano's teeth on edge. Most of the music in vogue there is the same that those pianos prattled in their innocent infancy; and there is something very pathetic about it when they go over it now, in their asthmatic second childhood, dropping a note here and there, where a tooth is gone.

We attended evening service at the stately Episcopal church on the hill, where were five or six hundred people, half of them white and the other half black, according to the usual Bermudian proportions; and all well dressed, — a thing which is also usual in Bermuda and to be confidently expected. There was good music, which we heard, and doubtless a good sermon, but there was a wonderful deal of coughing, and so only the high parts of the argument carried over it. As we came out, after service, I overheard one young girl say to another, —

"Why, you don't mean to say you pay duty on

gloves and laces! I only pay postage; have them done up and sent in the 'Boston Advertiser.'"

There are those who believe that the most difficult thing to create is a woman who can comprehend that it is wrong to smuggle; and that an impossible thing to create is a woman who will not smuggle, whether or no, when she gets a chance. But these may be errors.

We went wandering off toward the country, and were soon far down in the lonely black depths of a road that was roofed over with the dense foliage of a double rank of great cedars. There was no sound of any kind, there; it was perfectly still. And it was so dark that one could detect nothing but sombre outlines. We strode farther and farther down this tunnel, cheering the way with chat.

Presently the chat took this shape: "How insensibly the character of a people and of a government makes its impress upon a stranger, and gives him a sense of security or of insecurity without his taking deliberate thought upon the matter or asking anybody a question! We have been in this land half a day; we have seen none but honest faces; we have noted the British flag flying, which means efficient government and good order; so without inquiry we plunge unarmed and with perfect confidence into this dismal place, which in almost any other country would swarm with thugs and garroters —"

'Sh! What was that? Stealthy footsteps! Low voices! We gasp, we close up together, and wait. A vague shape glides out of the dusk and confronts us. A voice speaks — demands money!

"A shilling, gentlemen, if you please, to help build the new Methodist church."

Blessed sound! Holy sound! We contribute with thankful avidity to the new Methodist church, and are happy to think how lucky it was that those little colored Sunday-school scholars did not seize upon everything we had with violence, before we recovered from our momentary helpless condition. By the light of cigars we write down the names of weightier philanthropists than ourselves on the contribution-cards, and then pass on into the farther darkness, saying, What sort of a government do they call this, where they allow little black pious children, with contribution-cards, to plunge out upon peaceable strangers in the dark and scare them to death?

We prowled on several hours, sometimes by the seaside, sometimes inland, and finally managed to get lost, which is a feat that requires talent in Bermuda. I had on new shoes. They were No. 7's when I started, but were not more than 5's now, and still diminishing. I walked two hours in those shoes after that, before we reached home. Doubtless I could have the reader's sympathy for the asking. Many people have never had the headache or the toothache, and I am one of those myself; but everybody has worn tight shoes for two or three hours, and known the luxury of taking them off in a retired place and seeing his feet swell up and obscure the firmament. Once when I was a callow, bashful cub, I took a plain, unsentimental country girl to a comedy one night. I had known her a day; she seemed divine; I wore my new boots. At the end of the first half-hour she said, "Why do you fidget so with your feet?" I said, "Did I?" Then I put my attention there and kept still. At the end of another half-hour she said, "Why do you say, 'Yes, oh yes!' and 'Ha, ha, oh, certainly! very true!' to everything I say, when half the time those are entirely irrelevant answers?" I blushed, and explained that I had been a little absent-minded. At the end of another half-hour she said, "Please, why do you grin so steadfastly at vacancy, and yet look so sad?" I explained that I always did that when I was reflecting. An hour passed, and then she turned and contemplated me with her earnest eyes and said, "Why do you cry all the time?" I explained that very funny comedies always made me cry. At last human nature surrendered, and I secretly slipped my boots off. This was a mistake. I was not able to get them on any more. It was a rainy night; there were no omnibuses going our way; and as I walked home, burning up with shame, with the girl on one arm and my boots under the other, I was an object worthy of some compassion, — especially in those moments of martyrdom when I had to pass through the glare that fell upon the pavement from street lamps. Finally, this child of the forest said, "Where are your boots?" and being taken unprepared, I put a fitting finish to the follies of the evening with the stupid remark, "The higher classes do not wear them to the theatre."

The Reverend had been an army chaplain during the war, and while we were hunting for a road that would lead to Hamilton he told a story about two dying soldiers which interested me in spite of my feet. He said that in the Potomac hospitals rough pine coffins were furnished by government, but that it was not always possible to keep up with the demand; so, when a man died, if there was no coffin at hand he was buried without one. One night, late, two soldiers lay dying in a ward. A man came in with a coffin on his shoulder, and stood trying to make up his mind which of these two poor fellows would be likely to need it first. Both of them begged for it with their fading eyes, — they were past talking. Then one of them protruded a wasted hand from his blankets and made a feeble beckoning sign with the fingers, to signify, "Be a good fellow; put it under my bed, please." The man did it, and left. The lucky soldier painfully turned himself in his bed until he faced the other warrior, raised himself partly on his elbow, and began to work up a mysterious expression of some kind in his face. Gradually, irksomely, but surely and steadily, it developed, and at last it took definite form as a pretty successful *wink*. The sufferer fell back exhausted with his labor, but bathed in glory. Now entered a personal friend of No. 2, the despoiled soldier. No. 2 pleaded with him with eloquent eyes, till presently he understood, and removed the coffin from under No. 1's bed and put it under No. 2's. No. 2 indicated his joy, and made some more signs; the friend understood again, and put his arm under No. 2's shoulders and lifted him partly up. Then the dying hero turned the dim exultation of his eye upon No. 1, and began a slow and labored work with his hands; gradually he lifted one hand up toward his face; it grew weak and dropped back again; once more he made the effort, but failed again. He took a rest; he gathered all the remnant of his strength, and this time he slowly but surely carried his thumb to the side of his nose, spread the gaunt fingers wide in triumph, and dropped back dead. That picture sticks by me yet. The "situation" is unique.

The next morning, at what seemed a very early hour, the little white table-waiter appeared suddenly in my room and shot a single word out of himself: "Breakfast!"

This was a remarkable boy in many ways. He was about eleven years old; he had alert, intent black eyes; he was quick of movement; there was no hesitation, no uncertainty about him anywhere; there was a military decision in his lip, his manner, his speech, that was an astonishing thing to see in a little chap like him; he wasted no words; his answers always came so quick and brief that they seemed to be part of the question that had been asked instead of a reply to it. When he stood at table with his fly-brush, rigid, erect, his face set in a cast-iron gravity, he was a statue till he detected a dawning want in somebody's eye; then he pounced down, supplied it, and was instantly a statue again. When he was sent to the kitchen for anything, he marched upright till he got to the door; he turned hand-springs the rest of the way.

"Breakfast!"

I thought I would make one more effort to get some conversation out of this being.

"Have you called the Reverend, or are —"

"Yes s'r!"

"Is it early, or is —"

"Eight-five!"

"Do you have to do all the 'chores,' or is there somebody to give you a l—"

"Colored girl!"

"Is there only one parish in this island, or are there —"

"Eight!"

"Is the big church on the hill a parish church, or is it —"

"Chapel-of-ease!"

"Is taxation here classified into poll, parish, town, and —"

"Don't know!"

Before I could cudgel another question out of my head, he was below, hand-springing across the back yard. He had slid down the balusters, head-first. I gave up trying to provoke a discussion with him. The essential element of discussion had been left out of him; his answers were so final and exact that they did not leave a doubt to hang conversation on. I suspect that there is the making of a mighty man or a mighty rascal in this boy, — according to circumstances, — but they are going to apprentice him to a carpenter. It is the way the world uses its opportunities.

During this day and the next we took carriage drives about the island and over to the town of St. George's, fifteen or twenty miles away. Such hard, excellent roads to drive over are not to be found elsewhere out of Europe. An intelligent young colored man drove us, and acted as guide-book. In the edge of the town we saw five or six mountain cabbage palms (atrocious name!) standing in a straight row, and equidistant from each other. These were not the largest or the tallest trees I have ever seen, but they were the state liest, the most majestic. That row of them must be the nearest that nature has ever come to counterfeiting a colonnade. These trees are all the same height, say sixty feet; the trunks as gray as granite, with a very gradual and perfect taper; without sign of branch or knot or flaw; the surface not looking like bark, but like granite that has been dressed and not polished. Thus all the way up the diminishing shaft for fifty feet; then it begins to take the appearance of being closely wrapped, spool-fashion, with gray cord, or of having been turned in a lathe. Above this point there is an outward swell, and thence upwards, for six feet or more, the cylinder is a bright, fresh green, and is formed of wrappings like those of an ear of green Indian corn. Then comes the great, spraying palm plume, also green. Other palm-trees always lean out of the perpendicular, or have a curve in them. But the plumb-line could not detect a deflection in any individual of this stately row; they stand as straight as the colonnade of Baalbec; they have its great height, they have its gracefulness, they have its dignity; in moonlight or twilight, and shorn of their plumes, they would duplicate it.

The birds we came across in the country were singularly tame; even that wild creature, the quail, would pick around in the grass at ease while we inspected it and talked about it at leisure. A small bird of the canary species had to be stirred up with the butt-end of the whip before it would move, and then it moved only a couple of feet. It is said that even the suspicious flea is tame and sociable in Bermuda, and will allow himself to be caught and caressed without misgivings. This should be taken with allowance, for doubtless there is more or less brag about it. In San Francisco they used to claim that their native flea could kick a child over, as if it were a merit in a flea to be able to do that; as if the knowledge of it trumpeted abroad ought to entice immigration. Such a thing in nine cases out of ten would be almost sure to deter a thinking man from coming.

We saw no bugs or reptiles to speak of, and so I was thinking of saying in print, in a general way, that there were none at all; but one night after I had gone to bed, the Reverend came into my room carrying something, and asked, "Is this your boot?" I said it was, and he said he had met a spider going off with it. Next morning he stated that just at dawn the same spider raised his window and was coming in to get a shirt, but saw him and fled.

I inquired, "Did he get the shirt?"

"No."

"How did you know it was a shirt he was after?"

"I could see it in his eye."

We inquired around, but could hear of no Bermudian spider capable of doing these things. Citizens said that their largest spiders could not more than spread their legs over an ordinary saucer, and that they had always been considered honest. Here was testimony of a clergyman against the testimony of mere worldlings, — interested ones, too. On the whole, I judged it best to lock up my things.

Here and there on the country roads we found lemon, papaia, orange, lime, and fig trees; also several sorts of palms, among them the cocoa, the date, and the palmetto. We saw some bamboos forty feet high, with stems as thick as a man's arm. Jungles of the mangrove-tree stood up out of swamps, propped on their interlacing roots as upon a tangle of stilts. In dryer places the noble tamarind sent down its grateful cloud of shade. Here and there the blossomy tamarisk adorned the roadside. There was a curious gnarled and twisted black tree, without a single leaf on it. It might have passed itself off for a dead apple-tree but for the fact that it had a star-like, red-hot flower sprinkled sparsely over its person. It had the scattery red glow that a constellation might have when glimpsed through smoked glass. It is possible that our constellations have been so constructed as to be invisible through smoked glass; if this is so it is a great mistake.

We saw a tree that bears grapes, and just as calmly and unostentatiously as a vine would do it. We saw an India-rubber tree, but out of season, possibly, so there were no shoes on it, nor suspenders, nor anything that a person would properly expect to find there. This gave it an impressively fraudulent look. There was exactly one mahogany-tree on the island. I know this to be reliable, because I saw a man who said he had counted it many a time and could not be mistaken. He was a man with a hare lip and a pure heart, and everybody said he was as true as steel. Such men are all too few.

One's eye caught near and far the pink cloud of the oleander and the red blaze of the pomegranate blossom. In one piece of wild wood the morning-glory vines had wrapped the trees to their very tops, and decorated them all over with couples and clusters of great blue bells, — a fine and striking spectacle, at a little distance. But the dull cedar is everywhere, and its is the prevailing foliage. One does not appreciate how dull it is until the varnished, bright green attire of the infrequent lemon-tree pleasantly intrudes its contrast. In one thing Bermuda is eminently tropical, — was in May, at least, — the unbrilliant, slightly faded, unrejoicing look of the landscape. For forests arrayed in a blemishless magnificence of glowing green foliage that seems to exult in its own existence and can move the beholder to an enthusiasm that will make him either shout or cry, one must go to countries that have malignant winters.

We saw scores of colored farmers digging their crops of potatoes and onions, their wives and children helping, — entirely contented and comfortable, if looks go for anything. We never met a man, or woman, or child anywhere in this sunny island who seemed to be unprosperous, or discontented, or sorry about anything. This sort of monotony became very tiresome presently, and even something worse. The spectacle of an entire nation grovelling in contentment is an infuriating thing. We felt the lack of something in this community, — a vague, an undefinable, an elusive something, and yet a lack. But after considerable thought we made out what it was, — tramps. Let them go there, right now, in a body. It is utterly virgin soil. Passage is cheap. Every true patriot in America will help buy tickets. Whole armies of these excellent beings can be spared from our midst and our polls; they will find a delicious climate and a green, kind-hearted people. There are potatoes and onions for all, and a generous welcome for the first batch that arrives, and elegant graves for the second.

It was the Early Rose potato the people were digging. Later in the year they have another crop, which they call the Garnet. We buy their potatoes (retail) at fifteen dollars a barrel; and those colored farmers buy ours for a song, and live on them. Havana might exchange cigars with Connecticut in the same advantageous way, if she thought of it.

We passed a roadside grocery with a sign up, "Potatoes Wanted." An ignorant stranger, doubtless. He could not have gone thirty steps from his place without finding plenty of them.

In several fields the arrowroot crop was already sprouting. Bermuda used to make a vast annual profit out of this staple before fire-arms came into such general use.

The island is not large. Somewhere in the interior a man ahead of us had a very slow horse. I suggested that we had better go by him; but the driver said the man had but a little way to go. I waited to see, wondering how he could know. Presently the man did turn down another road. I asked, "How did you know he would?"

"Because I knew the man, and where he lived."

I asked him, satirically, if he knew everybody in the island; he answered, very simply, that he did. This gives a body's mind a good substantial grip on the dimensions of the place.

At the principal hotel in St. George's, a young girl, with a sweet, serious face, said we could not be furnished with dinner, because we had not been expected, and no preparation had been made. Yet it was still an hour before dinner time. We argued, she yielded not; we supplicated, she was serene. The hotel had not been expecting an inundation of two people, and so it seemed that we should have to go home dinnerless. I said we were not very hungry; a fish would do. My little maid answered, it was not the market-day for fish. Things began to look serious; but presently the boarder who sustained the hotel came in, and when the case was laid before him he was cheerfully willing to divide. So we had much pleasant chat at table about St. George's chief industry, the repairing of damaged ships; and in between we had a soup that had something in it that seemed to taste like the hereafter, but it proved to be only pepper of a particularly vivacious kind. And we had an iron-clad chicken that was deliciously cooked, but not in the right way. Baking was not the thing to convince his sort. He ought to have been put through a quartz mill until the "tuck" was taken out of him, and then boiled till we came again. We got a good deal of sport out of him, but not enough sustenance to leave the victory on our side. No matter; we had potatoes and a pie and a sociable good time. Then a ramble through the town, which is a quaint one, with interesting, crooked streets, and narrow, crooked lanes, with here and there a grain of dust. Here, as in Hamilton, the dwellings had Venetian blinds of a very sensible pattern. They were not double shutters, hinged at the sides, but a single broad shutter, hinged at the top; you push it outward, from the bottom, and fasten it at any angle required by the sun or desired by yourself.

All about the island one sees great white scars on the hill-slopes. These are dished spaces where the soil has been scraped off and the coral exposed and glazed with hard whitewash. Some of these are a quarter-acre in size. They catch and carry the rainfall to reservoirs; for the wells are few and poor, and there are no natural springs and no brooks.

They say that the Bermuda climate is mild and equable, with never any snow or ice, and that one may be very comfortable in spring clothing the year round, there. We had delightful and decided summer weather in May, with a flaming sun that permitted the thinnest of raiment, and yet there was a constant breeze; consequently we were never discomforted by heat. At four or five in the afternoon the mercury began to go down, and then it became necessary to change to thick garments. I went to St. George's in the morning clothed in the thinnest of linen, and reached home at five in the afternoon with two overcoats on. The nights are said to be always cool and bracing. We had mosquito nets, and the Reverend said the mosquitoes persecuted him a good deal. I often heard him slapping and banging at these imaginary creatures with as much zeal as if they had been real. There are no mosquitoes in the Bermudas in May.

The poet Thomas Moore spent several months in Bermuda more than seventy years ago. He was sent out to be registrar of the admiralty. I am not quite clear as to the function of a registrar of the admiralty of Bermuda, but I think it is his duty to keep a record of all the admirals born there. I will inquire into this. There was not much doing in admirals, and Moore got tired and went away. A reverently preserved souvenir of him is still one of the treasures of the islands. I gathered the idea, vaguely, that it was a jug, but was persistently thwarted in the twenty-two efforts I made to visit it. However, it was no matter, for I found afterwards that it was only a chair.

There are several "sights" in the Bermudas, of course, but they are easily avoided. This is a great advantage, — one cannot have it in Europe. Bermuda is the right country for a jaded man to "loaf" in. There are no harassments; the deep peace and quiet of the country sink into one's body and bones and give his conscience a rest, and chloroform the legion of invisible small devils that are always trying to whitewash his hair. A good many Americans go there about the first of March and remain until the early spring weeks have finished their villanies at home.

The Bermudians are hoping soon to have telegraphic communication with the world. But even after they shall have acquired this curse it will still be a good country to go to for a vacation, for there are charming little islets scattered about the enclosed sea where one could live secure from interruption. The telegraph boy would have to come in a boat, and one could easily kill him while he was making his landing.

We had spent four days in Bermuda, — three bright ones out of doors and one rainy one in the house, we being disappointed about getting a yacht for a sail; and now our furlough was ended, and we entered into the ship again and sailed homeward.

Among the passengers was a most lean and lank and forlorn invalid, whose weary look and patient eyes and sorrowful mien awoke every one's kindly interest and stirred every one's compassion. When he spoke — which was but seldom — there was a gentleness in his tones that made each hearer his friend. The second night of the voyage — we were all in the smoking cabin at the time — he drifted, little by little, into the general conversation. One thing brought on another, and so, in due course, he happened to fall into the biographical vein, and the following strange narrative was the result.

THE INVALID'S STORY.[1]

I seem sixty and married, but these effects are due to my condition and sufferings, for I am a bachelor,

[1] Left out of these "Rambling Notes," when originally published in the "Atlantic Monthly," because it was feared that the

and only forty-one. It will be hard for you to believe that I, who am now but a shadow, was a hale, hearty man two short years ago, — a man of iron, a very athlete! — yet such is the simple truth. But stranger still than this fact is the way in which I lost my health. I lost it through helping to take care of a box of guns on a two-hundred-mile railway journey one winter's night. It is the actual truth, and I will tell you about it.

I belong in Cleveland, Ohio. One winter's night, two years ago, I reached home just after dark, in a driving snow-storm, and the first thing I heard when I entered the house was that my dearest boyhood friend and schoolmate, John B. Hackett, had died the day before, and that his last utterance had been a desire that I would take his remains home to his poor old father and mother in Wisconsin. I was greatly shocked and grieved, but there was no time to waste in emotions; I must start at once. I took the card, marked "Deacon Levi Hackett, Bethlehem, Wisconsin," and hurried off through the whistling storm to the railway station. Arrived there I found the long white-pine box which had been described to me; I fastened the card to it with some tacks, saw it put safely aboard the express car, and then ran into the eating-room to provide myself with a sandwich and some cigars. When I returned, presently, there was my coffin-box *back again*, apparently, and a young fellow examining around it, with a card in his hand, and some tacks and

story was not true, and at that time there was no way of proving that it was not. — M. T.

a hammer! I was astonished and puzzled. He began to nail on his card, and I rushed out to the express car, in a good deal of a state of mind, to ask for an explanation. But no — there was my box, all right, in the express car; it had n't been disturbed. [The fact is that without my suspecting it a prodigious mistake had been made. I was carrying off a box of *guns* which that young fellow had come to the station to ship to a rifle company in Peoria, Illinois, and *he* had got my corpse!] Just then the conductor sung out "All aboard," and I jumped into the express car and got a comfortable seat on a bale of buckets. The expressman was there, hard at work, — a plain man of fifty, with a simple, honest, good-natured face, and a breezy, practical heartiness in his general style. As the train moved off a stranger skipped into the car and set a package of peculiarly mature and capable Limburger cheese on one end of my coffin-box — I mean my box of guns. That is to say, I know *now* that it was Limburger cheese, but at that time I never had heard of the article in my life, and of course was wholly ignorant of its character. Well, we sped through the wild night, the bitter storm raged on, a cheerless misery stole over me, my heart went down, down, down! The old expressman made a brisk remark or two about the tempest and the arctic weather, slammed his sliding doors to, and bolted them, closed his window down tight, and then went bustling around, here and there and yonder, setting things to rights, and all the time contentedly humming "Sweet By and By," in a low tone, and flatting a good deal. Presently I began to detect a most evil and searching odor stealing about on the frozen air. This depressed my spirits still more, because of course I attributed it to my poor departed friend. There was something infinitely saddening about his calling himself to my remembrance in this dumb pathetic way, so it was hard to keep the tears back. Moreover, it distressed me on account of the old expressman, who, I was afraid, might notice it. However, he went humming tranquilly on, and gave no sign; and for this I was grateful. Grateful, yes, but still uneasy; and soon I began to feel more and more uneasy every minute, for every minute that went by that odor thickened up the more, and got to be more and more gamey and hard to stand. Presently, having got things arranged to his satisfaction, the expressman got some wood and made up a tremendous fire in his stove. This distressed me more than I can tell, for I could not but feel that it was a mistake. I was sure that the effect would be deleterious upon my poor departed friend. Thompson — the expressman's name was Thompson, as I found out in the course of the night — now went poking around his car, stopping up whatever stray cracks he could find, remarking that it did n't make any difference what kind of a night it was outside, he calculated to make *us* comfortable, anyway. I said nothing, but I believed he was not choosing the right way. Meantime he was humming to himself just as before; and meantime, too, the stove was getting hotter and hotter, and the place closer and closer. I felt myself growing pale and qualmish, but grieved in silence and said nothing. Soon I noticed that the "Sweet By and By" was gradually fading out; next it ceased altogether, and there was an ominous stillness. After a few moments Thompson said, —

"Pfew! I reckon it ain't no cinnamon 't I 've loaded up thish-yer stove with!"

He gasped once or twice, then moved toward the cof— gun-box, stood over that Limburger cheese part of a moment, then came back and sat down near me, looking a good deal impressed. After a contemplative pause, he said, indicating the box with a gesture, —

"Friend of yourn?"

"Yes," I said with a sigh.

"He 's pretty ripe, *ain't* he!"

Nothing further was said for perhaps a couple of minutes, each being busy with his own thoughts; then Thompson said, in a low, awed voice, —

"Sometimes it 's uncertain whether they 're really gone or not, — *seem* gone, you know — body warm, joints limber — and so, although you *think* they 're gone, you don't really know. I 've had cases in my car. It 's perfectly awful, becuz *you* don't know what minute they 'll rise right up and look at you!" Then, after a pause, and slightly lifting his elbow toward the box, — "But *he* ain't in no trance! No, sir, I go bail for *him!*"

We sat some time, in meditative silence, listening to the wind and the roar of the train; then Thompson said, with a good deal of feeling, —

"Well-a-well, we 've all got to go, they ain't no getting around it. Man that is born of woman is of few days and far between, as Scriptur' says. Yes, you look at it any way you want to, it 's awful solemn and cu-r'us: they ain't *nobody* can get around it; *all's* got to go — just *everybody*, as you may say. One day you 're hearty and strong" — here he scrambled to his feet and broke a pane and stretched his nose out at it a moment or two, then sat down again while I struggled up and thrust my nose out at the same place, and this we kept on doing every now and then — "and next day he 's cut down like the grass, and the places which knowed him then knows him no more forever, as Scrip-tur' says. Yes-'ndeedy, it 's awful solemn and cur'us; but we 've all got to go, one time or another; they ain't no getting around it."

There was another long pause; then, —

"What did he die of?"

I said I did n't know.

"How long has he ben dead?"

It seemed judicious to enlarge the facts to fit the probabilities; so I said, —

"Two or three days."

But it did no good; for Thompson received it with an injured look which plainly said, "Two or three *years*, you mean." Then he went right along, placidly ignor-ing my statement, and gave his views at considerable length upon the unwisdom of putting off burials too long. Then he lounged off toward the box, stood a moment, then came back on a sharp trot and visited the broken pane, observing, —

"'T would 'a' ben a dum sight better, all around, if they 'd started him along last summer."

Thompson sat down and buried his face in his red silk handkerchief, and began to slowly sway and rock his body like one who is doing his best to endure the almost unendurable. By this time the fragrance — if you may call it fragrance — was just about suffocating, as near as you can come at it. Thompson's face was turning gray; I knew mine had n't any color left in it. By and by Thompson rested his forehead in his left hand, with his elbow on his knee, and sort of waved his red handkerchief towards the box with his other hand, and said, —

"I 've carried a many a one of 'em, — some of 'em considerable overdue, too, — but, lordy, he just lays over 'em all! — and does it *easy*. Cap., they was heliotrope to *him!*"

This recognition of my poor friend gratified me, in spite of the sad circumstances, because it had so much the sound of a compliment.

Pretty soon it was plain that something had got to be done. I suggested cigars. Thompson thought it was a good idea. He said, —

"Likely it 'll modify him some."

We puffed gingerly along for a while, and tried hard to imagine that things were improved. But it was n't any use. Before very long, and without any consultation, both cigars were quietly dropped from our nerveless fingers at the same moment. Thompson said, with a sigh, —

"No, Cap., it don't modify him worth a cent. Fact is, it makes him worse, becuz it appears to stir up his ambition. What do you reckon we better do, now?"

I was not able to suggest anything; indeed, I had to be swallowing and swallowing, all the time, and did not like to trust myself to speak. Thompson fell to maundering, in a desultory and low-spirited way, about the miserable experiences of this night; and he got to referring to my poor friend by various titles, — sometimes military ones, sometimes civil ones; and I noticed that as fast as my poor friend's effectiveness grew, Thompson promoted him accordingly, — gave him a bigger title. Finally he said, —

"I 've got an idea. Suppos'n' we buckle down to it and give the Colonel a bit of a shove towards t' other end of the car? — about ten foot, say. He would n't have so much influence, then, don't you reckon?"

I said it was a good scheme. So we took in a good fresh breath at the broken pane, calculating to hold it till we got through; then we went there and bent down over that deadly cheese and took a grip on the box. Thompson nodded "All ready," and then we threw ourselves forward with all our might; but Thompson slipped, and slumped down with his nose on the cheese, and his breath got loose. He gagged and gasped, and floundered up and made a break for the door, pawing the air and saying, hoarsely, "Don't hender me! — gimme the road! I 'm a-dying; gimme the road!" Out on the cold platform I sat down and held his head a while, and he revived. Presently he said, —

"Do you reckon we started the Gen'rul any?"

I said no; we had n't budged him.

"Well, then, *that* idea 's up the flume. We got to

think up something else. He 's suited wher' he is, I reckon; and if that 's the way he feels about it, and has made up his mind that he don't wish to be disturbed, you bet you he 's a-going to have his own way in the business. Yes, better leave him right wher' he is, long as he wants it so; becuz he holds all the trumps, don't you know, and so it stands to reason that the man that lays out to alter his plans for him is going to get left."

But we could n't stay out there in that mad storm; we should have frozen to death. So we went in again and shut the door, and began to suffer once more and take turns at the break in the window. By and by, as we were starting away from a station where we had stopped a moment Thompson pranced in cheerily, and exclaimed, —

"We 're all right, now! I reckon we 've got the Commodore this time. I judge I 've got the stuff here that 'll take the tuck out of him."

It was carbolic acid. He had a carboy of it. He sprinkled it all around everywhere; in fact he drenched everything with it, rifle-box, cheese, and all. Then we sat down, feeling pretty hopeful. But it was n't for long. You see the two perfumes began to mix, and then — well, pretty soon we made a break for the door; and out there Thompson swabbed his face with his bandanna and said in a kind of disheartened way, —

"It ain't no use. We can't buck agin *him*. He just utilizes everything we put up to modify him with, and gives it his own flavor and plays it back on us. Why, Cap., don't you know, it 's as much as a hundred times worse in there now than it was when he first got a-going. I never *did* see one of 'em warm up to his work so, and take such a dumnation interest in it. No, sir, I never did, as long as I 've ben on the road; and I 've carried a many a one of 'em, as I was telling you."

We went in again, after we were frozen pretty stiff; but my, we could n't *stay* in, now. So we just waltzed back and forth, freezing, and thawing, and stifling, by turns. In about an hour we stopped at another station; and as we left it Thompson came in with a bag, and said, —

"Cap., I 'm a-going to chance him once more, — just this once; and if we don't fetch him this time, the thing for us to do, is to just throw up the sponge and withdraw from the canvass. That 's the way *I* put it up."

He had brought a lot of chicken feathers, and dried apples, and leaf tobacco, and rags, and old shoes, and sulphur, and assafœtida, and one thing or another; and he piled them on a breadth of sheet iron in the middle of the floor, and set fire to them. When they got well started, I could n't see, myself, how even the corpse could stand it. All that went before was just simply poetry to that smell, — but mind you, the original smell stood up out of it just as sublime as ever, — fact is, these other smells just seemed to give it a better hold; and my, how rich it was! I did n't make these reflections there — there was n't time — made them on the platform. And breaking for the platform, Thompson got suffocated and fell; and before I got him dragged out, which I did by the collar, I was mighty near gone myself. When we revived, Thompson said dejectedly, —

"We got to stay out here, Cap. We got to do it. They ain't no other way. The Governor wants to travel alone, and he 's fixed so he can outvote us."

And presently he added, —

"And don't you know, we 're *pisoned.* It 's *our* last trip, you can make up your mind to it. Typhoid fever is what 's going to come of this. I feel it a-coming right now. Yes, sir, we 're elected, just as sure as you 're born."

We were taken from the platform an hour later, frozen and insensible, at the next station, and I went straight off into a virulent fever, and never knew anything again for three weeks. I found out, then, that I had spent that awful night with a harmless box of rifles and a lot of innocent cheese; but the news was too late to save *me;* imagination had done its work, and my health was permanently shattered; neither Bermuda nor any other land can ever bring it back to me. This is my last trip; I am on my way home to die.

We made the run home to New York quarantine in three days and five hours, and could have gone right along up to the city if we had had a health permit. But health permits are not granted after seven in the evening, partly because a ship cannot be inspected and overhauled with exhaustive thoroughness except in daylight, and partly because health officers are liable to catch cold if they expose themselves to the night air. Still, you can *buy* a permit after hours for five dollars extra, and the officer will do the inspecting next week. Our ship and passengers lay under expense and in humiliating captivity all night, under the very nose of the little official reptile who is supposed to protect New York from pestilence by his vigilant "inspections." This imposing rigor gave everybody a solemn and awful idea of the beneficent watchfulness of our government, and there were some who wondered if anything finer could be found in other countries.

In the morning we were all a-tiptoe to witness the intricate ceremony of inspecting the ship. But it was a disappointing thing. The health officer's tug ranged alongside for a moment, our purser handed the lawful three-dollar permit fee to the health officer's bootblack, who passed us a folded paper in a forked stick, and away we went. The entire "inspection" did not occupy thirteen seconds.

The health officer's place is worth a hundred thousand dollars a year to him. His system of inspection is perfect, and therefore cannot be improved on; but it seems to me that his system of collecting his fees might be amended. For a great ship to lie idle all night is a most costly loss of time; for her passengers to have to do the same thing works to them the same damage, with the addition of an amount of exasperation and bitterness of soul that the spectacle of that health officer's ashes on a shovel could hardly sweeten. Now why would it not be better and simpler to let the ships pass in unmolested, and the fees and permits be exchanged once a year by post?

THE FACTS CONCERNING THE RECENT CARNIVAL OF CRIME IN CONNECTICUT.

I WAS feeling blithe, almost jocund. I put a match to my cigar, and just then the morning's mail was handed in. The first superscription I glanced at was in a handwriting that sent a thrill of pleasure through and through me. It was Aunt Mary's; and she was the person I loved and honored most in all the world, outside of my own household. She had been my boyhood's idol; maturity, which is fatal to so many enchantments, had not been able to dislodge her from her pedestal; no, it had only justified her right to be there, and placed her dethronement permanently among the impossibilities. To show how strong her influence over me was, I will observe that long after everybody else's "*do*-stop-smoking" had ceased to affect me in the slightest degree, Aunt Mary could still stir my torpid conscience into faint signs of life when she touched upon the matter. But all things have their limit, in this world. A happy day came at last, when even Aunt Mary's words could no longer move me. I was not merely glad to see that day arrive; I was more than glad — I was grateful; for when its sun had set, the one alloy that was able to mar my enjoyment of my aunt's society was gone.

The remainder of her stay with us that winter was in every way a delight. Of course she pleaded with me just as earnestly as ever, after that blessed day, to quit my pernicious habit, but to no purpose whatever; the moment she opened the subject I at once became calmly, peacefully, contentedly indifferent — absolutely, adamantinely indifferent. Consequently the closing weeks of that memorable visit melted away as pleasantly as a dream, they were so freighted, for me, with tranquil satisfaction. I could not have enjoyed my pet vice more if my gentle tormentor had been a smoker herself, and an advocate of the practice. Well, the sight of her handwriting reminded me that I was getting very hungry to see her again. I easily guessed what I should find in her letter. I opened it. Good! just as I expected; she was coming! Coming this very day, too, and by the morning train; I might expect her any moment.

I said to myself, "I am thoroughly happy and content, now. If my most pitiless enemy could appear before me at this moment, I would freely right any wrong I may have done him."

Straightway the door opened, and a shrivelled, shabby dwarf entered. He was not more than two feet high. He seemed to be about forty years old. Every feature and every inch of him was a trifle out of shape; and so, while one could not put his finger upon any particular part and say, "This is a conspicuous deformity," the spectator perceived that this little person was a deformity as a whole, — a vague, general, evenly blended, nicely adjusted deformity. There was a foxlike cunning in the face and the sharp little eyes, and also alertness and malice. And yet, this vile bit of human rubbish seemed to bear a sort of remote and ill-defined resemblance to me! It was dully perceptible in the mean form, the countenance, and even the clothes, gestures, manner, and attitudes of the creature. He was a far-fetched, dim suggestion of a burlesque upon me, a caricature of me in little. One thing about him struck me forcibly, and most unpleasantly: he was covered all over with a fuzzy, greenish mould, such as one sometimes sees upon mildewed bread. The sight of it was nauseating.

He stepped along with a chipper air, and flung himself into a doll's chair in a very free and easy way, without waiting to be asked. He tossed his hat into the waste basket. He picked up my old chalk pipe from the floor, gave the stem a wipe or two on his knee, filled the bowl from the tobacco-box at his side, and said to me in a tone of pert command, —

"Gimme a match!"

I blushed to the roots of my hair; partly with indignation, but mainly because it somehow seemed to me that this whole performance was very like an exaggeration of conduct which I myself had sometimes been guilty of in my intercourse with familiar friends, — but never, never with strangers, I observed to myself. I wanted to kick the pygmy into the fire, but some incomprehensible sense of being legally and legitimately under his authority forced me to obey his order. He applied the match to the pipe, took a contemplative whiff or two, and remarked, in an irritatingly familiar way, —

"Seems to me it's devilish odd weather for this time of year."

I flushed again, and in anger and humiliation as before; for the language was hardly an exaggeration of some that I have uttered in my day, and moreover was delivered in a tone of voice and with an exasperating drawl that had the seeming of a deliberate travesty of my style. Now there is nothing I am quite so sensitive about as a mocking imitation of my drawling infirmity of speech. I spoke up sharply and said, —

"Look here, you miserable ash-cat! you will have to give a little more attention to your manners, or I will throw you out of the window!"

The manikin smiled a smile of malicious content and security, puffed a whiff of smoke contemptuously toward me, and said, with a still more elaborate drawl, —

"Come — go gently, now; don't put on *too* many airs with your betters."

This cool snub rasped me all over, but it seemed to subjugate me, too, for a moment. The pygmy contemplated me awhile with his weasel eyes, and then said, in a peculiarly sneering way, —

"You turned a tramp away from your door this morning."

I said crustily, —

"Perhaps I did, perhaps I did n't. How do *you* know?"

"Well, I know. It is n't any matter *how* I know."

"Very well. Suppose I *did* turn a tramp away from the door — what of it?"

"Oh, nothing; nothing in particular. Only you lied to him."

"I *did n't!* That is, I — "

"Yes, but you did; you lied to him."

I felt a guilty pang, — in truth I had felt it forty times before that tramp had travelled a block from my door, — but still I resolved to make a show of feeling slandered; so I said, —

"This is a baseless impertinence. I said to the tramp — "

"There — wait. You were about to lie again. *I* know what you said to him. You said the cook was gone down town and there was nothing left from breakfast. Two lies. You knew the cook was behind the door, and plenty of provisions behind *her*."

This astonishing accuracy silenced me; and it filled me with wondering speculations, too, as to how this cub could have got his information. Of course he could have culled the conversation from the tramp, but by what sort of magic had he contrived to find out about the concealed cook? Now the dwarf spoke again: —

"It was rather pitiful, rather small, in you to refuse to read that poor young woman's manuscript the other day, and give her an opinion as to its literary value; and she had come so far, too, and *so* hopefully. Now *was n't* it?"

I felt like a cur! And I had felt so every time the thing had recurred to my mind, I may as well confess. I flushed hotly and said, —

"Look here, have you nothing better to do than prowl around prying into other people's business? Did that girl tell you that?"

"Never mind whether she did or not. The main thing is, you did that contemptible thing. And you felt ashamed of it afterwards. Aha! you feel ashamed of it *now!*"

This with a sort of devilish glee. With fiery earnestness I responded, —

"I told that girl, in the kindest, gentlest way, that I could not consent to deliver judgment upon *any* one's manuscript, because an individual's verdict was worthless. It might underrate a work of high merit and lose it to the world, or it might overrate a trashy production and so open the way for its infliction upon the world. I said that the great public was the only tribunal competent to sit in judgment upon a literary effort, and therefore it must be best to lay it before that tribunal in the outset, since in the end it must stand or fall by that mighty court's decision any way."

"Yes, you said all that. So you did, you juggling, small-souled shuffler! And yet when the happy hopefulness faded out of that poor girl's face, when you saw her furtively slip beneath her shawl the scroll she had so patiently and honestly scribbled at, — so ashamed of her darling now, so proud of it before, — when you saw the gladness go out of her eyes and the tears come there, when she crept away so humbly who had come so —"

"Oh, peace! peace! peace! Blister your merciless tongue, have n't all these thoughts tortured me enough, without *your* coming here to fetch them back again?"

Remorse! remorse! It seemed to me that it would

eat the very heart out of me! And yet that small fiend only sat there leering at me with joy and contempt, and placidly chuckling. Presently he began to speak again. Every sentence was an accusation, and every accusation a truth. Every clause was freighted with sarcasm and derision, every slow-dropping word burned like vitriol. The dwarf reminded me of times when I had flown at my children in anger and punished them for faults which a little inquiry would have taught me that others, and not they, had committed. He reminded me of how I had disloyally allowed old friends to be traduced in my hearing, and been too craven to utter a word in their defence. He reminded me of many dishonest things which I had done; of many which I had procured to be done by children and other irresponsible persons; of some which I had planned, thought upon, and longed to do, and been kept from the performance by fear of consequences only. With exquisite cruelty he recalled to my mind, item by item, wrongs and unkindnesses I had inflicted and humiliations I had put upon friends since dead, "who died thinking of those injuries, maybe, and grieving over them," he added, by way of poison to the stab.

"For instance," said he, "take the case of your younger brother, when you two were boys together, many a long year ago. He always lovingly trusted in you with a fidelity that your manifold treacheries were not able to shake. He followed you about like a dog, content to suffer wrong and abuse if he might only be with you; patient under these injuries so long as it was your hand that inflicted them. The latest picture you have of him in health and strength must be such a comfort to you! You pledged your honor that if he would let you blindfold him no harm should come to him; and then, giggling and choking over the rare fun of the joke, you led him to a brook thinly glazed with ice, and pushed him in; and how you did laugh! Man, you will never forget the gentle, reproachful look he gave you as he struggled shivering out, if you live a thousand years! Oho! you see it now, you see it *now!*"

"Beast, I have seen it a million times, and shall see it a million more! and may you rot away piecemeal, and suffer till doomsday what I suffer now, for bringing it back to me again!"

The dwarf chuckled contentedly, and went on with his accusing history of my career. I dropped into a moody, vengeful state, and suffered in silence under the merciless lash. At last this remark of his gave me a sudden rouse: —

"Two months ago, on a Tuesday, you woke up, away in the night, and fell to thinking, with shame, about a peculiarly mean and pitiful act of yours toward a poor ignorant Indian in the wilds of the Rocky Mountains in the winter of eighteen hundred and —"

"Stop a moment, devil! Stop! Do you mean to tell me that even my very *thoughts* are not hidden from you?"

"It seems to look like that. Did n't you think the thoughts I have just mentioned?"

"If I did n't, I wish I may never breathe again!

8

Look here, friend — look me in the eye. Who *are* you?"

"Well, who do you think?"

"I think you are Satan himself. I think you are the devil."

"No."

"No? Then who *can* you be?"

"Would you really like to know?"

"*Indeed* I would."

"Well, I am your *Conscience!*"

In an instant I was in a blaze of joy and exultation. I sprang at the creature, roaring, —

"Curse you, I have wished a hundred million times that you were tangible, and that I could get my hands on your throat once! Oh, but I will wreak a deadly vengeance on —"

Folly! Lightning does not move more quickly than my Conscience did! He darted aloft so suddenly that in the moment my fingers clutched the empty air he was already perched on the top of the high book-case, with his thumb at his nose in token of derision. I flung the poker at him, and missed. I fired the boot-jack. In a blind rage I flew from place to place, and snatched and hurled any missile that came handy; the storm of books, inkstands, and chunks of coal gloomed the air and beat about the manikin's perch relentlessly, but all to no purpose; the nimble figure dodged every shot; and not only that, but burst into a cackle of sarcastic and triumphant laughter as I sat down exhausted. While I puffed and gasped with fatigue and excitement, my Conscience talked to this effect: —

"My good slave, you are curiously witless — no, I mean characteristically so. In truth, you are always consistent, always yourself, always an ass. Otherwise it must have occurred to you that if you attempted this murder with a sad heart and a heavy conscience, I would droop under the burdening influence instantly. Fool, I should have weighed a ton, and could not have budged from the floor; but instead, you are so cheerfully anxious to kill me that your conscience is as light as a feather; hence I am away up here out of your reach. I can almost respect a mere ordinary sort of fool; but *you* — pah!"

I would have given anything, then, to be heavy-hearted, so that I could get this person down from there and take his life, but I could no more be heavy-hearted over such a desire than I could have sorrowed over its accomplishment. So I could only look longingly up at my master, and rave at the ill-luck that denied me a heavy conscience the one only time that I had ever wanted such a thing in my life. By and by I got to musing over the hour's strange adventure, and of course my human curiosity began to work. I set myself to framing in my mind some questions for this fiend to answer. Just then one of my boys entered, leaving the door open behind him, and exclaimed, —

"My! what *has* been going on, here? The bookcase is all one riddle of — "

I sprang up in consternation, and shouted, —

"Out of this! Hurry! Jump! Fly! Shut the door! Quick, or my Conscience will get away!"

The door slammed to, and I locked it. I glanced up and was grateful, to the bottom of my heart, to see that my owner was still my prisoner. I said, —

"Hang you, I might have lost you! Children are the heedlessest creatures. But look here, friend, the boy did not seem to notice you at all; how is that?"

"For a very good reason. I am invisible to all but you."

I made mental note of that piece of information with a good deal of satisfaction. I could kill this miscreant now, if I got a chance, and no one would know it. But this very reflection made me so light-hearted that my Conscience could hardly keep his seat, but was like to float aloft toward the ceiling like a toy balloon. I said, presently, —

"Come, my Conscience, let us be friendly. Let us fly a flag of truce for a while. I am suffering to ask you some questions."

"Very well. Begin."

"Well, then, in the first place, why were you never visible to me before?"

"Because you never asked to see me before; that is, you never asked in the right spirit and the proper form before. You were just in the right spirit this time, and when you called for your most pitiless enemy I was that person by a very large majority, though you did not suspect it."

"Well, did that remark of mine turn you into flesh and blood?"

"No. It only made me visible to you. I am unsubstantial, just as other spirits are."

This remark prodded me with a sharp misgiving. If he was unsubstantial, how was I going to kill him? But I dissembled, and said persuasively, —

"Conscience, it is n't sociable of you to keep at such a distance. Come down and take another smoke."

This was answered with a look that was full of derision, and with this observation added: —

"Come where you can get at me and kill me? The invitation is declined with thanks."

"All right," said I to myself; "so it seems a spirit *can* be killed, after all; there will be one spirit lacking in this world, presently, or I lose my guess." Then I said aloud, —

"Friend — "

"There; wait a bit. I am not your friend, I am your enemy; I am not your equal, I am your master. Call me 'my lord,' if you please. You are too familiar."

"I don't like such titles. I am willing to call you *sir*. That is as far as — "

"We will have no argument about this. Just obey; that is all. Go on with your chatter."

"Very well, my lord, — since nothing but my lord will suit you, — I was going to ask you how long you will be visible to me?"

"Always!"

I broke out with strong indignation: "This is simply an outrage. That is what I think of it. You have dogged, and dogged, and *dogged* me, all the days of my life, invisible. That was misery enough; now to have such a looking thing as you tagging after me like another shadow all the rest of my days is an intolerable prospect. You have my opinion, my lord; make the most of it."

"My lad, there was never so pleased a conscience in this world as I was when you made me visible. It gives me an inconceivable advantage. *Now,* I can look you straight in the eye, and call you names, and leer at you, jeer at you, sneer at you; and *you* know what eloquence there is in visible gesture and expression, more especially when the effect is heightened by audible speech. I shall always address you henceforth in your o-w-n s-n-i-v-e-l-l-i-n-g d-r-a-w-l — baby!"

I let fly with the coal-hod. No result. My lord said, —

"Come, come! Remember the flag of truce!"

"Ah, I forgot that. I will try to be civil; and *you* try it, too, for a novelty. The idea of a *civil* conscience! It is a good joke; an excellent joke. All the consciences *I* have ever heard of were nagging, badgering, fault-finding, execrable savages! Yes; and always in a sweat about some poor little insignificant trifle or other — destruction catch the lot of them, *I* say! I would trade mine for the small-pox and seven kinds of consumption, and be glad of the chance. Now tell me, why *is* it that a conscience can't haul a man over the coals once, for an offence, and then let him alone? Why is it that it wants to keep on pegging at him, day and night and night and day, week in and week out, forever and ever, about the same old thing? There is no sense in that, and no reason in it. I think a conscience that will act like that is meaner than the very dirt itself."

"Well, *we* like it; that suffices."

"Do you do it with the honest intent to improve a man?"

That question produced a sarcastic smile, and this reply: —

"No, sir. Excuse me. We do it simply because it is 'business.' It is our trade. The *purpose* of it *is* to improve the man, but *we* are merely disinterested agents. We are appointed by authority, and have n't anything to say in the matter. We obey orders and leave the consequences where they belong. But I am willing to admit this much: we *do* crowd the orders a trifle when we get a chance, which is most of the time. We enjoy it. We are instructed to remind a man a few times of an error; and I don't mind acknowledging that we try to give pretty good measure. And when we get hold of a man of a peculiarly sensitive nature, oh, but we do haze him! I have known consciences to come all the way from China and Russia to see a person of that kind put through his paces, on a special occasion. Why, I knew a man of that sort who had accidentally crippled a mulatto baby; the news went abroad, and I wish you may never commit another sin if the consciences did n't flock from all over the earth to enjoy the fun and help his master exercise him. That man walked the floor in torture for forty-eight hours, without eating or sleeping, and then blew his brains out. The child was perfectly well again in three weeks."

"Well, you are a precious crew, not to put it too strong. I think I begin to see, now, why you have always been a trifle inconsistent with me. In your anxiety to get all the juice you can out of a sin, you make a man repent of it in three or four different ways. For instance, you found fault with me for lying to that tramp, and I suffered over that. But it was only yesterday that I told a tramp the square truth, to wit, that, it being regarded as bad citizenship to encourage vagrancy, I would give him nothing. What did you do *then?* Why, you made me say to myself, 'Ah, it would have been so much kinder and more blameless to ease him off with a little white lie, and send him away feeling that if he could not have bread, the gentle treatment was at least something to be grateful for!' Well, I suffered all day about *that.* Three days before, I had fed a tramp, and fed him freely, supposing it a virtuous act. Straight off you said, 'O false citizen, to have fed a tramp!' and I suffered as usual. I gave a tramp work; you objected to it, —*after* the contract was made, of course; you never speak up beforehand. Next, I *refused* a tramp work; you objected to *that.* Next, I proposed to kill a tramp; you kept me awake all night, oozing remorse at every pore. Sure I was going to be right *this* time, I sent the next tramp away with my benediction; and I wish you may live as long as I do, if you did n't make me smart all night again because I did n't kill him. Is there *any* way of satisfying that malignant invention which is called a conscience?"

"Ha, ha! this is luxury! Go on!"

"But come, now, answer me that question. *Is* there any way?"

"Well, none that I propose to tell *you*, my son. Ass! I don't care *what* act you may turn your hand to, I can straightway whisper a word in your ear and make you think you have committed a dreadful meanness. It is my *business* — and my joy — to make you repent of *every*thing you do. If I have fooled away any opportunities it was not intentional; I beg to assure you it was not intentional!"

"Don't worry; you have n't missed a trick that *I* know of. I never did a thing in all my life, virtuous or otherwise, that I did n't repent of within twenty-four hours. In church last Sunday I listened to a charity sermon. My first impulse was to give three hundred and fifty dollars; I repented of that and reduced it a hundred; repented of that and reduced it another hundred; repented of that and reduced it another hundred; repented of that and reduced the remaining fifty to twenty-five; repented of that and came down to fifteen; repented of that and dropped to two dollars and a half; when the plate came around at last, I repented once more and contributed ten cents. Well, when I got home, I did wish to goodness I had that ten cents back again! You never *did* let me get through a charity sermon without having something to sweat about."

"Oh, and I never shall, I never shall. You can always depend on me."

"I think so. Many and many 's the restless night I 've wanted to take you by the neck. If I could only get hold of you now!"

"Yes, no doubt. But I am not an ass; I am only

the saddle of an ass. But go on, go on. You entertain me more than I like to confess."

"I am glad of that. (You will not mind my lying a little, to keep in practice.) Look here; not to be too personal, I think you are about the shabbiest and most contemptible little shrivelled-up reptile that can be imagined. I am grateful enough that you are invisible to other people, for I should die with shame to be seen with such a mildewed monkey of a conscience as *you* are. Now if you were five or six feet high, and —"

"Oh, come! who is to blame?"

"*I* don't know."

"Why, you are; nobody else."

"Confound you, I was n't consulted about your personal appearance."

"I don't care, you had a good deal to do with it, nevertheless. When you were eight or nine years old, I was seven feet high, and as pretty as a picture."

"I wish you had died young! So you have grown the wrong way, have you?"

"Some of us grow one way and some the other. You had a large conscience once; if you 've a small conscience now, I reckon there are reasons for it. However, both of us are to blame, you and I. You see, you used to be conscientious about a great many things; morbidly so, I may say. It was a great many years ago. You probably do not remember it, now. Well, I took a great interest in my work, and I so enjoyed the anguish which certain pet sins of yours afflicted you with, that I kept pelting at you until I rather overdid the matter. You began to rebel. Of course I began to lose ground, then, and shrivel a little, — diminish in stature, get mouldy, and grow deformed. The more I weakened, the more stubbornly you fastened on to those particular sins; till at last the places on my person that represent those vices became as callous as shark skin. Take smoking, for instance. I played that card a little too long, and I lost. When people plead with you at this late day to quit that vice, that old callous place seems to enlarge and cover me all over like a shirt of mail. It exerts a mysterious, smothering effect; and presently I, your faithful hater, your devoted Conscience, go soùnd asleep! Sound? It is no name for it. I could n't hear it thunder at such a time. You have some few other vices — perhaps eighty, or maybe ninety — that affect me in much the same way."

"This is flattering; you must be asleep a good part of your time."

"Yes, of late years. I should be asleep *all* the time, but for the help I get."

"Who helps you?"

"Other consciences. Whenever a person whose conscience I am acquainted with tries to plead with you about the vices you are callous to, I get my friend to give his client a pang concerning some villany of his own, and that shuts off his meddling and starts him off to hunt personal consolation. My field of usefulness is about trimmed down to tramps, budding authoresses, and that line of goods, now; but don't you worry — I'll harry you on *them* while they last! Just you put your trust in me."

"I think I can. But if you had only been good enough to mention these facts some thirty years ago, I should have turned my particular attention to sin, and I think that by this time I should not only have had you pretty permanently asleep on the entire list of human vices, but reduced to the size of a homœopathic pill, at that. That is about the style of conscience *I* am pining for. If I only had you shrunk down to a homœopathic pill, and could get my hands on you, would I put you in a glass case for a keepsake? No, sir. I would give you to a yellow dog! That is where *you* ought to be — you and all your tribe. You are not fit to be in society, in my opinion. Now another question. Do you know a good many consciences in this section?"

"Plenty of them."

"I would give anything to see some of them! Could you bring them here? And would they be visible to me?"

"Certainly not."

"I suppose I ought to have known that, without asking. But no matter, you can describe them. Tell me about my neighbor Thompson's conscience, please."

"Very well. I know him intimately; have known him many years. I knew him when he was eleven feet high and of a faultless figure. But he is very rusty and tough and misshapen, now, and hardly ever interests himself about anything. As to his present size — well, he sleeps in a cigar box."

"Likely enough. There are few smaller, meaner men in this region than Hugh Thompson. Do you know Robinson's conscience?"

"Yes. He is a shade under four and a half feet high; used to be a blonde; is a brunette, now, but still shapely and comely."

"Well, Robinson is a good fellow. Do you know Tom Smith's conscience?"

"I have known him from childhood. He was thirteen inches high, and rather sluggish, when he was two years old — as nearly all of us are, at that age. He is thirty-seven feet high, now, and the stateliest figure in America. His legs are still racked with growing-pains, but he has a good time, nevertheless. Never sleeps. He is the most active and energetic member of the New England Conscience Club; is president of it. Night and day you can find him pegging away at Smith, panting with his labor, sleeves rolled up, countenance all alive with enjoyment. He has got his victim splendidly dragooned, now. He can make poor Smith imagine that the most innocent little thing he does is an odious sin; and then he sets to work and almost tortures the soul out of him about it."

"Smith is the noblest man in all this section, and the purest; and yet is always breaking his heart because he cannot be good! Only a conscience *could* find pleasure in heaping agony upon a spirit like that. Do you know my aunt Mary's conscience?"

"I have seen her at a distance, but am not acquainted with her. She lives in the open air altogether, because no door is large enough to admit her."

"I can believe that. Let me see. Do you know the conscience of that publisher who once stole some sketches of mine for a 'series' of his, and then left me to pay the law expenses I had to incur in order to choke him off?"

"Yes. He has a wide fame. He was exhibited, a month ago, with some other antiquities, for the benefit of a recent Member of the Cabinet's conscience, that was starving in exile. Tickets and fares were high, but I travelled for nothing by pretending to be the conscience of an editor, and got in for half price by representing myself to be the conscience of a clergyman. However, the publisher's conscience, which was to have been the main feature of the entertainment, was a failure — as an exhibition. He was there, but what of that? The management had provided a microscope with a magnifying power of only thirty thousand diameters, and so nobody got to see him, after all. There was great and general dissatisfaction, of course, but — "

Just here there was an eager footstep on the stair; I opened the door, and my aunt Mary burst into the room. It was a joyful meeting, and a cheery bombardment of questions and answers concerning family matters ensued. By and by my aunt said, —

"But I am going to abuse you a little now. You promised me, the day I saw you last, that you would look after the needs of the poor family around the corner as faithfully as I had done it myself. Well, I found out by accident that you failed of your promise. *Was* that right?"

In simple truth, I never had thought of that family a second time! And now such a splintering pang of guilt shot through me! I glanced up at my Conscience. Plainly, my heavy heart was affecting him. His body was drooping forward; he seemed about to fall from the book-case. My aunt continued: —

"And think how you have neglected my poor *protégée* at the almshouse, you dear, hard-hearted promise-breaker!" I blushed scarlet, and my tongue was tied. As the sense of my guilty negligence waxed sharper and stronger, my Conscience began to sway heavily back and forth; and when my aunt, after a little pause, said in a grieved tone, "Since you never once went to see her, maybe it will not distress you now to know that that poor child died, months ago, utterly friendless and forsaken!" my Conscience could no longer bear up under the weight of my sufferings, but tumbled headlong from his high perch and struck the floor with a dull, leaden thump. He lay there writhing with pain and quaking with apprehension, but straining every muscle in frantic efforts to get up. In a fever of expectancy I sprang to the door, locked it, placed my back against it, and bent a watchful gaze upon my struggling master. Already my fingers were itching to begin their murderous work.

"Oh, what *can* be the matter!" exclaimed my aunt, shrinking from me, and following with her frightened eyes the direction of mine. My breath was coming in short, quick gasps now, and my excitement was almost uncontrollable. My aunt cried out, —

"Oh, do not look so! You appall me! Oh, what can the matter be? What is it you see? Why do you stare so? Why do you work your fingers like that?"

"Peace, woman!" I said, in a hoarse whisper. "Look elsewhere; pay no attention to me; it is nothing — nothing. I am often this way. It will pass in a moment. It comes from smoking too much."

My injured lord was up, wild-eyed with terror, and trying to hobble toward the door. I could hardly breathe, I was so wrought up. My aunt wrung her hands, and said, —

"Oh, I knew how it would be; I knew it would come to this at last! Oh, I implore you to crush out that fatal habit while it may yet be time! You must not, you shall not be deaf to my supplications longer!" My struggling Conscience showed sudden signs of weariness! "Oh, promise me you will throw off this hateful slavery of tobacco!" My Conscience began to reel drowsily, and grope with his hands — enchanting spectacle! "I beg you, I beseech you, I implore you! Your reason is deserting you! There is madness in your eye! It flames with frenzy! Oh, hear me, hear me, and be saved! See, I plead with you on my very knees!" As she sank before me my Conscience reeled again, and then drooped languidly to the floor, blinking toward me a last supplication for mercy, with heavy eyes. "Oh, promise, or you are lost! Promise, and be redeemed! Promise! Promise and live!" With a long-drawn sigh my conquered Conscience closed his eyes and fell fast asleep!

With an exultant shout I sprang past my aunt, and in an instant I had my life-long foe by the throat. After so many years of waiting and longing, he was mine at last. I tore him to shreds and fragments. I rent the fragments to bits. I cast the bleeding rubbish into the fire, and drew into my nostrils the grateful incense of my burnt-offering. At last, and forever, my Conscience was dead!

I was a free man! I turned upon my poor aunt, who was almost petrified with terror, and shouted, —

"Out of this with your paupers, your charities, your reforms, your pestilent morals! You behold before you a man whose life-conflict is done, whose soul is at peace; a man whose heart is dead to sorrow, dead to suffering, dead to remorse; a man WITHOUT A CONSCIENCE! In my joy I spare you, though I could throttle you and never feel a pang! Fly!"

She fled. Since that day my life is all bliss. Bliss, unalloyed bliss. Nothing in all the world could persuade me to have a conscience again. I settled all my old outstanding scores, and began the world anew. I killed thirty-eight persons during the first two weeks — all of them on account of ancient grudges. I burned a dwelling that interrupted my view. I swindled a widow and some orphans out of their last cow, which is a very good one, though not thoroughbred, I believe. I have also committed scores of crimes, of various kinds, and have enjoyed my work exceedingly, whereas it would formerly have broken my heart and turned my hair gray, I have no doubt.

In conclusion I wish to state, by way of advertisement, that medical colleges desiring assorted tramps for scientific purposes, either by the gross, by cord measurement, or per ton, will do well to examine the lot in my cellar before purchasing elsewhere, as these were all selected and prepared by myself, and can be had at a low rate, because I wish to clear out my stock and get ready for the spring trade.

ABOUT MAGNANIMOUS-INCIDENT LITERATURE.

ALL my life, from boyhood up, I have had the habit of reading a certain set of anecdotes, written in the quaint vein of The World's ingenious Fabulist, for the lesson they taught me and the pleasure they gave me. They lay always convenient to my hand, and whenever I thought meanly of my kind I turned to them, and they banished that sentiment; whenever I felt myself to be selfish, sordid, and ignoble I turned to them, and they told me what to do to win back my self-respect. Many times I wished that the charming anecdotes had not stopped with their happy climaxes, but had continued the pleasing history of the several benefactors and beneficiaries. This wish rose in my breast so persistently that at last I determined to satisfy it by seeking out the sequels of those anecdotes myself. So I set about it, and after great labor and tedious research accomplished my task. I will lay the result before you, giving you each anecdote in its turn, and following it with its sequel as I gathered it through my investigations.

THE GRATEFUL POODLE.

One day a benevolent physician (who had read the books) having found a stray poodle suffering from a broken leg, conveyed the poor creature to his home, and after setting and bandaging the injured limb gave the little outcast its liberty again, and thought no more about the matter. But how great was his surprise, upon opening his door one morning, some days later, to find the grateful poodle patiently waiting there, and in its company another stray dog, one of whose legs, by some accident, had been broken. The kind physician at once relieved the distressed animal, nor did he forget to admire the inscrutable goodness and mercy of God, who had been willing to use so humble an instrument as the poor outcast poodle for the inculcating of, etc., etc., etc.

SEQUEL.

The next morning the benevolent physician found the two dogs, beaming with gratitude, waiting at his door, and with them two other dogs, — cripples. The cripples were speedily healed, and the four went their way, leaving the benevolent physician more overcome by pious wonder than ever. The day passed, the morning came. There at the door sat now the four reconstructed dogs, and with them four others requiring reconstruction. This day also passed, and another morning came; and now sixteen dogs, eight of them newly crippled, occupied the sidewalk, and the people were going around. By noon the broken legs were all set, but the pious wonder in the good physician's breast was beginning to get mixed with involuntary profanity. The sun rose once more, and exhibited thirty-two dogs, sixteen of them with broken legs, occupying the sidewalk and half of the street; the human spectators took up the rest of the room. The cries of the wounded, the songs of the healed brutes, and the comments of the on-looking citizens made great and inspiring cheer, but traffic was interrupted in that street. The good physician hired a couple of assistant surgeons and got through his benevolent work before dark, first taking the precaution to cancel his church membership, so that he might express himself with the latitude which the case required.

But some things have their limits. When once more the morning dawned, and the good physician looked out upon a massed and far-reaching multitude of clamorous and beseeching dogs, he said, "I might as well acknowledge it, I have been fooled by the books; they only tell the pretty part of the story, and then stop. Fetch me the shot-gun; this thing has gone along far enough."

He issued forth with his weapon, and chanced to step upon the tail of the original poodle, who promptly bit him in the leg. Now the great and good work which this poodle had been engaged in had engendered in him such a mighty and augmenting enthusiasm as to turn his weak head at last and drive him mad. A month later, when the benevolent physician lay in the death throes of hydrophobia, he called his weeping friends about him, and said, —

"Beware of the books. They tell but half of the story. Whenever a poor wretch asks you for help, and you feel a doubt as to what result may flow from your benevolence, give yourself the benefit of the doubt and kill the applicant."

And so saying he turned his face to the wall and gave up the ghost.

THE BENEVOLENT AUTHOR.

A poor and young literary beginner had tried in vain to get his manuscripts accepted. At last, when the horrors of starvation were staring him in the face, he laid his sad case before a celebrated author, beseeching his counsel and assistance. This generous man immediately put aside his own matters and proceeded to peruse one of the despised manuscripts. Having completed his kindly task, he shook the poor young man cordially by the hand, saying, "I perceive merit in this; come again to me on Monday." At the time specified, the celebrated author, with a sweet smile, but saying nothing, spread open a magazine which was damp from the press. What was the poor young man's astonishment to discover upon the printed page his own article. "How can I ever," said he, falling upon his knees and bursting into tears, "testify my gratitude for this noble conduct!" The celebrated author was the renowned Snodgrass; the poor young beginner thus rescued from obscurity and starvation was the afterwards equally renowned Snagsby. Let this pleasing incident admonish us to turn a charitable ear to all beginners that need help.

SEQUEL.

The next week Snagsby was back with five rejected manuscripts. The celebrated author was a little surprised, because in the books the young struggler had needed but one lift, apparently. However, he ploughed through these papers, removing unnecessary flowers and digging up some acres of adjective-stumps, and then succeeded in getting two of the articles accepted.

A week or so drifted by, and the grateful Snagsby arrived with another cargo. The celebrated author had felt a mighty glow of satisfaction within himself the first time he had successfully befriended the poor young struggler, and had compared himself with the generous people in the books with high gratification; but he was beginning to suspect now that he had struck upon something fresh in the noble-episode line. His enthusiasm took a chill. Still, he could not bear to repulse this struggling young author, who clung to him with such pretty simplicity and trustfulness.

Well, the upshot of it all was that the celebrated author presently found himself permanently freighted with the poor young beginner. All his mild efforts to unload his cargo went for nothing. He had to give daily counsel, daily encouragement; he had to keep on procuring magazine acceptances, and then revamping the manuscripts to make them presentable. When the young aspirant got a start at last, he rode into sudden fame by describing the celebrated author's private life with such a caustic humor and such minuteness of blistering detail that the book sold a prodigious edition, and broke the celebrated author's heart with mortification. With his latest gasp he said, "Alas, the books deceived me; they do not tell the whole story. Beware of the struggling young author, my friends. Whom God sees fit to starve, let not man presumptuously rescue to his own undoing."

THE GRATEFUL HUSBAND.

One day a lady was driving through the principal street of a great city with her little boy, when the horses took fright and dashed madly away, hurling the coachman from his box and leaving the occupants of the carriage paralyzed with terror. But a brave youth who was driving a grocery wagon threw himself before the plunging animals, and succeeded in arresting their flight at the peril of his own.[1] The grateful lady took his number, and upon arriving at her home she related the heroic act to her husband (who had read the books), who listened with streaming eyes to the moving recital, and who, after returning thanks, in conjunction with his restored loved ones, to Him who suffereth not even a sparrow to fall to the ground unnoticed, sent for the brave young person, and, placing a check for five hundred dollars in his hand, said, "Take this as a reward for your noble act, William Ferguson, and if ever you shall need a friend, remember that Thompson McSpadden has a grateful heart." Let us learn from this that a good deed cannot fail to benefit the doer, however humble he may be.

[1] This is probably a misprint. — M. T.

SEQUEL.

William Ferguson called the next week and asked Mr. McSpadden to use his influence to get him a higher employment, he feeling capable of better things than driving a grocer's wagon. Mr. McSpadden got him an under-clerkship at a good salary.

Presently William Ferguson's mother fell sick, and William — Well, to cut the story short, Mr. McSpadden consented to take her into his house. Before long she yearned for the society of her younger children; so Mary and Julia were admitted also, and little Jimmy, their brother. Jimmy had a pocket-knife, and he wandered into the drawing-room with it one day, alone, and reduced ten thousand dollars' worth of furniture to an indeterminable value in rather less than three quarters of an hour. A day or two later he fell downstairs and broke his neck, and seventeen of his family's relatives came to the house to attend the funeral. This made them acquainted, and they kept the kitchen occupied after that, and likewise kept the McSpaddens busy hunting up situations of various sorts for them, and hunting up more when they wore these out. The old woman drank a good deal and swore a good deal; but the grateful McSpaddens knew it was their duty to reform her, considering what her son had done for them, so they clave nobly to their generous task. William came often and got decreasing sums of money, and asked for higher and more lucrative employments, — which the grateful McSpadden more or less promptly procured for him. McSpadden consented also, after some demur, to fit William for college; but when the first vacation came and the hero requested to be sent to Europe for his health, the persecuted McSpadden rose against the tyrant and revolted. He plainly and squarely refused. William Ferguson's mother was so astounded that she let her gin-bottle drop, and her profane lips refused to do their office. When she recovered she said in a half-gasp, "Is this your gratitude? Where would your wife and boy be now, but for my son?"

William said, "Is this your gratitude? Did I save your wife's life or not? tell me that!"

Seven relations swarmed in from the kitchen and each said, "And this is his gratitude!"

William's sisters stared, bewildered, and said, "And this is his grat—" but were interrupted by their mother, who burst into tears and exclaimed, "To think that my sainted little Jimmy threw away his life in the service of such a reptile!"

Then the pluck of the revolutionary McSpadden rose to the occasion, and he replied with fervor, "Out of my house, the whole beggarly tribe of you! I was beguiled by the books, but shall never be beguiled again, — once is sufficient for me." And turning to William he shouted, "Yes, you did save my wife's life, and the next man that does it shall die in his tracks!"

Not being a clergyman, I place my text at the end of my sermon instead of at the beginning. Here it is, from Mr. Noah Brooks's Recollections of President Lincoln, in "Scribner's Monthly": —

"J. H. Hackett, in his part of Falstaff, was an actor who gave Mr. Lincoln great delight. With his usual desire to signify to others his sense of obligation, Mr. Lincoln wrote a genial little note to the actor, expressing his pleasure at witnessing his performance. Mr. Hackett, in reply, sent a book of some sort; perhaps it was one of his own authorship. He also wrote several notes to the President. One night, quite late, when the episode had passed out of my mind, I went to the White House in answer to a message. Passing into the President's office, I noticed, to my surprise, Hackett sitting in the anteroom as if waiting for an audience. The President asked me if any one was outside. On being told, he said, half sadly, 'Oh, I can't see him, I can't see him; I was in hopes he had gone away.' Then he added, 'Now this just illustrates the difficulty of having pleasant friends and acquaintances in this place. You know how I liked Hackett as an actor, and how I wrote to tell him so. He sent me that book, and there I thought the matter would end. He is a master of his place in the profession, I suppose, and well fixed in it; but just because we had a little friendly correspondence, such as any two men might have, he wants something. What do you suppose he wants?' I could not guess, and Mr. Lincoln added, 'Well, he wants to be consul to London. Oh, dear!'"

I will observe, in conclusion, that the William Ferguson incident occurred, and within my personal knowledge, — though I have changed the nature of the details, to keep William from recognizing himself in it.

All the readers of this article have in some sweet and gushing hour of their lives played the rôle of Magnanimous-Incident hero. I wish I knew how many there are among them who are willing to talk about that episode and like to be reminded of the consequences that flowed from it.

PUNCH, BROTHERS, PUNCH.

WILL the reader please to cast his eye over the following verses, and see if he can discover anything harmful in them?

> "Conductor, when you receive a fare,
> Punch in the presence of the passenjare!
> A blue trip slip for an eight-cent fare,
> A buff trip slip for a six-cent fare,
> A pink trip slip for a three-cent fare,
> Punch in the presence of the passenjare!
>
> CHORUS.
>
> Punch, brothers! punch with care!
> Punch in the presence of the passenjare!"

I came across these jingling rhymes in a newspaper, a little while ago, and read them a couple of times. They took instant and entire possession of me. All through breakfast they went waltzing through my brain; and when, at last, I rolled up my napkin, I could not tell whether I had eaten anything or not. I had carefully laid out my day's work the day before, — a thrilling tragedy in the novel which I am writing. I went to my den to begin my deed of blood. I took up my pen, but all I could get it to say was, "Punch in the presence of the passenjare." I fought hard for an hour, but it was useless. My head kept humming, "A blue trip slip for an eight-cent fare, a buff trip slip for a six-cent fare," and so on and so on, without peace or respite. The day's work was ruined — I could see that plainly enough. I gave up and drifted down town, and presently discovered that my feet were keeping time to that relentless jingle. When I could stand it no longer I altered my step. But it did no good; those rhymes accommodated themselves to the new step and went on harassing me just as before. I returned home, and suffered all the afternoon; suffered all through an unconscious and unrefreshing dinner; suffered, and cried, and jingled all through the evening; went to bed and rolled, tossed, and jingled right along, the same as ever; got up at midnight frantic, and tried to read; but there was nothing visible upon the whirling page except "Punch! punch in the presence of the passenjare." By sunrise I was out of my mind, and everybody marvelled and was distressed at the idiotic burden of my ravings, — "Punch! oh, punch! punch in the presence of the passenjare!"

Two days later, on Saturday morning, I arose, a tottering wreck, and went forth to fulfil an engagement with a valued friend, the Rev. Mr. ——, to walk to the Talcott Tower, ten miles distant. He stared at me, but asked no questions. We started. Mr. —— talked, talked, talked — as is his wont. I said nothing; I heard nothing. At the end of a mile, Mr. —— said, —

"Mark, are you sick? I never saw a man look so

haggard and worn and absent-minded. Say something; do!"

Drearily, without enthusiasm, I said: "Punch, brothers, punch with care! Punch in the presence of the passenjare!"

My friend eyed me blankly, looked perplexed, then said, —

"I do not think I get your drift, Mark. There does not seem to be any relevancy in what you have said, certainly nothing sad; and yet — maybe it was the way you *said* the words — I never heard anything that sounded so pathetic. What is —"

But I heard no more. I was already far away with my pitiless, heart-breaking "blue trip slip for an eight-cent fare, buff trip slip for a six-cent fare, pink trip slip for a three-cent fare; punch in the presence of the passenjare." I do not know what occurred during the other nine miles. However, all of a sudden Mr. —— laid his hand on my shoulder and shouted, —

"Oh, wake up! wake up! wake up! Don't sleep all day! Here we are at the Tower, man! I have talked myself deaf and dumb and blind, and never got a response. Just look at this magnificent autumn landscape! Look at it! look at it! Feast your eyes on it! You have travelled; you have seen boasted landscapes elsewhere. Come, now, deliver an honest opinion. What do you say to this?"

I sighed wearily, and murmured, —

"A buff trip slip for a six-cent fare, a pink trip slip for a three-cent fare, punch in the presence of the passenjare."

Rev. Mr. —— stood there, very grave, full of concern, apparently, and looked long at me; then he said, —

"Mark, there is something about this that I cannot understand. Those are about the same words you said before; there does not seem to be anything in them, and yet they nearly break my heart when you say them. Punch in the — how is it they go?"

I began at the beginning and repeated all the lines. My friend's face lighted with interest. He said, —

"Why, what a captivating jingle it is! It is almost music. It flows along so nicely. I have nearly caught the rhymes myself. Say them over just once more, and then I'll have them, sure."

I said them over. Then Mr. —— said them. He made one little mistake, which I corrected. The next time and the next he got them right. Now a great burden seemed to tumble from my shoulders. That torturing jingle departed out of my brain, and a grateful sense of rest and peace descended upon me. I was light-hearted enough to sing; and I did sing for half an hour, straight along, as we went jogging homeward. Then my freed tongue found blessed speech again, and the pent talk of many a weary hour began to gush and flow. It flowed on and on, joyously, jubilantly, until the fountain was empty and dry. As I wrung my friend's hand at parting, I said, —

"Have n't we had a royal good time! But now I remember, you have n't said a word for two hours. Come, come, out with something!"

The Rev. Mr. —— turned a lack-lustre eye upon

me, drew a deep sigh, and said, without animation, without apparent consciousness, —

"Punch, brothers, punch with care! Punch in the presence of the passenjare!"

A pang shot through me as I said to myself, "Poor fellow, poor fellow! *he* has got it, now."

I did not see Mr. —— for two or three days after that. Then, on Tuesday evening, he staggered into my presence and sank dejectedly into a seat. He was pale, worn; he was a wreck. He lifted his faded eyes to my face and said, —

"Ah, Mark, it was a ruinous investment that I made in those heartless rhymes. They have ridden me like a nightmare, day and night, hour after hour, to this very moment. Since I saw you I have suffered the torments of the lost. Saturday evening I had a sudden call, by telegraph, and took the night train for Boston. The occasion was the death of a valued old friend who had requested that I should preach his funeral sermon. I took my seat in the cars and set myself to framing the discourse. But I never got beyond the opening paragraph; for then the train started and the car-wheels began their 'clack, clack — clack-clack-clack! clack, clack — clack-clack-clack!' and right away those odious rhymes fitted themselves to that accompaniment. For an hour I sat there and set a syllable of those rhymes to every separate and distinct clack the car-wheels made. Why, I was as fagged out, then, as if I had been chopping wood all day. My skull was splitting with headache. It seemed to me that I must go mad if I sat there any longer; so I undressed and went to bed. I stretched myself out in my berth, and — well, you know what the result was. The thing went right along, just the same. 'Clack-clack-clack, a blue trip slip, clack-clack-clack, for an eight-cent fare; clack-clack-clack, a buff trip slip, clack-clack-clack, for a six-cent fare, and so on, and so on, and so on — *punch*, in the presence of the passenjare!' Sleep? Not a single wink! I was almost a lunatic when I got to Boston. Don't ask me about the funeral. I did the best I could, but every solemn individual sentence was meshed and tangled and woven in and out with 'Punch, brothers, punch with care, punch in the presence of the passenjare.' And the most distressing thing was that my *delivery* dropped into the undulating rhythm of those pulsing rhymes, and I could actually catch absent-minded people nodding *time* to the swing of it with their stupid heads. And, Mark, you may believe it or not, but before I got through, the entire assemblage were placidly bobbing their heads in solemn unison, mourners, undertaker, and all. The moment I had finished, I fled to the anteroom in a state bordering on frenzy. Of course it would be my luck to find a sorrowing and aged maiden aunt of the deceased there, who had arrived from Springfield too late to get into the church. She began to sob, and said, —

"'Oh, oh, he is gone, he is gone, and I did n't see him before he died!'

"'Yes!' I said, 'he *is* gone, he *is* gone, he *is* gone — oh, *will* this suffering never cease!'

"'*You* loved him, then! Oh, you too loved him!'

"'Loved him! Loved *who?*'

"'Why, my poor George! my poor nephew!'

"'Oh — *him!* Yes — oh, yes, yes. Certainly — certainly. Punch — punch — oh, this misery will kill me!'

"'Bless you! bless you, sir, for these sweet words! *I*, too, suffer in this dear loss. Were you present during his last moments?'

"'Yes! I — *whose* last moments?'

"'*His.* The dear departed's.'

"'Yes! Oh, yes — yes — *yes!* I suppose so, I think so, *I* don't know! Oh, certainly — I was there — *I* was there!'

"'Oh, what a privilege! what a precious privilege! And his last words — oh, tell me, tell me his last words! What did he say?'

"'He said — he said — oh, my head, my head, my head! He said — he said — he never said *any*thing but Punch, punch, *punch* in the presence of the passenjare! Oh, leave me, madam! In the name of all that is generous, leave me to my madness, my misery, my despair! — a buff trip slip for a six-cent fare, a pink trip slip for a three-cent fare — endu-rance *can* no fur-ther go! — PUNCH in the presence of the passenjare!"'

My friend's hopeless eyes rested upon mine a pregnant minute, and then he said impressively, —

"Mark, you do not say anything. You do not offer me any hope. But, ah me, it is just as well — it is just as well. You could not do me any good. The time has long gone by when words could comfort me. Something tells me that my tongue is doomed to wag forever to the jigger of that remorseless jingle. There — there it is coming on me again : a blue trip slip for an eight-cent fare, a buff trip slip for a —"

Thus murmuring faint and fainter, my friend sank into a peaceful trance and forgot his sufferings in a blessed respite.

How did I finally save him from the asylum? I took him to a neighboring university and made him discharge the burden of his persecuting rhymes into the eager ears of the poor, unthinking students. How is it with *them*, now? The result is too sad to tell. Why did I write this article? It was for a worthy, even a noble, purpose. It was to warn you, reader, if you should come across those merciless rhymes, to avoid them — avoid them as you would a pestilence!

A CURIOUS EXPERIENCE.

THIS is the story which the Major told me, as nearly as I can recall it: —

In the winter of 1862–3, I was commandant of Fort Trumbull, at New London, Conn. Maybe our life there was not so brisk as life at "the front"; still it was brisk enough, in its way — one's brains did n't cake together there for lack of something to keep them stirring. For one thing, all the Northern atmosphere at that time was thick with mysterious rumors — rumors to the effect that rebel spies were flitting everywhere, and getting ready to blow up our Northern forts, burn our hotels, send infected clothing into our towns, and all that sort of thing. You remember it. All this had a tendency to keep us awake, and knock the traditional dulness out of garrison life. Besides, ours was a recruiting station — which is the same as saying we had n't any time to waste in dozing, or dreaming, or fooling around. Why, with all our watchfulness, fifty per cent of a day's recruits would leak out of our hands and give us the slip the same night. The bounties were so prodigious that a recruit could pay a sentinel three or four hundred dollars to let him escape, and still have enough of his bounty-money left to constitute a fortune for a poor man. Yes, as I said before, our life was not drowsy.

Well, one day I was in my quarters alone, doing some writing, when a pale and ragged lad of fourteen or fifteen entered, made a neat bow, and said, —

"I believe recruits are received here?"

"Yes."

"Will you please enlist me, sir?"

"Dear me, no! You are too young, my boy, and too small."

A disappointed look came into his face, and quickly deepened into an expression of despondency. He turned slowly away, as if to go; hesitated, then faced me again, and said, in a tone which went to my heart, —

"I have no home, and not a friend in the world. If you *could* only enlist me!"

But of course the thing was out of the question, and I said so as gently as I could. Then I told him to sit down by the stove and warm himself, and added, —

"You shall have something to eat presently. You are hungry?"

He did not answer; he did not need to; the gratitude in his big soft eyes was more eloquent than any words could have been. He sat down by the stove, and I went on writing. Occasionally I took a furtive glance at him. I noticed that his clothes and shoes, although soiled and damaged, were of good style and material. This fact was suggestive. To it I added the facts that his voice was low and musical; his eyes deep and melancholy; his carriage and address gentlemanly; evidently the poor chap was in trouble. As a result, I was interested.

However, I became absorbed in my work, by and by, and forgot all about the boy. I don't know how long this lasted; but, at length, I happened to look up. The boy's back was toward me, but his face was turned in such a way that I could see one of his cheeks — and down that cheek a rill of noiseless tears was flowing.

"God bless my soul!" I said to myself; "I forgot the poor rat was starving." Then I made amends for my brutality by saying to him, "Come along, my lad; you shall dine with *me;* I am alone to-day."

He gave me another of those grateful looks, and a happy light broke in his face. At the table he stood with his hand on his chair-back until I was seated, then seated himself. I took up my knife and fork and — well, I simply held them, and kept still; for the boy had inclined his head and was saying a silent grace. A thousand hallowed memories of home and my childhood poured in upon me, and I sighed to think how far I had drifted from religion and its balm for hurt minds, its comfort and solace and support.

As our meal progressed, I observed that young Wicklow — Robert Wicklow was his full name — knew what to do with his napkin; and — well, in a word, I observed that he was a boy of good breeding; never mind the details. He had a simple frankness, too, which won upon me. We talked mainly about himself, and I had no difficulty in getting his history out of him. When he spoke of his having been born and reared in Louisiana, I warmed to him decidedly, for I had spent some time down there. I knew all the "coast" region of the Mississippi, and loved it, and had not been long enough away from it for my interest in it to begin to pale. The very names that fell from his lips sounded good to me, — so good that I steered the talk in directions that would bring them out. Baton Rouge, Plaquemine, Donaldsonville, Sixty-mile Point, Bonnet-Carre, the Stock-Landing, Carrollton, the Steamship Landing, the Steamboat Landing, New Orleans, Tchoupitoulas Street, the Esplanade, the Rue des Bons Enfants, the St. Charles Hotel, the Tivoli Circle, the Shell Road, Lake Pontchartrain; and it was particularly delightful to me to hear once more of the "R. E. Lee," the "Natchez," the "Eclipse," the "General Quitman," the "Duncan F. Kenner," and other old familiar steamboats. It was almost as good as being back there, these names so vividly reproduced in my mind the look of the things they stood for. Briefly, this was little Wicklow's history: —

When the war broke out, he and his invalid aunt and his father were living near Baton Rouge, on a great and rich plantation which had been in the family for fifty years. The father was a Union man. He was persecuted in all sorts of ways, but clung to his principles. At last, one night, masked men burned his mansion down, and the family had to fly for their lives. They were hunted from place to place, and learned all there was to know about poverty, hunger, and distress.

The invalid aunt found relief at last: misery and exposure killed her; she died in an open field, like a tramp, the rain beating upon her and the thunder booming overhead. Not long afterward, the father was captured by an armed band; and while the son begged and pleaded, the victim was strung up before his face. [At this point a baleful light shone in the youth's eyes, and he said, with the manner of one who talks to himself: "If I cannot be enlisted, no matter — I shall find a way — I shall find a way."] As soon as the father was pronounced dead, the son was told that if he was not out of that region within twenty-four hours, it would go hard with him. That night he crept to the riverside and hid himself near a plantation landing. By and by the "Duncan F. Kenner," stopped there, and he swam out and concealed himself in the yawl that was dragging at her stern. Before daylight the boat reached the Stock-Landing, and he slipped ashore. He walked the three miles which lay between that point and the house of an uncle of his in Good-Children Street, in New Orleans, and then his troubles were over for the time being. But this uncle was a Union man, too, and before very long he concluded that he had better leave the South. So he and young Wicklow slipped out of the country on board a sailing vessel, and in due time reached New York. They put up at the Astor House. Young Wicklow had a good time of it for a while, strolling up and down Broadway, and observing the strange Northern sights; but in the end a change came, — and not for the better. The uncle had been cheerful at first, but now he began to look troubled and despondent; moreover, he became moody and irritable; talked of money giving out, and no way to get more, — "not enough left for one, let alone two." Then, one morning, he was missing — did not come to breakfast. The boy inquired at the office, and was told that the uncle had paid his bill the night before and gone away — to Boston, the clerk believed, but was not certain.

The lad was alone and friendless. He did not know what to do, but concluded he had better try to follow and find his uncle. He went down to the steamboat landing; learned that the trifle of money in his pocket would not carry him to Boston; however, it would carry him to New London; so he took passage for that port, resolving to trust to Providence to furnish him means to travel the rest of the way. He had now been wandering about the streets of New London three days and nights, getting a bite and a nap here and there for charity's sake. But he had given up at last; courage and hope were both gone. If he could enlist, nobody could be more thankful; if he could not get in as a soldier, could n't he be a drummer-boy? Ah, he would work *so* hard to please, and would be so grateful!

Well, there 's the history of young Wicklow, just as he told it to me, barring details. I said, —

"My boy, you 're among friends, now, — don't you be troubled any more." How his eyes glistened! I called in Sergeant John Rayburn, — he was from Hartford; lives in Hartford yet; maybe you know him, — and said, "Rayburn, quarter this boy with the musicians. I am going to enroll him as a drummer-boy, and I want you to look after him and see that he is well treated."

Well, of course, intercourse between the commandant of the post and the drummer-boy came to an end, now; but the poor little friendless chap lay heavy on my heart, just the same. I kept on the lookout, hoping to see him brighten up and begin to be cheery and gay; but no, the days went by, and there was no change. He associated with nobody; he was always absent-minded, always thinking; his face was always sad. One morning Rayburn asked leave to speak to me privately. Said he, —

"I hope I don't offend, sir; but the truth is, the musicians are in such a sweat it seems as if somebody 's *got* to speak."

"Why, what is the trouble?"

"It 's the Wicklow boy, sir. The musicians are down on him to an extent you can't imagine."

"Well, go on, go on. What has he been doing?"

"Prayin', sir."

"Praying!"

"Yes, sir; the musicians have n't any peace of their life for that boy's prayin'. First thing in the morning he 's at it; noons he 's at it; and nights — well, *nights* he just lays into 'em like all possessed! Sleep? Bless you, they *can't* sleep: he 's got the floor, as the sayin' is, and then when he once gets his supplication-mill agoin', there just simply ain't any let-up *to* him. He starts in with the band-master, and he prays for him; next he takes the head bugler, and he prays for him; next the bass drum, and he scoops *him* in; and so on, right straight through the band, givin' them all a show, and takin' that amount of interest in it which would make you think he thought he warn't but a little while for this world, and believed he could n't be happy in heaven without he had a brass band along, and wanted to pick 'em out for himself, so he could depend on 'em to do up the national tunes in a style suitin' to the place. Well, sir, heavin' boots at him don't have no effect; it 's dark in there; and, besides, he don't pray fair, anyway, but kneels down behind the big drum; so it don't make no difference if they *rain* boots at him, *he* don't give a dern — warbles right along, same as if it was applause. They sing out, 'Oh, dry up!' 'Give us a rest!' 'Shoot him!' 'Oh, take a walk!' and all sorts of such things. But what of it? It don't phaze him. *He* don't mind it." After a pause: "Kind of a good little fool, too; gits up in the mornin' and carts all that stock of boots back, and sorts 'em out and sets each man's pair where they belong. And they 've been throwed at him so much now, that he knows every boot in the band, — can sort 'em out with his eyes shut."

After another pause, which I forbore to interrupt, —

"But the roughest thing about it is, that when he 's done prayin', — when he ever *does* get done, — he pipes up and begins to *sing*. Well, you know what a honey kind of a voice he 's got when he talks; you know how it would persuade a cast-iron dog to come down off of a doorstep and lick his hand. Now if you 'll take my word for it, sir, it ain't a circumstance to his singin'!

Flute music is harsh to that boy's singin'. Oh, he just gurgles it out so soft and sweet and low, there in the dark, that it makes you think you are in heaven."

"What is there 'rough' about that?"

"Ah, that 's just it, sir. You hear him sing

"'Just as I am — poor, wretched, blind,'

— just you hear him sing that, once, and see if you don't melt all up and the water come into your eyes! I don't care *what* he sings, it goes plum straight home to you — it goes deep down to where you *live* — and it fetches you every time! Just you hear him sing: —

"'Child of sin and sorrow, filled with dismay,
Wait not till to-morrow, yield thee to-day;
Grieve not that love
Which, from above' —

and so on. It makes a body feel like the wickedest, ungratefulest brute that walks. And when he sings them songs of his about home, and mother, and childhood, and old memories, and things that 's vanished, and old friends dead and gone, it fetches everything before your face that you 've ever loved and lost in all your life — and it 's just beautiful, it 's just divine to listen to, sir — but, Lord, Lord, the heart-break of it! The band — well, they all cry — every rascal of them blubbers, and don't try to hide it, either; and first you know, that very gang that 's been slammin' boots at that boy will skip out of their bunks all of a sudden, and rush over in the dark and hug him! Yes, they do — and slobber all over him, and call him pet names, and beg him to forgive them. And just at that time, if a regiment was to offer to hurt a hair of that cub's head, they 'd go for that regiment, if it was a whole army corps!"

Another pause.

"Is that all?" said I.

"Yes, sir."

"Well, dear me, what is the complaint? What do they want done?"

"Done? Why, bless you, sir, they want you to stop him from *singin'*."

"What an idea! You said his music was divine."

"That 's just it. It 's *too* divine. Mortal man can't stand it. It stirs a body up so; it turns a body inside out; it racks his feelin's all to rags; it makes him feel bad and wicked, and not fit for any place but perdition. It keeps a body in such an everlastin' state of repentin', that nothin' don't taste good and there ain't no comfort in life. And then the *cryin'*, you see — every mornin' they are ashamed to look one another in the face."

"Well, this is an odd case, and a singular complaint. So they really want the singing stopped?"

"Yes, sir, that is the idea. They don't wish to ask too much; they would like powerful well to have the prayin' shut down on, or leastways trimmed off around the edges; but the main thing 's the singin.' If they can only get the singin' choked off, they think they can stand the prayin', rough as it is to be bullyragged so much that way."

I told the sergeant I would take the matter under consideration. That night I crept into the musicians' quarters and listened. The sergeant had not overstated the case. I heard the praying voice pleading in the dark; I heard the execrations of the harassed men; I heard the rain of boots whiz through the air, and bang and thump around the big drum. The thing touched me, but it amused me, too. By and by, after an impressive silence, came the singing. Lord, the pathos of it, the enchantment of it! Nothing in the world was ever so sweet, so gracious, so tender, so holy, so moving. I made my stay very brief; I was beginning to experience emotions of a sort not proper to the commandant of a fortress.

Next day I issued orders which stopped the praying and singing. Then followed three or four days which were so full of bounty-jumping excitements and irritations that I never once thought of my drummer-boy. But now comes Sergeant Rayburn, one morning, and says, —

"That new boy acts mighty strange, sir."

"How?"

"Well, sir, he 's all the time writing."

"Writing? What does he write — letters?"

"I don't know, sir; but whenever he 's off duty, he is always poking and nosing around the fort, all by himself, — blest if I think there 's a hole or corner in it he has n't been into, — and every little while he outs with pencil and paper and scribbles something down."

This gave me a most unpleasant sensation. I wanted to scoff at it, but it was not a time to scoff at *anything* that had the least suspicious tinge about it. Things were happening all around us, in the North, then, that warned us to be always on the alert, and always suspecting. I recalled to mind the suggestive fact that this boy was from the South, — the extreme South, Louisiana, — and the thought was not of a reassuring nature, under the circumstances. Nevertheless, it cost me a pang to give the orders which I now gave to Rayburn. I felt like a father who plots to expose his own child to shame and injury. I told Rayburn to keep quiet, bide his time, and get me some of those writings whenever he could manage it without the boy's finding it out. And I charged him not to do anything which might let the boy discover that he was being watched. I also ordered that he allow the lad his usual liberties, but that he be followed at a distance when he went out into the town.

During the next two days, Rayburn reported to me several times. No success. The boy was still writing, but he always pocketed his paper with a careless air whenever Rayburn appeared in his vicinity. He had gone twice to an old deserted stable in the town, remained a minute or two, and come out again. One could not pooh-pooh these things — they had an evil look. I was obliged to confess to myself that I was getting uneasy. I went into my private quarters and sent for my second in command, — an officer of intelligence and judgment, son of General James Watson Webb. He was surprised and troubled. We had a long talk over the matter, and came to the conclusion that it would be worth while to institute a secret search. I determined to take charge of that myself. So I had myself called at two in the morning; and, pretty soon after, I was in the musicians' quarters, crawling along the floor on my stomach among the snorers. I reached my slumbering waif's bunk at last, without disturbing anybody, captured his clothes and kit, and crawled stealthily back again. When I got to my own quarters, I found Webb there, waiting and eager to know the result. We made search immediately. The clothes were a disappointment. In the pockets we found blank paper and a pencil; nothing else, except a jack-knife and such queer odds and ends and useless trifles as boys hoard and value. We turned to the kit hopefully. Nothing there but a rebuke for us! — a little Bible with this written on the fly-leaf: "Stranger, be kind to my boy, for his mother's sake."

I looked at Webb — he dropped his eyes; he looked at me — I dropped mine. Neither spoke. I put the book reverently back in its place. Presently Webb got up and went away, without remark. After a little I nerved myself up to my unpalatable job, and took the plunder back to where it belonged, crawling on my stomach as before. It seemed the peculiarly appropriate attitude for the business I was in. I was most honestly glad when it was over and done with.

About noon next day Rayburn came, as usual, to report. I cut him short. I said, —

"Let this nonsense be dropped. We are making a bugaboo out of a poor little cub who has got no more harm in him than a hymn-book."

The sergeant looked surprised, and said, —

"Well, you know it was your orders, sir, and I 've got some of the writing."

"And what does it amount to? How did you get it?"

"I peeped through the key-hole, and see him writing. So when I judged he was about done, I made a sort of a little cough, and I see him crumple it up and throw it in the fire, and look all around to see if anybody was coming. Then he settled back as comfortable and careless as anything. Then I comes in, and passes the time of day pleasantly, and sends him of an errand. He never looked uneasy, but went right along. It was a coal-fire and new-built; the writing had gone over behind a chunk, out of sight; but I got it out; there it is; it ain't hardly scorched, you see."

I glanced at the paper and took in a sentence or two. Then I dismissed the sergeant and told him to send Webb to me. Here is the paper in full:—

"FORT TRUMBULL, the 8th.

"COLONEL,—I was mistaken as to the calibre of the three guns I ended my list with. They are 18-pounders; all the rest of the armament is as I stated. The garrison remains as before reported, except that the two light infantry companies that were to be detached for service at the front are to stay here for the present—can't find out for how long, just now, but will soon. We are satisfied that, all things considered, matters had better be postponed un—"

There it broke off—there is where Rayburn coughed and interrupted the writer. All my affection for the boy, all my respect for him and charity for his forlorn condition, withered in a moment under the blight of this revelation of cold-blooded baseness.

But never mind about that. Here was business, — business that required profound and immediate attention, too. Webb and I turned the subject over and over, and examined it all around. Webb said, —

"What a pity he was interrupted! Something is going to be postponed until — when? And what *is* the something? Possibly he would have mentioned it, the pious little reptile!"

"Yes," I said, "we have missed a trick. And who is '*we*,' in the letter? Is it conspirators inside the fort or outside?"

That "we" was uncomfortably suggestive. However, it was not worth while to be guessing around that, so we proceeded to matters more practical. In the first place, we decided to double the sentries and keep the strictest possible watch. Next, we thought of calling Wicklow in and making him divulge everything; but that did not seem wisest until other methods should fail. We must have some more of the writings; so we began to plan to that end. And now we had an idea: Wicklow never went to the post-office, — perhaps the deserted stable was his post-office. We sent for my confidential clerk — a young German named Sterne, who was a sort of natural detective — and told him all about the case, and ordered him to go to work on it. Within the hour we got word that Wicklow was writing again. Shortly afterward, word came that he had asked leave to go out into the town. He was detained awhile, and meantime Sterne hurried off and concealed himself in the stable. By and by he saw Wicklow saunter in, look about him, then hide something under some rubbish in a corner, and take leisurely leave again. Sterne pounced upon the hidden article — a letter — and brought it to us. It had no superscription and no signature. It repeated what we had already read, and then went on to say: —

"We think it best to postpone till the two companies are gone. I mean the four inside think so; have not communicated with the others — afraid of attracting attention. I say four because we have lost two; they had hardly enlisted and got inside when they were shipped off to the front. It will be absolutely necessary to have two in their places. The two that went were the brothers from Thirty-mile Point. I have something of the greatest importance to reveal, but must not trust it to this method of communication; will try the other."

"The little scoundrel!" said Webb; "who *could* have supposed he was a spy? However, never mind about that; let us add up our particulars, such as they are, and see how the case stands to date. First, we've got a rebel spy in our midst, whom we know; secondly, we've got three more in our midst whom we don't know; thirdly, these spies have been introduced among us through the simple and easy process of enlisting as soldiers in the Union army — and evidently two of them have got sold at it, and been shipped off to the front; fourthly, there are assistant spies 'outside' — number indefinite; fifthly, Wicklow has very important matter which he is afraid to communicate by the 'present method' — will 'try the other.' That is the case, as it now stands. Shall we collar Wicklow and make him confess? Or shall we catch the person who removes the letters from the stable and make *him* tell? Or shall we keep still and find out more?"

We decided upon the last course. We judged that we did not need to proceed to summary measures now, since it was evident that the conspirators were likely to wait till those two light infantry companies were out of the way. We fortified Sterne with pretty ample powers, and told him to use his best endeavors to find out Wicklow's "other method" of communication. We meant to play a bold game; and to this end we proposed to keep the spies in an unsuspecting state as long as possible. So we ordered Sterne to return to the stable immediately, and, if he found the coast clear, to conceal Wicklow's letter where it was before, and leave it there for the conspirators to get.

The night closed down without further event. It was cold and dark and sleety, with a raw wind blowing; still I turned out of my warm bed several times during the night, and went the rounds in person, to see that all was right and that every sentry was on the alert. I always found them wide awake and watchful; evidently whispers of mysterious dangers had been floating about, and the doubling of the guards had been a kind of indorsement of those rumors. Once, toward morning, I encountered Webb, breasting his way against the bitter wind, and learned then that he, also, had been the rounds several times to see that all was going right.

Next day's events hurried things up somewhat. Wicklow wrote another letter; Sterne preceded him to the stable and saw him deposit it; captured it as soon as Wicklow was out of the way, then slipped out and followed the little spy at a distance, with a detective in plain clothes at his own heels, for we thought it judicious to have the law's assistance handy in case of need. Wicklow went to the railway station, and waited around till the train from New York came in, then stood scanning the faces of the crowd as they poured out of the cars. Presently an aged gentleman, with green goggles and a cane, came limping along, stopped in Wicklow's neighborhood, and began to look about him expectantly. In an instant Wicklow darted forward, thrust an envelope into his hand, then glided away and disappeared in the throng. The next instant Sterne had snatched the letter; and as he hurried past the detective, he said: "Follow the old gentleman — don't lose sight of him." Then Sterne skurried out with the crowd, and came straight to the fort.

We sat with closed doors, and instructed the guard outside to allow no interruption.

First we opened the letter captured at the stable. It read as follows: —

"HOLY ALLIANCE, — Found, in the usual gun, commands from the Master, left there last night, which set aside the instructions heretofore received from the subordinate quarter. Have left in the gun the usual indication that the commands reached the proper hand — "

Webb, interrupting: "Is n't the boy under constant surveillance now?"

I said yes; he had been under strict surveillance ever since the capturing of his former letter.

"Then how could he put anything into a gun, or take anything out of it, and not get caught?"

"Well," I said, "I don't like the look of that very well."

"I don't, either," said Webb. "It simply means that there are conspirators among the very sentinels. Without their connivance in some way or other, the thing could n't have been done."

I sent for Rayburn, and ordered him to examine the batteries and see what he could find. The reading of the letter was then resumed: —

"The new commands are peremptory, and require that the MMMM shall be FFFFF at 3 o'clock to-morrow morning. Two hundred will arrive, in small parties, by train and otherwise, from various directions, and will be at appointed place at right time. I will distribute the sign to-day. Success is apparently sure, though something must have got out, for the sentries have been doubled, and the chiefs went the rounds last night several times. W. W. comes from southerly to-day and will receive secret orders — by the other method. All six of you must be in 166 at sharp 2 A. M. You will find B. B. there, who will give you detailed instructions. Password same as last time, only reversed — put first syllable last and last syllable first. REMEMBER XXXX. Do not forget. Be of good heart; before the next sun rises you will be heroes; your fame will be permanent; you will have added a deathless page to history. Amen."

"Thunder and Mars," said Webb, "but we are getting into mighty hot quarters, as I look at it!"

I said there was no question but that things were beginning to wear a most serious aspect. Said I, —

"A desperate enterprise is on foot, that is plain enough. To-night is the time set for it, — that, also, is plain. The exact nature of the enterprise — I mean the manner of it — is hidden away under those blind bunches of M's and F's, but the end and aim, I judge, is the surprise and capture of the post. We must move quick and sharp now. I think nothing can be gained by continuing our clandestine policy as regards Wicklow. We *must* know, and as soon as possible, too, where '166' is located, so that we can make a descent upon the gang there at 2 A. M.; and doubtless the quickest way to get that information will be to force it out of that boy. But first of all, and before we make any important move, I must lay the facts before the War Department, and ask for plenary powers."

The despatch was prepared in cipher to go over the wires; I read it, approved it, and sent it along.

We presently finished discussing the letter which was under consideration, and then opened the one which had been snatched from the lame gentleman. It contained nothing but a couple of perfectly blank sheets of note-paper! It was a chilly check to our hot eagerness and expectancy. We felt as blank as the paper, for a moment, and twice as foolish. But it was for a moment only; for, of course, we immediately afterward thought of "sympathetic ink." We held the paper close to the fire and watched for the characters to come out, under the influence of the heat; but nothing appeared but some faint tracings, which we could make nothing of. We then called in the surgeon, and sent him off with orders to apply every test he was acquainted with till he got the right one, and report the contents of the letter to me the instant he brought them to the surface. This check was a confounded annoyance, and we naturally chafed under the delay; for we had fully expected to get out of that letter some of the most important secrets of the plot.

Now appeared Sergeant Rayburn, and drew from his pocket a piece of twine string about a foot long, with three knots tied in it, and held it up.

"I got it out of a gun on the water-front," said he. "I took the tompions out of all the guns and examined close; this string was the only thing that was in any gun."

So this bit of string was Wicklow's "sign" to signify that the "Master's" commands had not miscarried. I ordered that every sentinel who had served near that gun during the past twenty-four hours be put in confinement at once and separately, and not allowed to communicate with any one without my privity and consent.

A telegram now came from the Secretary of War. It read as follows: —

> "Suspend *habeas corpus*. Put town under martial law. Make necessary arrests. Act with vigor and promptness. Keep the Department informed."

We were now in shape to go to work. I sent out and had the lame gentleman quietly arrested and as quietly brought into the fort; I placed him under guard, and forbade speech to him or from him. He was inclined to bluster at first, but he soon dropped that.

Next came word that Wicklow had been seen to give something to a couple of our new recruits; and that, as soon as his back was turned, these had been seized and confined. Upon each was found a small bit of paper, bearing these words and signs in pencil: —

> EAGLE'S THIRD FLIGHT.
>
> REMEMBER XXXX.
>
> 166.

In accordance with instructions, I telegraphed to the Department, in cipher, the progress made, and also described the above ticket. We seemed to be in a strong enough position now to venture to throw off the mask as regarded Wicklow; so I sent for him. I also sent for and received back the letter written in sympathetic ink, the surgeon accompanying it with the information that thus far it had resisted his tests, but that there were others he could apply when I should be ready for him to do so.

Presently Wicklow entered. He had a somewhat worn and anxious look, but he was composed and easy, and if he suspected anything it did not appear in his face or manner. I allowed him to stand there a moment or two, then I said pleasantly, —

"My boy, why do you go to that old stable so much?"

He answered, with simple demeanor and without embarrassment, —

"Well, I hardly know, sir; there is n't any particular reason, except that I like to be alone, and I amuse myself there."

"You amuse yourself there, do you?"

"Yes, sir," he replied, as innocently and simply as before.

"Is that all you do there?"

"Yes, sir," he said, looking up with childlike wonderment in his big soft eyes.

"You are *sure?*"

"Yes, sir, sure."

After a pause, I said, —

"Wicklow, why do you write so much?"

"I? I do not write much, sir."

"You don't?"

"No, sir. Oh, if you mean scribbling, I *do* scribble some, for amusement."

"What do you do with your scribblings?"

"Nothing, sir — throw them away."

"Never send them to anybody?"

"No, sir."

I suddenly thrust before him the letter to the "Colonel." He started slightly, but immediately composed himself. A slight tinge spread itself over his cheek.

"How came you to send *this* piece of scribbling, then?"

"I nev-never meant any harm, sir."

"Never meant any harm! You betray the armament and condition of the post, and mean no harm by it?"

He hung his head and was silent.

"Come, speak up, and stop lying. Whom was this letter intended for?"

He showed signs of distress, now; but quickly collected himself, and replied, in a tone of deep earnestness, —

"I will tell you the truth, sir — the whole truth. The letter was never intended for anybody at all. I wrote it only to amuse myself. I see the error and foolishness of it, now, — but it is the only offence, sir, upon my honor."

"Ah, I am glad of that. It is dangerous to be writing such letters. I hope you are sure this is the only one you wrote?"

"Yes, sir, perfectly sure."

His hardihood was stupefying. He told that lie with as sincere a countenance as any creature ever wore. I waited a moment to soothe down my rising temper, and then said, —

"Wicklow, jog your memory now, and see if you can help me with two or three little matters which I wish to inquire about."

"I will do my very best, sir."

"Then, to begin with — who is 'the Master'?"

It betrayed him into darting a startled glance at our faces, but that was all. He was serene again in a moment, and tranquilly answered, —

"I do not know, sir."

"You do not know?"

"I do not know."

"You are *sure* you do not know?"

He tried hard to keep his eyes on mine, but the strain was too great; his chin sunk slowly toward his breast and he was silent; he stood there nervously fumbling with a button, an object to command one's pity, in spite of his base acts. Presently I broke the stillness with the question, —

"Who are the 'Holy Alliance'?"

His body shook visibly, and he made a slight random gesture with his hands, which to me was like the appeal of a despairing creature for compassion. But he made no sound. He continued to stand with his face bent toward the ground. As we sat gazing at him, waiting for him to speak, we saw the big tears begin to roll down his cheeks. But he remained silent. After a little, I said, —

"You must answer me, my boy, and you must tell me the truth. Who are the Holy Alliance?"

He wept on in silence. Presently I said, somewhat sharply, —

"Answer the question!"

He struggled to get command of his voice; and then, looking up appealingly, forced the words out between his sobs, —

"Oh, have pity on me, sir! I cannot answer it, for I do not know."

"What!"

"Indeed, sir, I am telling the truth. I never have

heard of the Holy Alliance till this moment. On my honor, sir, this is so."

"Good heavens! Look at this second letter of yours; there, do you see those words, '*Holy Alliance*'? What do you say now?"

He gazed up into my face with the hurt look of one upon whom a great wrong has been wrought, then said, feelingly, —

"This is some cruel joke, sir; and how could they play it upon me, who have tried all I could to do right, and have never done harm to anybody? Some one has counterfeited my hand; I never wrote a line of this; I have never seen this letter before!"

"Oh, you unspeakable liar! Here, what do you say to *this?*" — and I snatched the sympathetic-ink letter from my pocket and thrust it before his eyes.

His face turned white! — as white as a dead person's. He wavered slightly in his tracks, and put his hand against the wall to steady himself. After a moment he asked, in so faint a voice that it was hardly audible, —

"Have you — read it?"

Our faces must have answered the truth before my lips could get out the false "yes," for I distinctly saw the courage come back into that boy's eyes. I waited for him to say something, but he kept silent. So at last I said, —

"Well, what have you to say as to the revelations in this letter?"

He answered, with perfect composure, —

"Nothing, except that they are entirely harmless and innocent; they can hurt nobody."

I was in something of a corner now, as I could n't disprove his assertion. I did not know exactly how to proceed. However, an idea came to my relief, and I said, —

"You are sure you know nothing about the Master and the Holy Alliance, and did not write the letter which you say is a forgery?"

"Yes, sir — sure."

I slowly drew out the knotted twine string and held it up without speaking. He gazed at it indifferently, then looked at me inquiringly. My patience was sorely taxed. However, I kept my temper down, and said in my usual voice, —

"Wicklow, do you see this?"

"Yes, sir."

"What is it?"

"It seems to be a piece of string."

"*Seems?* It *is* a piece of string. Do you recognize it?"

"No, sir," he replied, as calmly as the words could be uttered.

His coolness was perfectly wonderful! I paused now for several seconds, in order that the silence might add impressiveness to what I was about to say; then I rose and laid my hand on his shoulder, and said gravely, —

"It will do you no good, poor boy, none in the world. This sign to the 'Master,' this knotted string, found in one of the guns on the water-front — "

"Found *in* the gun! Oh, no, no, no! do not say *in* the gun, but in a crack in the tompion! — it *must* have been in the crack!" and down he went on his knees and clasped his hands and lifted up a face that was pitiful to see, so ashy it was, and so wild with terror.

"No, it was *in* the gun."

"Oh, something has gone wrong! My God, I am lost!" and he sprang up and darted this way and that, dodging the hands that were put out to catch him, and doing his best to escape from the place. But of course escape was impossible. Then he flung himself on his knees again, crying with all his might, and clasped me around the legs; and so he clung to me and begged and pleaded, saying, "Oh, have pity on me! Oh, be merciful to me! Do not betray me; they would not spare my life a moment! Protect me, save me. I will confess everything!"

It took us some time to quiet him down and modify his fright, and get him into something like a rational frame of mind. Then I began to question him, he answering humbly, with downcast eyes, and from time to time swabbing away his constantly flowing tears.

"So you are at heart a rebel?"

"Yes, sir."

"And a spy?"

"Yes, sir."

"And have been acting under distinct orders from outside?"

"Yes, sir."

"Willingly?"

"Yes, sir."

"*Gladly*, perhaps?"

"Yes, sir; it would do no good to deny it. The South is my country; my heart is Southern, and it is all in her cause."

"Then the tale you told me of your wrongs and the persecution of your family was made up for the occasion?"

"They — they told me to say it, sir."

"And you would betray and destroy those who pitied and sheltered you. Do you comprehend how base you are, you poor misguided thing?"

He replied with sobs only.

"Well, let that pass. To business. Who is the 'Colonel,' and where is he?"

He began to cry hard, and tried to beg off from answering. He said he would be killed if he told. I threatened to put him in the dark cell and lock him up if he did not come out with the information. At the same time I promised to protect him from all harm if he made a clean breast. For all answer, he closed his mouth firmly and put on a stubborn air which I could not bring him out of. At last I started with him; but a single glance into the dark cell converted him. He broke into a passion of weeping and supplicating, and declared he would tell everything.

So I brought him back, and he named the "Colonel," and described him particularly. Said he would be found at the principal hotel in the town, in citizen's dress. I had to threaten him again, before he would describe and name the "Master." Said the Master would be found at No. 15 Bond Street, New York, passing under the name of R. F. Gaylord. I telegraphed name and description to the chief of police of the metropolis, and asked that Gaylord be arrested and held till I could send for him.

"Now," said I, "it seems that there are several of the conspirators 'outside,' — presumably in New London. Name and describe them."

He named and described three men and two women, — all stopping at the principal hotel. I sent out quietly, and had them and the "Colonel" arrested and confined in the fort.

"Next, I want to know all about your three fellow-conspirators who are here in the fort."

He was about to dodge me with a falsehood, I thought; but I produced the mysterious bits of paper which had been found upon two of them, and this had a salutary effect upon him. I said we had possession of two of the men, and he must point out the third. This frightened him badly, and he cried out, —

"Oh, please don't make me; he would kill me on the spot!"

I said that that was all nonsense; I would have somebody near by to protect him, and, besides, the men should be assembled without arms. I ordered all the raw recruits to be mustered, and then the poor trembling little wretch went out and stepped along down the line, trying to look as indifferent as possible. Finally he spoke a single word to one of the men, and before he had gone five steps the man was under arrest.

As soon as Wicklow was with us again, I had those three men brought in. I made one of them stand forward, and said, —

"Now, Wicklow, mind, not a shade's divergence from the exact truth. Who is this man, and what do you know about him?"

Being "in for it," he cast consequences aside, fastened his eyes on the man's face, and spoke straight along without hesitation, — to the following effect.

"His real name is George Bristow. He is from New Orleans; was second mate of the coast-packet 'Capitol,' two years ago; is a desperate character, and has served two terms for manslaughter, — one for killing a deck-hand named Hyde with a capstan-bar, and one for killing a roustabout for refusing to heave the lead, which is no part of a roustabout's business. He is a spy, and was sent here by the Colonel, to act in that capacity. He was third mateof the 'St. Nicholas,' when she blew up in the neighborhood of Memphis, in '58, and came near being lynched for robbing the dead and wounded while they were being taken ashore in an empty wood-boat."

And so forth and so on — he gave the man's biography in full. When he had finished, I said to the man, —

"What have you to say to this?"

"Barring your presence, sir, it is the infernalest lie that ever was spoke!"

I sent him back into confinement, and called the others forward in turn. Same result. The boy gave a detailed history of each, without ever hesitating for a word or a fact; but all I could get out of either rascal was the indignant assertion that it was all a lie. They would confess nothing. I returned them to captivity, and brought out the rest of my prisoners, one by one. Wicklow told all about them — what towns in the South they were from, and every detail of their connection with the conspiracy.

But they all denied his facts, and not one of them confessed a thing. The men raged, the women cried. According to their stories, they were all innocent people from out West, and loved the Union above all things in this world. I locked the gang up, in disgust, and fell to catechising Wicklow once more.

"Where is No. 166, and who is B. B?"

But *there* he was determined to draw the line. Neither coaxing nor threats had any effect upon him. Time was flying — it was necessary to institute sharp measures. So I tied him up a-tiptoe by the thumbs. As the pain increased, it wrung screams from him which were almost more than I could bear. But I held my ground, and pretty soon he shrieked out, —

"Oh, *please* let me down, and I will tell!"

"No — you'll tell *before* I let you down."

Every instant was agony to him, now, so out it came:—

"No. 166, Eagle Hotel!" — naming a wretched tavern down by the water, a resort of common laborers, 'longshoremen, and less reputable folk.

So I released him, and then demanded to know the object of the conspiracy.

"To take the fort to-night," said he, doggedly and sobbing.

"Have I got all the chiefs of the conspiracy?"

"No. You've got all except those that are to meet at 166."

"What does 'Remember XXXX' mean?"

No reply.

"What is the password to No. 166?"

No reply.

"What do those bunches of letters mean,—'FFFFF' and 'MMMM'? Answer! or you will catch it again."

"I never *will* answer! I will die first. Now do what you please."

"Think what you are saying, Wicklow. Is it final?"

He answered steadily, and without a quiver in his voice, —

"It is final. As sure as I love my wronged country and hate everything this Northern sun shines on, I will die before I will reveal those things."

I triced him up by the thumbs again. When the agony was full upon him, it was heart-breaking to hear the poor thing's shrieks, but we got nothing else out of him. To every question he screamed the same reply: "I can die, and I *will* die; but I will never tell."

Well, we had to give it up. We were convinced that he certainly would die rather than confess. So we took him down and imprisoned him, under strict guard.

Then for some hours we busied ourselves with sending telegrams to the War Department, and with making preparations for a descent upon No. 166.

It was stirring times, that black and bitter night. Things had leaked out, and the whole garrison was on the alert. The sentinels were trebled, and nobody could move, outside or in, without being brought to a stand with a musket levelled at his head. However, Webb and I were less concerned now than we had previously been, because of the fact that the conspiracy must necessarily be in a pretty crippled condition, since so many of its principals were in our clutches.

I determined to be at No. 166 in good season, capture and gag B. B., and be on hand for the rest when they arrived. At about a quarter past one in the morning I crept out of the fortress with half a dozen stalwart and gamy U. S. regulars at my heels — and the boy Wicklow, with his hands tied behind him. I told him we were going to No. 166, and that if I found he had lied again and was misleading us, he would have to show us the right place or suffer the consequences.

We approached the tavern stealthily and reconnoitred. A light was burning in the small bar-room, the rest of the house was dark. I tried the front door; it yielded, and we softly entered, closing the door behind us. Then we removed our shoes, and I led the way to the bar-room. The German landlord sat there, asleep in his chair. I woke him gently, and told him to take off his boots and precede us; warning him at the same time to utter no sound. He obeyed without a murmur, but evidently he was badly frightened. I ordered him to lead the way to 166. We ascended two or three flights of stairs as softly as a file of cats; and then, having arrived near the farther end of a long hall, we came to a door through the glazed transom of which we could discern the glow of a dim light from within. The landlord felt for me in the dark and whispered me that that was 166. I tried the door — it was locked on the inside. I whispered an order to one of my biggest soldiers; we set our ample shoulders to the door and with one heave we burst it from its hinges. I caught a half-glimpse of a figure in a bed — saw its head dart toward the candle; out went the light, and we were in pitch darkness. With one big bound I lit on that bed and pinned its occupant down with my knees. My prisoner struggled fiercely, but I got a grip on his throat with my left hand, and that was a good assistance to my knees in holding him down. Then straightway I snatched out my revolver, cocked it, and laid the cold barrel warningly against his cheek.

"Now somebody strike a light!" said I. "I 've got him safe."

It was done. The flame of the match burst up. I looked at my captive, and, by George, it was a young woman!

I let go and got off the bed, feeling pretty sheepish. Everybody stared stupidly at his neighbor. Nobody had any wit or sense left, so sudden and overwhelming had been the surprise. The young woman began to cry, and covered her face with the sheet. The landlord said, meekly, —

"My daughter, she has been doing something that is not right, *nicht wahr?*"

"Your daughter? Is she your daughter?"

"Oh, yes, she is my daughter. She is just to-night come home from Cincinnati a little bit sick."

"Confound it, that boy has lied again. This is not the right 166; this is not B. B. Now, Wicklow, you will find the correct 166 for us, or — hello! where is that boy?"

Gone, as sure as guns! And, what is more, we failed to find a trace of him. Here was an awkward predicament. I cursed my stupidity in not tying him to one of the men; but it was of no use to bother about that now. What should I do in the present circumstances? — that was the question. That girl *might* be B. B., after all. I did not believe it, but still it would not answer to take unbelief for proof. So I finally put my men in a vacant room across the hall from 166, and told them to capture anybody and everybody that approached the girl's room, and to keep the landlord with them, and under strict watch, until further orders. Then I hurried back to the fort to see if all was right there yet.

Yes, all was right. And all remained right. I stayed up all night to make sure of that. Nothing happened. I was unspeakably glad to see the dawn come again, and be able to telegraph the Department that the Stars and Stripes still floated over Fort Trumbull.

An immense pressure was lifted from my breast. Still I did not relax vigilance, of course, nor effort either; the case was too grave for that. I had up my prisoners, one by one, and harried them by the hour, trying to get them to confess, but it was a failure. They only gnashed their teeth and tore their hair, and revealed nothing.

About noon came tidings of my missing boy. He

had been seen on the road, tramping westward, some eight miles out, at six in the morning. I started a cavalry lieutenant and a private on his track at once. They came in sight of him twenty miles out. He had climbed a fence and was wearily dragging himself across a slushy field toward a large old-fashioned mansion in the edge of a village. They rode through a bit of woods, made a detour, and closed up on the house from the opposite side; then dismounted and skurried into the kitchen. Nobody there. They slipped into the next room, which was also unoccupied; the door from that room into the front or sitting room was open. They were about to step through it when they heard a low voice; it was somebody praying. So they halted reverently, and the lieutenant put his head in and saw an old man and an old woman kneeling in a corner of that sitting-room. It was the old man that was praying, and just as he was finishing his prayer, the Wicklow boy opened the front door and stepped in. Both of those old people sprang at him and smothered him with embraces, shouting, —

"Our boy! our darling! God be praised. The lost is found! He that was dead is alive again!"

Well, sir, what do you think! That young imp was born and reared on that homestead, and had never been five miles away from it in all his life, till the fortnight before he loafed into my quarters and gulled me with that maudlin yarn of his! It's as true as gospel. That old man was his father — a learned old retired clergyman; and that old lady was his mother.

Let me throw in a word or two of explanation concerning that boy and his performances. It turned out that he was a ravenous devourer of dime novels and sensation-story papers — therefore, dark mysteries and gaudy heroisms were just in his line. Then he had read newspaper reports of the stealthy goings and comings of rebel spies in our midst, and of their lurid purposes and their two or three startling achievements, till his imagination was all aflame on that subject. His constant comrade for some months had been a Yankee youth of much tongue and lively fancy, who had served for a couple of years as "mud clerk" (that is, subordinate purser) on certain of the packet-boats plying between New Orleans and points two or three hundred miles up the Mississippi — hence his easy facility in handling the names and other details pertaining to that region. Now I had spent two or three months in that part of the country before the war; and I knew just enough about it to be easily taken in by that boy, whereas a born Louisianian would probably have caught him tripping before he had talked fifteen minutes. Do you know the reason he said he would rather die than explain certain of his treasonable enigmas? Simply because he *could n't* explain them! — they had no meaning; he had fired them out of his imagination without forethought or afterthought; and so, upon sudden call, he was n't able to invent an explanation of them. For instance, he could n't reveal what was hidden in the "sympathetic ink" letter, for the ample reason that there was n't anything hidden in it; it was blank paper only. He had n't put anything into a gun, and had never intended to — for his letters were all written to imaginary persons, and when he hid one in the stable he always removed the one he had put there the day before; so he was not acquainted with that knotted string, since he was seeing it for the first time when I showed it to him; but as soon as I had let him find out where it came from, he straightway adopted it, in his romantic fashion, and got some fine effects out of it. He invented Mr. "Gaylord;" there was n't any 15 Bond Street, just then — it had been pulled down three months before. He invented the "Colonel;" he invented the glib histories of those unfortunates whom I captured and confronted with him; he invented "B. B.;" he even invented No. 166, one may say, for he did n't know there *was* such a number in the Eagle Hotel until we went there. He stood ready to invent anybody or anything whenever it was wanted. If I called for "outside" spies, he promptly described strangers whom he had seen at the hotel, and whose names he had happened to hear. Ah, he lived in a gorgeous, mysterious, romantic world during those few stirring days, and I think it was *real* to him, and that he enjoyed it clear down to the bottom of his heart.

But he made trouble enough for us, and just no end of humiliation. You see, on account of him we had fifteen or twenty people under arrest and confinement in the fort, with sentinels before their doors. A lot of the captives were soldiers and such, and to them I did n't have to apologize; but the rest were first-class citizens, from all over the country, and no amount of apologies was sufficient to satisfy them. They just fumed and raged and made no end of trouble! And those two ladies, — one was an Ohio Congressman's wife, the other a Western bishop's sister, — well, the scorn and ridicule and angry tears they poured out on me made up a keepsake that was likely to make me remember them for a considerable time, — and I shall. That old lame gentleman with the goggles was a college president from Philadelphia, who had come up to attend his nephew's funeral. He had never seen young Wicklow before, of course. Well, he not only missed the funeral, and got jailed as a rebel spy, but Wicklow had stood up there in my quarters and coldly described him as a counterfeiter, nigger-trader, horse-thief, and fire-bug from the most notorious rascal-nest in Galveston; and this was a thing which that poor old gentleman could n't seem to get over at all.

And the War Department! But, O my soul, let 's draw the curtain over that part!

NOTE. — I showed my manuscript to the Major, and he said: "Your unfamiliarity with military matters has betrayed you into some little mistakes. Still, they are picturesque ones — let them go; military men will smile at them, the rest won't detect them. You have got the main facts of the history right, and have set them down just about as they occurred." — M. T.

THE GREAT REVOLUTION IN PITCAIRN.

LET me refresh the reader's memory a little. Nearly a hundred years ago the crew of the British ship "Bounty" mutinied, set the captain and his officers adrift upon the open sea, took possession of the ship, and sailed southward. They procured wives for themselves among the natives of Tahiti, then proceeded to a lonely little rock in mid-Pacific, called Pitcairn's Island, wrecked the vessel, stripped her of everything that might be useful to a new colony, and established themselves on shore.

Pitcairn's is so far removed from the track of commerce that it was many years before another vessel touched there. It had always been considered an uninhabited island; so when a ship did at last drop its anchor there, in 1808, the captain was greatly surprised to find the place peopled. Although the mutineers had fought among themselves, and gradually killed each other off until only two or three of the original stock remained, these tragedies had not occurred before a number of children had been born; so in 1808 the island had a population of twenty-seven persons. John

Adams, the chief mutineer, still survived, and was to live many years yet, as governor and patriarch of the flock. From being mutineer and homicide, he had turned Christian and teacher, and his nation of twenty-seven persons was now the purest and devoutest in Christendom. Adams had long ago hoisted the British flag and constituted his island an appanage of the British crown.

To-day the population numbers ninety persons, — sixteen men, nineteen women, twenty-five boys, and thirty girls, — all descendants of the mutineers, all bearing the family names of those mutineers, and all speaking English, and English only. The island stands high up out of the sea, and has precipitous walls. It is about three quarters of a mile long, and in places is as much as half a mile wide. Such arable land as it affords is held by the several families, according to a division made many years ago. There is some live stock, — goats, pigs, chickens, and cats; but no dogs, and no large animals. There is one church building, — used also as a capitol, a school-house, and a public library. The title of the governor has been, for a generation or two, "Magistrate and Chief Ruler, in subordination to her Majesty the Queen of Great Britain." It was his province to *make* the laws, as well as execute them. His office was elective; everybody over seventeen years old had a vote, — no matter about the sex.

The sole occupations of the people were farming and fishing; their sole recreation, religious services. There has never been a shop in the island, nor any money. The habits and dress of the people have always been primitive, and their laws simple to puerility. They have lived in a deep Sabbath tranquillity, far from the world and its ambitions and vexations, and neither knowing nor caring what was going on in the mighty empires that lie beyond their limitless ocean solitudes. Once in three or four years a ship touched there, moved them with aged news of bloody battles, devastating epidemics, fallen thrones, and ruined dynasties, then traded them some soap and flannel for some yams and bread-fruit, and sailed away, leaving them to retire into their peaceful dreams and pious dissipations once more.

On the 8th of last September, Admiral de Horsey, commander-in-chief of the British fleet in the Pacific, visited Pitcairn's Island, and speaks as follows in his official report to the admiralty: —

"They have beans, carrots, turnips, cabbages, and a little maize; pineapples, fig-trees, custard apples, and oranges; lemons and cocoa-nuts. Clothing is obtained alone from passing ships, in barter for refreshments. There are no springs on the island, but as it rains generally once a month they have plenty of water, although at times, in former years, they have suffered from drought. No alcoholic liquors, except for medicinal purposes, are used, and a drunkard is unknown. . . . .

"The necessary articles required by the islanders are best shown by those we furnished in barter for refreshments: namely, flannel, serge, drill, half-boots, combs, tobacco, and soap. They also stand much in need of maps and slates for their school, and tools of any kind are most acceptable. I caused them to be supplied from the public stores with a union-jack for display on the arrival of ships, and a pit saw, of which they were greatly in need. This, I trust, will meet the approval of their lordships. If the munificent people of England were only aware of the wants of this most deserving little colony, they would not long go unsupplied. . . . .

"Divine service is held every Sunday at 10.30 A. M. and at 3 P. M., in the house built and used by John Adams for that purpose until he died in 1829. It is conducted strictly in accordance with the liturgy of the Church of England, by Mr. Simon Young, their selected pastor, who is much respected. A Bible class is held every Wednesday, when all who conveniently can, attend. There is also a general meeting for prayer on the first Friday in every month. Family prayers are said in every house the first thing in the morning and the last thing in the evening, and no food is partaken of without asking God's blessing before and afterwards. Of these islanders' religious attributes no one can speak without deep respect. A people whose greatest pleasure and privilege is to commune in prayer with their God, and to join in hymns of praise, and who are, moreover, cheerful, diligent, and probably freer from vice than any other community, need no priest among them."

Now I come to a sentence in the admiral's report which he dropped carelessly from his pen, no doubt, and never gave the matter a second thought. He little imagined what a freight of tragic prophecy it bore! This is the sentence: —

"One stranger, an American, has settled on the island, — *a doubtful acquisition.*"

A doubtful acquisition indeed! Captain Ormsby, in the American ship "Hornet," touched at Pitcairn's nearly four months after the admiral's visit, and from the facts which he gathered there we now know all about that American. Let us put these facts together, in historical form. The Ámerican's name was Butterworth Stavely. As soon as he had become well acqainted with all the people, — and this took but a few days, of course, — he began to ingratiate himself with them by all the arts he could command. He became exceedingly popular, and much looked up to; for one of the first things he did was to forsake his worldly way of life, and throw all his energies into religion. He was always reading his Bible, or praying, or singing hymns, or asking blessings. In prayer, no one had such "liberty" as he, no one could pray so long or so well.

At last, when he considered the time to be ripe, he began secretly to sow the seeds of discontent among the people. It was his deliberate purpose, from the beginning, to subvert the government, but of course he kept that to himself for a time. He used different arts with different individuals. He awakened dissatisfaction in one quarter by calling attention to the shortness of the Sunday services; he argued that there should be three three-hour services on Sunday instead of only two. Many had secretly held this opinion before; they now privately banded themselves into a party to work for it. He showed certain of the women that they were not allowed sufficient voice in the prayer-meetings; thus another party was formed. No weapon was beneath his notice; he even descended to the children, and awoke discontent in their breasts because — as *he* discovered for them — they had not enough Sunday-school. This created a third party.

Now, as the chief of these parties, he found himself the strongest power in the community. So he proceeded to his next move, — a no less important one than the impeachment of the chief magistrate, James Russell Nickoy; a man of character and ability, and possessed of great wealth, he being the owner of a house with a parlor to it, three acres and a half of yam land, and the only boat in Pitcairn's, a whale-boat; and, most unfortunately, a pretext for this impeachment offered itself at just the right time. One of the earliest and most precious laws of the island was the law against trespass. It was held in great reverence, and was regarded as the palladium of the people's liberties. About thirty years ago an important case came before the courts under this law, in this wise: a chicken belonging to Elizabeth Young (aged, at that time, fifty-eight, a daughter of John Mills, one of the mutineers of the "Bounty") trespassed upon the grounds of Thursday October Christian (aged twenty-nine, a grandson of Fletcher Christian, one of the mutineers). Christian killed the chicken. According to the law, Christian could keep the chicken; or, if he preferred, he could restore its remains to the owner, and receive damages in "produce" to an amount equivalent to the waste and injury wrought by the trespasser. The court records set forth that "the said Christian aforesaid did deliver the aforesaid remains to the said Elizabeth Young, and did demand one bushel of yams in satisfaction of the damage done." But Elizabeth Young considered the demand exorbitant; the parties could not agree; therefore Christian brought suit in the courts. He lost his case in the justice's court; at least, he was awarded only a half-peck of yams, which he considered insufficient, and in the nature of a defeat. He appealed. The case lingered several years in an ascending grade of courts, and always resulted in decrees sustaining the original verdict; and finally the thing got into the supreme court, and there it stuck for twenty years. But last summer, even the supreme court managed to arrive at a decision at last. Once more the orginal verdict was sustained. Christian then said he was satisfied; but Stavely was present, and whispered to him and to his lawyer, suggesting, "as a mere form," that the original law be exhibited, in order to make sure that it still existed. It seemed an odd idea, but an ingenious one. So the demand was made. A messenger was sent to the magistrate's house; he presently returned with the tidings that it had disappeared from among the state archives.

The court now pronounced its late decision void, since it had been made under a law which had no actual existence.

Great excitement ensued, immediately. The news swept abroad over the whole island that the palladium of the public liberties was lost, — may be treasonably destroyed. Within thirty minutes almost the entire nation were in the court-room, — that is to say, the church. The impeachment of the chief magistrate followed, upon Stavely's motion. The accused met his misfortune with the dignity which became his great office. He did not plead, or even argue: he offered the simple defence that he had not meddled with the missing law; that he had kept the state archives in the same candle-box that had been used as their depository from the beginning; and that he was innocent of the removal or destruction of the lost document.

But nothing could save him; he was found guilty of misprision of treason, and degraded from his office, and all his property was confiscated.

The lamest part of the whole shameful matter was the *reason* suggested by his enemies for his destruction of the law, to wit: that he did it to favor Christian, because Christian was his cousin! Whereas Stavely was the only individual in the entire nation who was *not* his cousin. The reader must remember that all of these people are the descendants of half a dozen men; that the first children intermarried together and bore grandchildren to the mutineers; that these grandchildren intermarried; after them, great and great-great-grandchildren intermarried: so that to-day everybody is blood kin to everybody. Moreover, the relationships are wonderfully, even astoundingly, mixed up and complicated. A stranger, for instance, says to an islander, —

"You speak of that young woman as your cousin; a while ago you called her your aunt."

"Well, she *is* my aunt, and my cousin too. And

also my step-sister, my niece, my fourth cousin, my thirty-third cousin, my forty-second cousin, my great-aunt, my grandmother, my widowed sister-in-law, — and next week she will be my wife."

So the charge of nepotism against the chief magistrate was weak. But no matter; weak or strong, it suited Stavely. Stavely was immediately elected to the vacant magistracy; and, oozing reform from every pore, he went vigorously to work. In no long time religious services raged everywhere and unceasingly. By command, the second prayer of the Sunday morning service, which had customarily endured some thirty-five or forty minutes, and had pleaded for the world, first by continent and then by national and tribal detail, was extended to an hour and a half, and made to include supplications in behalf of the possible peoples in the several planets. Everybody was pleased with this; everybody said, "Now *this* is something *like*." By command, the usual three-hour sermons were doubled in length. The nation came in a body to testify their gratitude to the new magistrate. The old law forbidding cooking on the Sabbath was extended to the prohibition of eating, also. By command, Sunday-school was privileged to spread over into the week. The joy of all classes was complete. In one short month the new magistrate had become the people's idol!

The time was ripe for this man's next move. He began, cautiously at first, to poison the public mind against England. He took the chief citizens aside, one by one, and conversed with them on this topic. Presently he grew bolder, and spoke out. He said the nation owed it to itself, to its honor, to its great traditions, to rise in its might and throw off "this galling English yoke."

But the simple islanders answered, —

"We had not noticed that it galled. How does it gall? England sends a ship once in three or four years to give us soap and clothing, and things which we sorely need and gratefully receive; but she never troubles us; she lets us go our own way."

"She lets you go your own way! So slaves have felt and spoken in all the ages! This speech shows how fallen you are, how base, how brutalized, you have become, under this grinding tyranny! What! has all manly pride forsaken you? Is liberty nothing? Are you content to be a mere appendage to a foreign and hateful sovereignty, when you might rise up and take your rightful place in the august family of nations, great, free, enlightened, independent, the minion of no sceptred master, but the arbiter of your own destiny, and a voice and a power in decreeing the destinies of your sister-sovereignties of the world?"

Speeches like this produced an effect by and by. Citizens began to feel the English yoke; they did not know exactly how or whereabouts they felt it, but they were perfectly certain they did feel it. They got to grumbling a good deal, and chafing under their chains, and longing for relief and release. They presently fell to hating the English flag, that sign and symbol of their nation's degradation; they ceased to glance up at it as they passed the capitol, but averted their eyes and grated their teeth; and one morning, when it was found trampled into the mud at the foot of the staff, they left it there, and no man put his hand to it to hoist it again. A certain thing which was sure to happen sooner or later happened now. Some of the chief citizens went to the magistrate by night, and said, —

"We can endure this hated tyranny no longer. How can we cast it off?"

"By a *coup d'état.*"

"How?"

"A coup d'état. It is like this: everything is got ready, and at the appointed moment I, as the official head of the nation, publicly and solemnly proclaim its independence, and absolve it from allegiance to any and all other powers whatsoever."

"That sounds simple and easy. We can do that right away. Then what will be the next thing to do?"

"Seize all the defences and public properties of all kinds, establish martial law, put the army and navy on a war footing, and proclaim the empire!"

This fine programme dazzled these innocents. They said, —

"This is grand, — this is splendid; but will not England resist?"

"Let her. This rock is a Gibraltar."

"True. But about the empire? Do we *need* an empire, and an emperor?"

"What you *need*, my friends, is unification. Look at Germany; look at Italy. They are unified. Unification is the thing. It makes living dear. That constitutes progress. We must have a standing army, and a navy. Taxes follow, as a matter of course. All these things summed up make grandeur. With unification and grandeur, what more can you want? Very well, — only the empire can confer these boons."

So on the 8th day of December Pitcairn's Island was proclaimed a free and independent nation; and on the same day the solemn coronation of Butterworth I., emperor of Pitcairn's Island, took place, amid great rejoicings and festivities. The entire nation, with the exception of fourteen persons, mainly little children, marched past the throne in single file, with banners and music, the procession being upwards of ninety feet long; and some said it was as much as three quarters of a minute passing a given point. Nothing like it had ever been seen in the history of the island before. Public enthusiasm was measureless.

Now straightway imperial reforms began. Orders of nobility were instituted. A minister of the navy was appointed, and the whale-boat put in commission. A minister of war was created, and ordered to proceed at once with the formation of a standing army. A first lord of the treasury was named, and commanded to get up a taxation scheme, and also open negotiations for treaties, offensive, defensive, and commercial, with foreign powers. Some generals and admirals were appointed; also some chamberlains, some equerries in waiting, and some lords of the bed-chamber.

At this point all the material was used up. The Grand Duke of Galilee, minister of war, complained that all the sixteen grown men in the empire had been given great offices, and consequently would not consent to serve in the ranks; wherefore his standing army was at a stand-still. The Marquis of Ararat, minister of the navy, made a similar complaint. He said he was willing to steer the whale-boat himself, but he *must* have somebody to man her.

The emperor did the best he could in the circumstances: he took all the boys above the age of ten years away from their mothers, and pressed them into the army, thus constructing a corps of seventeen privates, officered by one lieutenant-general and two major-generals. This pleased the minister of war, but procured the enmity of all the mothers in the land; for they said their precious ones must now find bloody graves in the fields of war, and he would be answerable for it. Some of the more heart-broken and inappeasable among them lay constantly in wait for the emperor and threw yams at him, unmindful of the body-guard.

On account of the extreme scarcity of material, it was found necessary to require the Duke of Bethany, postmaster-general, to pull stroke-oar in the navy, and thus sit in the rear of a noble of lower degree, namely, Viscount Canaan, lord-justice of the common pleas. This turned the Duke of Bethany into a tolerably open malcontent and a secret conspirator, — a thing which the emperor foresaw, but could not help.

Things went from bad to worse. The emperor raised Nancy Peters to the peerage on one day, and married her the next, notwithstanding, for reasons of state, the cabinet had strenuously advised him to marry Emmeline, eldest daughter of the Archbishop of Bethlehem. This caused trouble in a powerful quarter, — the church. The new empress secured the support and friendship of two thirds of the thirty-six grown women in the nation by absorbing them into her court as maids of honor; but this made deadly enemies of the remaining twelve. The families of the maids of honor soon began to rebel, because there was now nobody at home to keep house. The twelve snubbed women refused to enter the imperial kitchen as servants; so the empress had to require the Countess of Jericho and other great court dames to fetch water, sweep the palace, and perform other menial and equally distasteful services. This made bad blood in that department.

Everybody fell to complaining that the taxes levied for the support of the army, the navy, and the rest of the imperial establishment were intolerably burdensome, and were reducing the nation to beggary. The emperor's reply — "Look at Germany; look at Italy. Are you better than they? and have n't you unification?" — did not satisfy them. They said, "People can't *eat* unification, and we are starving. Agriculture has ceased. Everybody is in the army, everybody is in the navy, everybody is in the public service, standing around in a uniform, with nothing whatever to do, nothing to eat, and nobody to till the fields — "

"Look at Germany; look at Italy. It is the same there. Such is unification, and there 's no other way to get it, — no other way to keep it after you 've got it," said the poor emperor always.

But the grumblers only replied, "We can't *stand* the taxes, — we can't *stand* them."

Now right on top of this the cabinet reported a national debt amounting to upwards of forty-five dollars, — half a dollar to every individual in the nation. And they proposed to fund something. They had heard that this was always done in such emergencies. They proposed duties on exports; also on imports. And they wanted to issue bonds; also paper money, redeemable in yams and cabbages in fifty years. They said the pay of the army and of the navy and of the whole governmental machine was far in arrears, and unless something was done, and done immediately, national bankruptcy must ensue, and possibly insurrection and revolution. The emperor at once resolved upon a high-handed measure, and one of a nature never before heard of in Pitcairn's Island. He went in state to the church on Sunday morning, with the army at his back, and commanded the minister of the treasury to take up a collection.

That was the feather that broke the camel's back. First one citizen, and then another, rose and refused to submit to this unheard-of outrage, — and each refusal was followed by the immediate confiscation of the malcontent's property. This vigor soon stopped the refusals, and the collection proceeded amid a sullen and ominous silence. As the emperor withdrew with the troops, he said, "I will teach you who is master here." Several persons shouted, "Down with unification!" They were at once arrested and torn from the arms of their weeping friends by the soldiery.

But in the mean time, as any prophet might have foreseen, a Social Democrat had been developed. As the emperor stepped into the gilded imperial wheelbarrow at the church door, the social democrat stabbed at him fifteen or sixteen times with a harpoon, but fortunately with such a peculiarly social democratic unprecision of aim as to do no damage.

That very night the convulsion came. The nation rose as one man, — though forty-nine of the revolutionists were of the other sex. The infantry threw down their pitchforks; the artillery cast aside their cocoa-nuts; the navy revolted; the emperor was seized, and bound hand and foot in his palace. He was very much depressed. He said, —

"I freed you from a grinding tyranny; I lifted you up out of your degradation, and made you a nation among nations; I gave you a strong, compact, centralized government; and, more than all, I gave you the blessing of blessings, — unification. I have done all this, and my reward is hatred, insult, and these bonds. Take me; do with me as ye will. I here resign my crown and all my dignities, and gladly do I release myself from their too heavy burden. For your sake I took them up; for your sake I lay them down. The imperial jewel is no more; now bruise and defile as ye will the useless setting."

By a unanimous voice the people condemned the ex-emperor and the social democrat to perpetual banishment from church services, or to perpetual labor as galley-slaves in the whale-boat, — whichever they might prefer. The next day the nation assembled again, and rehoisted the British flag, reinstated the British tyranny, reduced the nobility to the condition of commoners again, and then straightway turned their diligent attention to the weeding of the ruined and neglected yam patches, and the rehabilitation of the old useful industries and the old healing and solacing pieties. The ex-emperor restored the lost trespass law, and explained that he had stolen it, — not to injure any one, but to further his political projects. Therefore the nation gave the late chief magistrate his office again, and also his alienated property.

Upon reflection, the ex-emperor and the social democrat chose perpetual banishment from religious services, in preference to perpetual labor as galley-slaves "*with* perpetual religious services," as they phrased it; wherefore the people believed that the poor fellows' troubles had unseated their reason, and so they judged it best to confine them for the present. Which they did.

Such is the history of Pitcairn's "doubtful acquisition."

MRS. McWILLIAMS AND THE LIGHTNING.

WELL, sir, — continued Mr. McWilliams, for this was not the beginning of his talk, — the fear of lightning is one of the most distressing infirmities a human being can be afflicted with. It is mostly confined to women; but now and then you find it in a little dog, and sometimes in a man. It is a particularly distressing infirmity, for the reason that it takes the sand out of a person to an extent which no other fear can, and it can't be *reasoned* with, and neither can it be shamed out of a person. A woman who could face the very devil himself — or a mouse — loses her grip and goes all to pieces in front of a flash of lightning. Her fright is something pitiful to see.

Well, as I was telling you, I woke up, with that smothered and unlocatable cry of "Mortimer! Mortimer!" wailing in my ears; and as soon as I could scrape my faculties together I reached over in the dark and then said, —

"Evangeline, is that you calling? What is the matter? Where are you?"

"Shut up in the boot-closet. You ought to be ashamed to lie there and sleep so, and such an awful storm going on."

"Why, how *can* one be ashamed when he is asleep? It is unreasonable; a man *can't* be ashamed when he is asleep, Evangeline."

"You never try, Mortimer, — you know very well you never try."

I caught the sound of muffled sobs.

That sound smote dead the sharp speech that was on my lips, and I changed it to —

"I 'm sorry, dear, — I 'm truly sorry. I never meant to act so. Come back and — "

"MORTIMER!"

"Heavens! what is the matter, my love?"

"Do you mean to say you are in that bed yet?"

"Why, of course."

"Come out of it instantly. I should think you would take some *little* care of your life, for *my* sake and the children's, if you will not for your own."

"But my love — "

"Don't talk to me, Mortimer. You *know* there is no place so dangerous as a bed, in such a thunder-storm as this, — all the books say that; yet there you would lie, and deliberately throw away your life, — for goodness knows what, unless for the sake of arguing and arguing, and — "

"But, confound it, Evangeline, I 'm *not* in the bed, *now*. I 'm — "

[Sentence interrupted by a sudden glare of lightning, followed by a terrified little scream from Mrs. McWilliams and a tremendous blast of thunder.]

"There! You see the result. Oh, Mortimer, how *can* you be so profligate as to swear at such a time as this?"

"I *did n't* swear. And that *was n't* a result of it, any way. It would have come, just the same, if I had n't said a word; and you know very well, Evangeline, — at least you ought to know, — that when the atmosphere is charged with electricity — "

"Oh, yes, now argue it, and argue it, and argue it! — I don't see how you can act so, when you *know* there is not a lightning-rod on the place, and your poor wife and children are absolutely at the mercy of Providence. What *are* you doing? — lighting a match at such a time as this! Are you stark mad?"

"Hang it, woman, where's the harm? The place is as dark as the inside of an infidel, and — "

"Put it out! put it out instantly! Are you determined to sacrifice us all? You *know* there is nothing attracts lightning like a light. [*Fzt! — crash! boom — boloom-boom-boom!*] Oh, just hear it! Now you see what you've done!"

"No, I *don't* see what I've done. A match may attract lightning, for all I know, but it don't *cause* lightning, — I'll go odds on that. And it did n't attract it worth a cent this time; for if that shot was levelled at my match, it was blessed poor marksmanship, — about an average of none out of a possible million, I should say. Why, at Dollymount, such marksmanship as that — "

"For shame, Mortimer! Here we are standing right in the very presence of death, and yet in so solemn a moment you are capable of using such language as that. If you have no desire to — Mortimer!"

"Well?"

"Did you say your prayers to-night?"

"I — I — meant to, but I got to trying to cipher out how much twelve times thirteen is, and —"

[*Fzt! — boom-berroom-boom! bumble-umble bang*-SMASH!]

"Oh, we are lost, beyond all help! How *could* you neglect such a thing at such a time as this?"

"But it *was n't* 'such a time as this.' There was n't a cloud in the sky. How could *I* know there was going to be all this rumpus and pow-wow about a little slip like that? And I don't think it 's just fair for you to make so much out of it, any way, seeing it happens so seldom; I have n't missed before since I brought on that earthquake, four years ago."

"MORTIMER! How you talk! Have you forgotten the yellow fever?"

"My dear, you are always throwing up the yellow fever to me, and I think it is perfectly unreasonable. You can't even send a telegraphic message as far as Memphis without relays, so how is a little devotional slip of mine going to carry so far? I 'll *stand* the earthquake, because it was in the neighborhood; but I 'll be hanged if I 'm going to be responsible for every blamed —"

[*Fzt!* — BOOM *beroom*-boom! boom! — BANG!]

"Oh, dear, dear, dear! I *know* it struck something, Mortimer. We never shall see the light of another day; and if it will do you any good to remember, when we are gone, that your dreadful language — *Mortimer!*"

"WELL! What now?"

"Your voice sounds as if— Mortimer, are you actually standing in front of that open fireplace?"

"That is the very crime I am committing."

"Get away from it, this moment. You do seem determined to bring destruction on us all. Don't you *know* that there is no better conductor for lightning than an open chimney? *Now* where have you got to?"

"I 'm here by the window."

"Oh, for pity's sake, have you lost your mind? Clear out from there, this moment. The very children in arms know it is fatal to stand near a window in a thunder-storm. Dear, dear, I know I shall never see the light of another day. Mortimer?"

"Yes?"

"What is that rustling?"

"It 's me."

"What are you doing?"

"Trying to find the upper end of my pantaloons."

"Quick! throw those things away! I do believe you would deliberately put on those clothes at such a time as this; yet you know perfectly well that *all* authorities agree that woollen stuffs attract lightning. Oh, dear, dear, it is n't sufficient that one's life must be in peril from natural causes, but you must do everything you can possibly think of to augment the danger. Oh, *don't* sing! What *can* you be thinking of?"

"Now where 's the harm in it?"

"Mortimer, if I have told you once, I have told you a hundred times, that singing causes vibrations in the atmosphere which interrupt the flow of the electric fluid, and — What on *earth* are you opening that door for?"

"Goodness gracious, woman, is there any harm in *that?*"

"*Harm?* There's *death* in it. Anybody that has given this subject any attention knows that to create a draught is to invite the lightning. You have n't half shut it; shut it *tight*, — and do hurry, or we are all destroyed. Oh, it is an awful thing to be shut up with a lunatic at such a time as this. Mortimer, what *are* you doing?"

"Nothing. Just turning on the water. This room is smothering hot and close. I want to bathe my face and hands."

"You have certainly parted with the remnant of your mind! Where lightning strikes any other substance once, it strikes water fifty times. Do turn it off. Oh, dear, I am sure that nothing in this world can save us. It does seem to me that — Mortimer, what was that?"

"It was a da— it was a picture. Knocked it down."

"Then you are close to the wall! I never heard of such imprudence! Don't you *know* that there's no better conductor for lightning than a wall? Come away from there! And you came as near as anything to swearing, too. Oh, how can you be so desperately wicked, and your family in such peril? Mortimer, did you order a feather bed, as I asked you to do?"

"No. Forgot it."

"Forgot it! It may cost you your life. If you had a feather bed, now, and could spread it in the middle of the room and lie on it, you would be perfectly safe. Come in here, — come quick, before you have a chance to commit any more frantic indiscretions."

I tried, but the little closet would not hold us both with the door shut, unless we could be content to smother. I gasped awhile, then forced my way out. My wife called out, —

"Mortimer, something *must* be done for your preservation. Give me that German book that is on the end of the mantel-piece, and a candle; but don't light it; give me a match; I will light it in here. That book has some directions in it."

I got the book, — at cost of a vase and some other brittle things; and the madam shut herself up with her candle. I had a moment's peace; then she called out, —

"Mortimer, what was that?"

"Nothing but the cat."

"The cat! Oh, destruction! Catch her, and shut her up in the wash-stand. Do be quick, love; cats are *full* of electricity. I just know my hair will turn white with this night's awful perils."

I heard the muffled sobbings again. But for that, I should not have moved hand or foot in such a wild enterprise in the dark.

However, I went at my task, — over chairs, and against all sorts of obstructions, all of them hard ones, too, and most of them with sharp edges, — and at last

I got kitty cooped up in the commode, at an expense of over four hundred dollars in broken furniture and shins. Then these muffled words came from the closet: —

"It says the safest thing is to stand on a chair in the middle of the room, Mortimer; and the legs of the chair must be insulated, with non-conductors. That is, you must set the legs of the chair in glass tumblers. [*Fzt! — boom — bang! — smash!*] Oh, hear that! Do hurry, Mortimer, before you are struck."

I managed to find and secure the tumblers. I got the last four, — broke all the rest. I insulated the chair legs, and called for further instructions.

"Mortimer, it says, 'Während eines Gewitters entferne man Metalle, wie z. B., Ringe, Uhren, Schlüssel, etc., von sich und halte sich auch nicht an solchen Stellen auf, wo viele Metalle bei einander liegen, oder mit andern Körpern verbunden sind, wie an Herden, Oefen, Eisengittern u. dgl.' What does that mean, Mortimer? Does it mean that you must keep metals *about* you, or keep them *away* from you?"

"Well, I hardly know. It appears to be a little mixed. All German advice is more or less mixed. However, I think that that sentence is mostly in the dative case, with a little genitive and accusative sifted in, here and there, for luck; so I reckon it means that you must keep some metals *about* you."

"Yes, that must be it. It stands to reason that it is. They are in the nature of lightning-rods, you know. Put on your fireman's helmet, Mortimer; that is mostly metal."

I got it and put it on, — a very heavy and clumsy and uncomfortable thing on a hot night in a close room. Even my night-dress seemed to be more clothing than I strictly needed.

"Mortimer, I think your middle ought to be protected. Won't you buckle on your militia sabre, please?"

I complied.

"Now, Mortimer, you ought to have some way to protect your feet. Do please put on your spurs."

I did it, — in silence, — and kept my temper as well as I could.

"Mortimer, it says, 'Das Gewitter läuten ist sehr gefährlich, weil die Glocke selbst, sowie der durch das Läuten veranlasste Luftzug und die Höhe des Thurmes den Blitz anziehen könnten.' Mortimer, does that mean that it is dangerous not to ring the church bells during a thunder-storm?"

"Yes, it seems to mean that, — if that is the past participle of the nominative case singular, and I reckon it is. Yes, I think it means that on account of the height of the church tower and the absence of *Luftzug* it would be very dangerous (*sehr gefährlich*) not to ring the bells in time of a storm; and moreover, don't you see, the very wording — "

"Never mind that, Mortimer; don't waste the precious time in talk. Get the large dinner-bell; it is right there in the hall. Quick, Mortimer dear; we are almost safe. Oh, dear, I do believe we are going to be saved, at last!"

Our little summer establishment stands on top of a

high range of hills, overlooking a valley. Several farm-houses are in our neighborhood, — the nearest some three or four hundred yards away.

When I, mounted on the chair, had been clanging that dreadful bell a matter of seven or eight minutes, our shutters were suddenly torn open from without, and a brilliant bull's-eye lantern was thrust in at the window, followed by a hoarse inquiry : —

"What in the nation is the matter here?"

The window was full of men's heads, and the heads were full of eyes that stared wildly at my night-dress and my warlike accoutrements.

I dropped the bell, skipped down from the chair in confusion, and said, —

"There is nothing the matter, friends, — only a little discomfort on account of the thunder-storm. I was trying to keep off the lightning."

"Thunder-storm? Lightning? Why, Mr. McWilliams, have you lost your mind? It is a beautiful starlight night; there has been no storm."

I looked out, and I was so astonished I could hardly speak for a while. Then I said, —

"I do not understand this. We distinctly saw the glow of the flashes through the curtains and shutters, and heard the thunder."

One after another of those people lay down on the ground to laugh, — and two of them died. One of the survivors remarked, —

"Pity you did n't think to open your blinds and look over to the top of the high hill yonder. What you heard was cannon; what you saw was the flash.

You see, the telegraph brought some news, just at midnight: Garfield 's nominated, — and that 's what 's the matter!"

Yes, Mr. Twain, as I was saying in the beginning (said Mr. McWilliams), the rules for preserving people against lightning are so excellent and so innumerable that the most incomprehensible thing in the world to me is how anybody ever manages to get struck.

So saying, he gathered up his satchel and umbrella, and departed; for the train had reached his town.

ON THE DECAY OF THE ART OF LYING.

ESSAY, FOR DISCUSSION, READ AT A MEETING OF THE HISTORICAL AND ANTIQUARIAN CLUB OF HARTFORD, AND OFFERED FOR THE THIRTY-DOLLAR PRIZE. NOW FIRST PUBLISHED.[1]

OBSERVE, I do not mean to suggest that the *custom* of lying has suffered any decay or interruption, — no, for the Lie, as a Virtue, a Principle, is eternal; the Lie, as a recreation, a solace, a refuge in time of need, the fourth Grace, the tenth Muse, man's best and surest friend, is immortal, and cannot perish from the earth while this Club remains. My complaint simply concerns the decay of the *art* of lying. No high-minded man, no man of right feeling, can contemplate the lumbering and slovenly lying of the present day without grieving to see a noble art so prostituted. In this veteran presence I naturally enter upon this theme with diffidence; it is like an old maid trying to teach nursery matters to the mothers in Israel. It would not become me to criticise you, gentlemen, who are nearly all my elders — and my

[1] Did not take the prize.

superiors, in this thing — and so, if I should here and there *seem* to do it, I trust it will in most cases be more in a spirit of admiration than of fault-finding; indeed if this finest of the fine arts had everywhere received the attention, encouragement, and conscientious practice and development which this Club has devoted to it, I should not need to utter this lament, or shed a single tear. I do not say this to flatter: I say it in a spirit of just and appreciative recognition. [It had been my intention, at this point, to mention names and give illustrative specimens, but indications observable about me admonished me to beware of particulars and confine myself to generalities.]

No fact is more firmly established than that lying is a necessity of our circumstances, — the deduction that it is then a Virtue goes without saying. No virtue can reach its highest usefulness without careful and diligent cultivation, — therefore, it goes without saying, that this one ought to be taught in the public schools — at the fireside — even in the newspapers. What chance has the ignorant, uncultivated liar against the educated expert? What chance have I against Mr. Per— against a lawyer? *Judicious* lying is what the world needs. I sometimes think it were even better and safer not to lie at all than to lie injudiciously. An awkward, unscientific lie is often as ineffectual as the truth.

Now let us see what the philosophers say. Note that venerable proverb: Children and fools *always* speak the truth. The deduction is plain, — adults and wise persons *never* speak it. Parkman, the historian, says, "The principle of truth may itself be carried into an absurdity." In another place in the same chapter he says, "The saying is old that truth should not be spoken at all times; and those whom a sick conscience worries into habitual violation of the maxim are imbeciles and nuisances." It is strong language, but true. None of us could *live* with an habitual truth-teller; but thank goodness none of us has to. An habitual truth-teller is simply an impossible creature; he does not exist; he never has existed. Of course there are people who *think* they never lie, but it is not so, — and this ignorance is one of the very things that shame our so-called civilization. Everybody lies — every day; every hour; awake; asleep; in his dreams; in his joy; in his mourning; if he keeps his tongue still, his hands, his feet, his eyes, his attitude, will convey deception — and purposely. Even in sermons — but that is a platitude.

In a far country where I once lived the ladies used to go around paying calls, under the humane and kindly pretence of wanting to see each other; and when they returned home, they would cry out with a glad voice, saying, "We made sixteen calls and found fourteen of them out," — not meaning that they found out anything against the fourteen, — no, that was only a colloquial phrase to signify that they were not at home, — and their manner of saying it expressed their lively satisfaction in that fact. Now their pretence of wanting to see the fourteen — and the other two whom they had been less lucky with — was that commonest and mildest form of lying which is sufficiently described as a deflection from the truth. Is it justifiable? Most certainly. It is beautiful, it is noble; for its object is, *not* to reap profit, but to convey a pleasure to the sixteen. The iron-souled truth-monger would plainly manifest, or even utter the fact that he did n't want to see those people, — and he would be an ass, and inflict a totally unnecessary pain. And next, those ladies in that far country — but never mind, they had a thousand pleasant ways of lying, that grew out of gentle impulses, and were a credit to their intelligence and an honor to their hearts. Let the particulars go.

The men in that far country were liars, every one. Their mere howdy-do was a lie, because *they* did n't care how you did, except they were undertakers. To the ordinary inquirer you lied in return; for you made no conscientious diagnosis of your case, but answered at random, and usually missed it considerably. You lied to the undertaker, and said your health was failing — a wholly commendable lie, since it cost you nothing and pleased the other man. If a stranger called and interrupted you, you said with your hearty tongue, "I 'm glad to see you," and said with your heartier soul, "I wish you were with the cannibals and it was dinner time." When he went, you said regretfully, "*Must* you go?" and followed it with a "Call again;" but you did no harm, for you did not deceive anybody nor inflict any hurt, whereas the truth would have made you both unhappy.

I think that all this courteous lying is a sweet and loving art, and should be cultivated. The highest perfection of politeness is only a beautiful edifice, built, from the base to the dome, of graceful and gilded forms of charitable and unselfish lying.

What I bemoan is the growing prevalence of the brutal truth. Let us do what we can to eradicate it. An injurious truth has no merit over an injurious lie. Neither should ever be uttered. The man who speaks an injurious truth lest his soul be not saved if he do otherwise, should reflect that that sort of a soul is not strictly worth saving. The man who tells a lie to help a poor devil out of trouble, is one of whom the angels doubtless say, "Lo, here is an heroic soul who casts his own welfare into jeopardy to succor his neighbor's; let us exalt this magnanimous liar."

An injurious lie is an uncommendable thing; and so, also, and in the same degree, is an injurious truth, — a fact which is recognized by the law of libel.

Among other common lies, we have the *silent* lie, — the deception which one conveys by simply keeping still and concealing the truth. Many obstinate truth-mongers indulge in this dissipation, imagining that if they *speak* no lie, they lie not at all. In that far country where I once lived, there was a lovely spirit, a lady whose impulses were always high and pure, and whose character answered to them. One day I was there at dinner, and remarked, in a general way, that we are all liars. She was amazed, and said, "Not *all?*" It was before Pinafore's time, so I did not make the response which would naturally follow in our day, but frankly said, "Yes, *all* — we are all liars; there are no exceptions." She looked almost offended, and said,

"Why, do you include *me?*" "Certainly," I said, "I think you even rank as an expert." She said, "Sh— sh! the children!" So the subject was changed in deference to the children's presence, and we went on talking about other things. But as soon as the young people were out of the way, the lady came warmly back to the matter and said, "I have made it the rule of my life to never tell a lie; and I have never departed from it in a single instance." I said, "I don't mean the least harm or disrespect, but really you have been lying like smoke ever since I've been sitting here. It has caused me a good deal of pain, because I am not used to it." She required of me an instance — just a single instance. So I said, —

"Well, here is the unfilled duplicate of the blank which the Oakland hospital people sent to you by the hand of the sick-nurse when she came here to nurse your little nephew through his dangerous illness. This blank asks all manner of questions as to the conduct of that sick-nurse: 'Did she ever sleep on her watch? Did she ever forget to give the medicine?' and so forth and so on. You are warned to be very careful and explicit in your answers, for the welfare of the service requires that the nurses be promptly fined or otherwise punished for derelictions. You told me you were perfectly delighted with that nurse — that she had a thousand perfections and only one fault: you found you never could depend on her wrapping Johnny up half sufficiently while he waited in a chilly chair for her to rearrange the warm bed. You filled up the duplicate of this paper, and sent it back to the hospital by the hand of the nurse. How did you answer this question, — 'Was the nurse at any time guilty of a negligence which was likely to result in the patient's taking cold?' Come — everything is decided by a bet here in California: ten dollars to ten cents you lied when you answered that question." She said, "I did n't; *I left it blank!*" "Just so — you have told a *silent* lie; you have left it to be inferred that you had no fault to find in that matter." She said, "Oh, was that a lie? And how *could* I mention her one single fault, and she so good? — it would have been cruel." I said, "One ought always to lie, when one can do good by it; your impulse was right, but your judgment was crude; this comes of unintelligent practice. Now observe the result of this inexpert deflection of yours. You know Mr. Jones's Willie is lying very low with scarlet fever; well, your recommendation was so enthusiastic that that girl is there nursing him, and the worn-out family have all been trustingly sound asleep for the last fourteen hours, leaving their darling with full confidence in those fatal hands, because you, like young George Washington, have a reputa— However, if you are not going to have anything to do, I will come around to-morrow and we 'll attend the funeral together, for of course you 'll naturally feel a peculiar interest in Willie's case, — as personal a one, in fact, as the undertaker."

But that was all lost. Before I was half-way through she was in a carriage and making thirty miles an hour toward the Jones mansion to save what was left of Willie and tell all she knew about the deadly nurse.

All of which was unnecessary, as Willie was n't sick; I had been lying myself. But that same day, all the same, she sent a line to the hospital which filled up the neglected blank, and stated the *facts*, too, in the squarest possible manner.

Now, you see, this lady's fault was *not* in lying, but only in lying injudiciously. She should have told the truth, *there*, and made it up to the nurse with a fraudulent compliment further along in the paper. She could have said, "In one respect this sick-nurse is perfection, — when she is on watch, she never snores." Almost any little pleasant lie would have taken the sting out of that troublesome but necessary expression of the truth.

Lying is universal — we *all* do it; we all *must* do it. Therefore, the wise thing is for us diligently to train ourselves to lie thoughtfully, judiciously; to lie with a good object, and not an evil one; to lie for others' advantage, and not our own; to lie healingly, charitably, humanely, not cruelly, hurtfully, maliciously; to lie gracefully and graciously, not awkwardly and clumsily; to lie firmly, frankly, squarely, with head erect, not haltingly, tortuously, with pusillanimous mien, as being ashamed of our high calling. Then shall we be rid of the rank and pestilent truth that is rotting the land; then shall we be great and good and beautiful, and worthy dwellers in a world where even benign Nature habitually lies, except when she promises execrable weather. Then — But I am but a new and feeble student in this gracious art; I cannot instruct *this* Club.

Joking aside, I think there is much need of wise examination into what sorts of lies are best and wholesomest to be indulged, seeing we *must* all lie and *do* all lie, and what sorts it may be best to avoid, — and this is a thing which I feel I can confidently put into the hands of this experienced Club, — a ripe body, who may be termed, in this regard, and without undue flattery, Old Masters.

THE CANVASSER'S TALE.

POOR, sad-eyed stranger! There was that about his humble mien, his tired look, his decayed-gentility clothes, that almost reached the mustard-seed of charity that still remained, remote and lonely, in the empty vastness of my heart, notwithstanding I observed a portfolio under his arm, and said to myself, Behold, Providence hath delivered his servant into the hands of another canvasser.

Well, these people always get one interested. Before I well knew how it came about, this one was telling me his history, and I was all attention and sympathy. He told it something like this: —

My parents died, alas, when I was a little, sinless child. My uncle Ithuriel took me to his heart and reared me as his own. He was my only relative in the wide world; but he was good and rich and generous. He reared me in the lap of luxury. I knew no want that money could satisfy.

In the fulness of time I was graduated, and went with two of my servants — my chamberlain and my valet — to travel in foreign countries. During four years I flitted upon careless wing amid the beauteous gardens of the distant strand, if you will permit this form of speech in one whose tongue was ever attuned to poesy; and indeed I so speak with confidence, as one unto his kind, for I perceive by your eyes that you too, sir, are gifted with the divine inflation. In those far lands I revelled in the ambrosial food that fructifies the soul, the mind, the heart. But of all things, that which most appealed to my inborn æsthetic taste was the prevailing custom there, among the rich, of making collections of elegant and costly rarities, dainty *objets de vertu*, and in an evil hour I tried to uplift my uncle Ithuriel to a plane of sympathy with this exquisite employment.

I wrote and told him of one gentleman's vast collection of shells; another's noble collection of meerschaum pipes; another's elevating and refining collection of undecipherable autographs; another's priceless collection of old china; another's enchanting collection of postage stamps, — and so forth and so on. Soon my letters yielded fruit. My uncle began to look about for something to make a collection of. You may know, perhaps, how fleetly a taste like this dilates. His soon became a raging fever, though I knew it not. He began to neglect his great pork .business; presently he wholly retired and turned an elegant leisure into a rabid search for curious things. His wealth was vast, and he spared it not. First he tried cow-bells. He made a collection which filled five large *salons*, and comprehended all the different sorts of cow-bells that ever had been contrived, save one. That one — an antique, and the only specimen extant — was possessed by another collector. My uncle offered enormous sums for it, but the gentleman would not sell. Doubtless you know what necessarily resulted. A true collector attaches no value to a collection that is not complete. His great heart breaks, he sells his hoard, he turns his mind to some field that seems unoccupied.

Thus did my uncle. He next tried brickbats. After piling up a vast and intensely interesting collection, the former difficulty supervened; his great heart broke again; he sold out his soul's idol to the retired brewer who possessed the missing brick. Then he tried flint hatchets and other implements of Primeval Man, but by and by discovered that the factory where they were made was supplying other collectors as well as himself. He tried Aztec inscriptions and stuffed whales — another failure, after incredible labor and expense. When his collection seemed at last perfect, a stuffed whale arrived from Greenland and an Aztec inscription from the Cundurango regions of Central America that made all former specimens insignificant. My uncle hastened to secure these noble gems. He got the stuffed whale, but another collector got the inscription. A real Cundurango, as possibly you know, is a possession of such supreme value that, when once a collector gets it, he will rather part with his family than with it. So my uncle sold out, and saw his darlings go forth, never more to return; and his coal-black hair turned white as snow in a single night.

Now he waited, and thought. He knew another disappointment might kill him. He was resolved that he would choose things next time that no other man was collecting. He carefully made up his mind, and once more entered the field — this time to make a collection of echoes.

"Of what?" said I.

Echoes, sir. His first purchase was an echo in Georgia that repeated four times; his next was a six-repeater in Maryland; his next was a thirteen-repeater in Maine; his next was a nine-repeater in Kansas; his next was a twelve-repeater in Tennessee, which he got cheap, so to speak, because it was out of repair, a portion of the crag which reflected it having tumbled down. He believed he could repair it at a cost of a few thousand dollars, and, by increasing the elevation with masonry, treble the repeating capacity; but the architect who undertook the job had never built an echo before, and so he utterly spoiled this one. Before he meddled with it, it used to talk back like a mother-in-law, but now it was only fit for the deaf and dumb asylum. Well, next he bought a lot of cheap little double-barrelled echoes, scattered around over various States and Territories; he got them at twenty per cent off by taking the lot. Next he bought a perfect Gatling gun of an echo in Oregon, and it cost a fortune, I can tell you. You may know, sir, that in the echo market the scale of prices is cumulative, like the carat-scale in diamonds; in fact, the same phraseology is used. A single-carat echo is worth but ten dollars over and above the value of the land it is on; a two-carat or double-barrelled echo is worth thirty dollars; a five-carat is worth nine hundred and fifty; a ten-carat is worth thirteen thousand. My uncle's Oregon echo, which he called the Great Pitt Echo, was a twenty-two carat gem, and cost two hundred and sixteen thousand dollars — they threw the land in, for it was four hundred miles from a settlement.

Well, in the mean time my path was a path of roses. I was the accepted suitor of the only and lovely daughter of an English earl, and was beloved to distraction. In that dear presence I swam in seas of bliss. The family were content, for it was known that I was sole heir to an uncle held to be worth five millions of dollars. However, none of us knew that my uncle had become a collector, at least in anything more than a small way, for æsthetic amusement.

Now gathered the clouds above my unconscious head. That divine echo, since known throughout the world as the Great Koh-i-noor, or Mountain of Repetitions, was discovered. It was a sixty-five-carat gem. You could utter a word and it would talk back at you for fifteen minutes, when the day was otherwise quiet. But behold, another fact came to light at the same time: another echo-collector was in the field. The two rushed to make the peerless purchase. The property consisted of a couple of small hills with a shallow swale between, out yonder among the back settlements of New York State. Both men arrived on the ground at the same time, and neither knew the other was there. The echo was not all owned by one man; a person by the name of Williamson Bolivar Jarvis owned the east hill, and a person by the name of Harbison J. Bledso owned the west hill; the swale between was the dividing line. So while my uncle was buying

Jarvis's hill for three million two hundred and eighty-five thousand dollars, the other party was buying Bledso's hill for a shade over three million.

Now, do you perceive the natural result? Why, the noblest collection of echoes on earth was forever and ever incomplete, since it possessed but the one half of the king echo of the universe. Neither man was content with this divided ownership, yet neither would sell to the other. There were jawings, bickerings, heart-burnings. And at last, that other collector, with a malignity which only a collector can ever feel toward a man and a brother, proceeded to cut down his hill!

You see, as long as he could not have the echo, he was resolved that nobody should have it. He would remove his hill, and then there would be nothing to reflect my uncle's echo. My uncle remonstrated with him, but the man said, "I own one end of this echo; I choose to kill my end; you must take care of your own end yourself."

Well, my uncle got an injunction put on him. The other man appealed and fought it in a higher court. They carried it on up, clear to the Supreme Court of the United States. It made no end of trouble there. Two of the judges believed that an echo was personal property, because it was impalpable to sight and touch, and yet was purchasable, salable, and consequently taxable; two others believed that an echo was real estate, because it was manifestly attached to the land, and was not removable from place to place; other of the judges contended that an echo was not property at all.

It was finally decided that the echo was property; that the hills were property; that the two men were separate and independent owners of the two hills, but tenants in common in the echo; therefore defendant was at full liberty to cut down his hill, since it belonged solely to him, but must give bonds in three million dollars as indemnity for damages which might result to my uncle's half of the echo. This decision also debarred my uncle from using defendant's hill to reflect his part of the echo, without defendant's consent; he must use only his own hill; if his part of the echo would not go, under these circumstances, it was sad, of course, but the court could find no remedy. The court also debarred defendant from using my uncle's hill to reflect *his* end of the echo, without consent. You see the grand result! Neither man would give consent, and so that astonishing and most noble echo had to cease from its great powers; and since that day that magnificent property is tied up and unsalable.

A week before my wedding day, while I was still swimming in bliss and the nobility were gathering from far and near to honor our espousals, came news of my uncle's death, and also a copy of his will, making me his sole heir. He was gone; alas, my dear benefactor was no more. The thought surcharges my heart even at this remote day. I handed the will to the earl; I could not read it for the blinding tears. The earl read it; then he sternly said, "Sir, do you call this wealth?—but doubtless you do in your inflated country. Sir, you are left sole heir to a vast collection of echoes — if a thing can be called a collection that is scattered far and wide over the huge length and breadth of the American continent; sir, this is not all; you are head and ears in debt; there is not an echo in the lot but has a mortgage on it; sir, I am not a hard man, but I must look to my child's interest; if you had but one echo which you could honestly call your own, if you had but one echo which was free from incumbrance, so that you could retire to it with my child, and by humble, painstaking industry, cultivate and improve it, and thus wrest from it a maintenance, I would not say you nay; but I cannot marry my child to a beggar. Leave his side, my darling; go, sir; take your mortgage-ridden echoes and quit my sight forever."

My noble Celestine clung to me in tears, with loving arms, and swore she would willingly, nay, gladly marry me, though I had not an echo in the world. But it could not be. We were torn asunder, she to pine and die within the twelvemonth, I to toil life's long journey sad and lone, praying daily, hourly, for that release which shall join us together again in that dear realm where the wicked cease from troubling and the weary are at rest. Now, sir, if you will be so kind as to look at these maps and plans in my portfolio, I am sure I can sell you an echo for less money than any man in the trade. Now this one, which cost my uncle ten dollars, thirty years ago, and is one of the sweetest things in Texas, I will let you have for —

"Let me interrupt you," I said. "My friend, I have not had a moment's respite from canvassers this day.

I have bought a sewing-machine which I did not want; I have bought a map which is mistaken in all its details; I have bought a clock which will not go; I have bought a moth poison which the moths prefer to any other beverage; I have bought no end of useless inventions, and now I have had enough of this foolishness. I would not have one of your echoes if you were even to give it to me. I would not let it stay on the place. I always hate a man that tries to sell me echoes. You see this gun? Now take your collection and move on; let us not have bloodshed."

But he only smiled a sad, sweet smile, and got out some more diagrams. You know the result perfectly well, because you know that when you have once opened the door to a canvasser, the trouble is done and you have got to suffer defeat.

I compromised with this man at the end of an intolerable hour. I bought two double-barrelled echoes in good condition, and he threw in another, which he said was not salable because it only spoke German. He said, "She was a perfect polyglot once, but somehow her palate got down."

AN ENCOUNTER WITH AN INTERVIEWER.

THE nervous, dapper, "peart" young man took the chair I offered him, and said he was connected with the "Daily Thunderstorm," and added, —

"Hoping it 's no harm, I 've come to interview you."

"Come to what?"

"*Interview* you."

"Ah! I see. Yes — yes. Um! Yes — yes."

I was not feeling bright that morning. Indeed, my powers seemed a bit under a cloud. However, I went to the bookcase, and when I had been looking six or seven minutes, I found I was obliged to refer to the young man. I said, —

"How do you spell it?"

"Spell what?"

"Interview."

"Oh my goodness! what do you want to spell it for?"

"I don't want to spell it; I want to see what it means."

"Well, this is astonishing, I must say. *I* can tell you what it means, if you — if you —"

"Oh, all right! That will answer, and much obliged to you, too."

"In, *in*, ter, *ter*, *in*ter— "

"Then you spell it with an *I?*"

"Why, certainly!"

"Oh, that is what took me so long."

"Why, my *dear* sir, what did *you* propose to spell it with?"

"Well, I—I—hardly know. I had the Unabridged, and I was ciphering around in the back end, hoping I might tree her among the pictures. But it's a very old edition."

"Why, my friend, they would n't have a *picture* of it in even the latest e— My dear sir, I beg your pardon, I mean no harm in the world, but you do not look as—as—intelligent as I had expected you would. No harm—I mean no harm at all."

"Oh, don't mention it! It has often been said, and by people who would not flatter and who could have no inducement to flatter, that I am quite remarkable in that way. Yes—yes; they always speak of it with rapture."

"I can easily imagine it. But about this interview. You know it is the custom, now, to interview any man who has become notorious."

"Indeed, I had not heard of it before. It must be very interesting. What do you do it with?"

"Ah, well—well—well—this is disheartening. It *ought* to be done with a club in some cases; but customarily it consists in the interviewer asking questions and the interviewed answering them. It is all the rage now. Will you let me ask you certain questions calculated to bring out the salient points of your public and private history?"

"Oh, with pleasure, — with pleasure. I have a very bad memory, but I hope you will not mind that. That is to say, it is an irregular memory, — singularly irregular. Sometimes it goes in a gallop, and then again it will be as much as a fortnight passing a given point. This is a great grief to me."

"Oh, it is no matter, so you will try to do the best you can."

"I will. I will put my whole mind on it."

"Thanks. Are you ready to begin?"

"Ready."

Q. How old are you?

A. Nineteen, in June.

Q. Indeed! I would have taken you to be thirty-five or six. Where were you born?

A. In Missouri.

Q. When did you begin to write?

A. In 1836.

Q. Why, how could that be, if you are only nineteen now?

A. I don't know. It does seem curious, somehow.

Q. It does, indeed. Whom do you consider the most remarkable man you ever met?

A. Aaron Burr.

Q. But you never could have met Aaron Burr, if you are only nineteen years —

A. Now, if you know more about me than I do, what do you ask me for?

Q. Well, it was only a suggestion; nothing more. How did you happen to meet Burr?

A. Well, I happened to be at his funeral one day, and he asked me to make less noise, and —

Q. But, good heavens! if you were at his funeral, he must have been dead; and if he was dead, how could he care whether you made a noise or not?

A. I don't know. He was always a particular kind of a man that way.

Q. Still, I don't understand it at all. You say he spoke to you, and that he was dead.

A. I did n't say he was dead.

Q. But was n't he dead?

A. Well, some said he was, some said he was n't.

Q. What did you think?

A. Oh, it was none of my business! It was n't any of my funeral.

Q. Did you — However, we can never get this matter straight. Let me ask about something else. What was the date of your birth?

A. Monday, October 31st, 1693.

Q. What! Impossible! That would make you a hundred and eighty years old. How do you account for that?

A. I don't account for it at all.

Q. But you said at first you were only nineteen, and now you make yourself out to be one hundred and eighty. It is an awful discrepancy.

A. Why, have you noticed that? (Shaking hands.) Many a time it has seemed to me like a discrepancy, but somehow I could n't make up my mind. How quick you notice a thing!

Q. Thank you for the compliment, as far as it goes. Had you, or have you, any brothers or sisters?

A. Eh! I — I — I think so — yes — but I don't remember.

Q. Well, that is the most extraordinary statement I ever heard!

A. Why, what makes you think that?

Q. How could I think otherwise? Why, look here! Who is this a picture of on the wall? Is n't that a brother of yours?

A. Oh! yes, yes, yes! Now you remind me of it; that *was* a brother of mine. That 's William — *Bill* we called him. Poor old Bill!

Q. Why? Is he dead, then?

A. Ah! well, I suppose so. We never could tell. There was a great mystery about it.

Q. That is sad, very sad. He disappeared, then?

A. Well, yes, in a sort of general way. We buried him.

Q. *Buried* him! *Buried* him, without knowing whether he was dead or not?

A. Oh, no! Not that. He was dead enough.

Q. Well, I confess that I can't understand this. If you buried him, and you knew he was dead —

A. No! no! We only thought he was.

Q. Oh, I see! He came to life again?

A. I bet he did n't.

Q. Well, I never heard anything like this. *Somebody* was dead. *Somebody* was buried. Now, where was the mystery?

A. Ah! that 's just it! That 's it exactly. You see,

we were twins, — defunct and I, — and we got mixed in the bath-tub when we were only two weeks old, and one of us was drowned. But we did n't know which. Some think it was Bill. Some think it was me.

Q. Well, that *is* remarkable. What do *you* think?

A. Goodness knows! I would give whole worlds to know. This solemn, this awful mystery has cast a gloom over my whole life. But I will tell you a secret now, which I never have revealed to any creature before. One of us had a peculiar mark — a large mole on the back of his left hand; that was *me. That child was the one that was drowned!*

Q. Very well, then, I don't see that there is any mystery about it, after all.

A. You don't? Well, *I* do. Anyway, I don't see how they could ever have been such a blundering lot as to go and bury the wrong child. But, 'sh! — don't mention it where the family can hear of it. Heaven knows they have heart-breaking troubles enough without adding this.

Q. Well, I believe I have got material enough for the present, and I am very much obliged to you for the pains you have taken. But I was a good deal interested in that account of Aaron Burr's funeral. Would you mind telling me what particular circumstance it was that made you think Burr was such a remarkable man?

A. Oh! it was a mere trifle! Not one man in fifty would have noticed it at all. When the sermon was over, and the procession all ready to start for the cemetery, and the body all arranged nice in the hearse, he said he wanted to take a last look at the scenery, and so he *got up and rode with the driver.*

Then the young man reverently withdrew. He was very pleasant company, and I was sorry to see him go.

PARIS NOTES.[1]

THE Parisian travels but little, he knows no language but his own, reads no literature but his own, and consequently he is pretty narrow and pretty self-sufficient. However, let us not be too sweeping; there are Frenchmen who know languages not their own: these are the waiters. Among the rest, they know English; that is, they know it on the European plan, — which is to say, they can speak it, but can't understand it. They easily make themselves understood, but it is next to impossible to word an English sentence in such a way as to enable them to comprehend it. They think they comprehend it; they pretend they do; but they don't. Here is a conversation which I had with one of these beings; I wrote it down at the time, in order to have it exactly correct.

I. These are fine oranges. Where are they grown?

He. More? Yes, I will bring them.

I. No, do not bring any more; I only want to know where they are from — where they are raised.

[1] Crowded out of "A Tramp Abroad" to make room for more vital statistics. — M. T.

He. Yes? (with imperturbable mien, and rising inflection.)

I. Yes. Can you tell me what country they are from?

He. Yes? (blandly, with rising inflection.)

I. (disheartened). They are very nice.

He. Good night. (Bows, and retires, quite satisfied with himself.)

That young man could have become a good English scholar by taking the right sort of pains, but he was French, and would n't do that. How different is the case with our people; they utilize every means that offers. There are some alleged French Protestants in Paris, and they built a nice little church on one of the great avenues that lead away from the Arch of Triumph, and proposed to listen to the correct thing, preached in the correct way, there, in their precious French tongue, and be happy. But their little game does not succeed. Our people are always there ahead of them, Sundays, and take up all the room. When the minister gets up to preach, he finds his house full of devout foreigners, each ready and waiting, with his little book in his hand, — a morocco-bound Testament, apparently. But only apparently; it is Mr. Bellows's admirable and exhaustive little French-English dictionary, which in look and binding and size is just like a Testament, — and those people are there to study French. The building has been nicknamed "The Church of the Gratis French Lesson."

These students probably acquire more language than general information, for I am told that a French sermon is like a French speech, — it never names an historical event, but only the date of it; if you are not up in dates, you get left. A French speech is something like this: —

"Comrades, citizens, brothers, noble parts of the only sublime and perfect nation, let us not forget that the 21st January cast off our chains; that the 10th August relieved us of the shameful presence of foreign spies; that the 5th September was its own justification before Heaven and humanity; that the 18th Brumaire contained the seeds of its own punishment; that the 14th July was the mighty voice of liberty proclaiming the resurrection, the new day, and inviting the oppressed peoples of the earth to look upon the divine face of France and live; and let us here record our everlasting curse against the man of the 2d December, and declare in thunder tones, the native tones of France, that but for him there had been no 17th March in history, no 12th October, no 19th January, no 22d April, no 16th November, no 30th September, no 2d July, no 14th February, no 29th June, no 15th August, no 31st May, — that but for him, France, the pure, the grand, the peerless, had had a serene and vacant almanac to-day!"

I have heard of one French sermon which closed in this odd yet eloquent way: —

"My hearers, we have sad cause to remember the man of the 13th January. The results of the vast crime of the 13th January have been in just proportion to the magnitude of the act itself. But for it there had been no 30th November, — sorrowful spectacle!

The grisly deed of the 16th June had not been done but for it, nor had the man of the 16th June known existence; to it alone the 3d September was due, also the fatal 12th October. Shall we, then, be grateful for the 13th January, with its freight of death for you and me and all that breathe? Yes, my friends, for it gave us also that which had never come but for it, and it alone, — the blessed 25th December."

It may be well enough to explain, though in the case of many of my readers this will hardly be necessary. The man of the 13th January is Adam; the crime of that date was the eating of the apple; the sorrowful spectacle of the 30th November was the expulsion from Eden; the grisly deed of the 16th June was the murder of Abel; the act of the 3d September was the beginning of the journey to the land of Nod; the 12th day of October, the last mountain-tops disappeared under the flood. When you go to church in France, you want to take your almanac with you, — annotated.

LEGEND OF SAGENFELD, IN GERMANY.[1]

I.

MORE than a thousand years ago this small district was a kingdom, — a little bit of a kingdom, a sort of dainty little toy kingdom, as one might say. It was far removed from the jealousies, strifes, and turmoils of that old warlike day, and so its life was a simple life, its people a gentle and guileless race; it lay always in a deep dream of peace, a soft Sabbath tranquillity; there was no malice, there was no envy, there was no ambition, consequently there were no heart-burnings, there was no unhappiness in the land.

In the course of time the old king died and his little son Hubert came to the throne. The people's love for him grew daily; he was so good and so pure and so noble, that by and by this love became a passion, almost a worship. Now at his birth the soothsayers had diligently studied the stars and found something written in that shining book to this effect: —

[1] Left out of "A Tramp Abroad" because its authenticity seemed doubtful, and could not at that time be proved. — M. T.

In Hubert's fourteenth year a pregnant event will happen; the animal whose singing shall sound sweetest in Hubert's ear shall save Hubert's life. So long as the king and the nation shall honor this animal's race for this good deed, the ancient dynasty shall not fail of an heir, nor the nation know war or pestilence or poverty. But beware an erring choice!

All through the king's thirteenth year but one thing was talked of by the soothsayers, the statesmen, the little parliament, and the general people. That one thing was this: How is the last sentence of the prophecy to be understood? What goes before seems to mean that the saving animal will choose *itself*, at the proper time; but the closing sentence seems to mean that the *king* must choose beforehand, and say what singer among the animals pleases him best, and that if he choose wisely the chosen animal will save his life, his dynasty, his people, but that if he should make "an erring choice" — beware!

By the end of the year there were as many opinions about this matter as there had been in the beginning; but a majority of the wise and the simple were agreed that the safest plan would be for the little king to make choice beforehand, and the earlier the better. So an edict was sent forth commanding all persons who owned singing creatures to bring them to the great hall of the palace in the morning of the first day of the new year. This command was obeyed. When everything was in readiness for the trial, the king made his solemn entry with the great officers of the crown, all clothed in their robes of state. The king mounted his golden throne and prepared to give judgment. But he presently said, —

"These creatures all sing at once; the noise is unendurable; no one can choose in such a turmoil. Take them all away, and bring back one at a time."

This was done. One sweet warbler after another charmed the young king's ear and was removed to make way for another candidate. The precious minutes slipped by; among so many bewitching songsters he found it hard to choose, and all the harder because the promised penalty for an error was so terrible that it unsettled his judgment and made him afraid to trust his own ears. He grew nervous and his face showed distress. His ministers saw this, for they never took their eyes from him a moment. Now they began to say in their hearts, —

"He has lost courage — the cool head is gone — he will err — he and his dynasty and his people are doomed!"

At the end of an hour the king sat silent awhile, and then said, —

"Bring back the linnet."

The linnet trilled forth her jubilant music. In the midst of it the king was about to uplift his sceptre in sign of choice, but checked himself and said, —

"But let us be sure. Bring back the thrush; let them sing together."

The thrush was brought, and the two birds poured out their marvels of song together. The king wavered, then his inclination began to settle and strengthen —

one could see it in his countenance. Hope budded in the hearts of the old ministers, their pulses began to beat quicker, the sceptre began to rise slowly, when —

There was a hideous interruption! It was a sound like this — just at the door: —

"Waw . . . . . . *he!* — waw . . . . . . *he!* — waw-he! waw-he! — waw-he!"

Everybody was sorely startled — and enraged at himself for showing it.

The next instant the dearest, sweetest, prettiest little peasant maid of nine years came tripping in, her brown eyes glowing with childish eagerness; but when she saw that august company and those angry faces she stopped and hung her head and put her poor coarse apron to her eyes. Nobody gave her welcome, none pitied her. Presently she looked up timidly through her tears, and said, —

"My lord the king, I pray you pardon me, for I meant no wrong. I have no father and no mother, but I have a goat and a donkey, and they are all in all to me. My goat gives me the sweetest milk, and when my dear good donkey brays it seems to me there is no music like to it. So when my lord the king's jester said the sweetest singer among all the animals should save the crown and nation, and moved me to bring him here — "

All the court burst into a rude laugh, and the child fled away crying, without trying to finish her speech. The chief minister gave a private order that she and her disastrous donkey be flogged beyond the precincts of the palace and commanded to come within them no more.

Then the trial of the birds was resumed. The two birds sang their best, but the sceptre lay motionless in the king's hand. Hope died slowly out in the breasts of all. An hour went by; two hours; still no decision. The day waned to its close, and the waiting multitudes outside the palace grew crazed with anxiety and apprehension. The twilight came on, the shadows fell deeper and deeper. The king and his court could no longer see each other's faces. No one spoke — none called for lights. The great trial had been made; it had failed; each and all wished to hide their faces from the light and cover up their deep trouble in their own hearts.

Finally — hark! A rich, full strain of the divinest melody streamed forth from a remote part of the hall, — the nightingale's voice!

"Up!" shouted the king, "let all the bells make proclamation to the people, for the choice is made and we have not erred. King, dynasty, and nation are saved. From henceforth let the nightingale be honored throughout the land forever. And publish it among all the people that whosoever shall insult a nightingale, or injure it, shall suffer death. The king hath spoken."

All that little world was drunk with joy. The castle and the city blazed with bonfires all night long, the people danced and drank and sang, and the triumphant clamor of the bells never ceased.

From that day the nightingale was a sacred bird.

Its song was heard in every house; the poets wrote its praises; the painters painted it; its sculptured image adorned every arch and turret and fountain and public building. It was even taken into the king's councils; and no grave matter of state was decided until the soothsayers had laid the thing before the state nightingale and translated to the ministry what it was that the bird had sung about it.

II.

The young king was very fond of the chase. When the summer was come he rode forth with hawk and hound, one day, in a brilliant company of his nobles. He got separated from them, by and by, in a great forest, and took what he imagined a near cut, to find them again; but it was a mistake. He rode on and on, hopefully at first, but with sinking courage finally. Twilight came on, and still he was plunging through a lonely and unknown land. Then came a catastrophe. In the dim light he forced his horse through a tangled thicket overhanging a steep and rocky declivity. When horse and rider reached the bottom, the former had a broken neck and the latter a broken leg. The poor little king lay there suffering agonies of pain, and each hour seemed a long month to him. He kept his ear strained to hear any sound that might promise hope of rescue; but he heard no voice, no sound of horn or bay of hound. So at last he gave up all hope, and said, "Let death come, for come it must."

Just then the deep, sweet song of a nightingale swept across the still wastes of the night.

"Saved!" the king said. "Saved! It is the sacred bird, and the prophecy is come true. The gods themselves protected me from error in the choice."

He could hardly contain his joy; he could not word his gratitude. Every few moments, now, he thought he caught the sound of approaching succor. But each time it was a disappointment; no succor came. The dull hours drifted on. Still no help came, — but still the sacred bird sang on. He began to have misgivings about his choice, but he stifled them. Toward dawn the bird ceased. The morning came, and with it thirst and hunger; but no succor. The day waxed and waned. At last the king cursed the nightingale.

Immediately the song of the thrush came from out the wood. The king said in his heart, "This was the true bird — my choice was false — succor will come now."

But it did not come. Then he lay many hours insensible. When he came to himself, a linnet was singing. He listened — with apathy. His faith was gone. "These birds," he said, "can bring no help; I and my house and my people are doomed." He turned him about to die; for he was grown very feeble from hunger and thirst and suffering, and felt that his end was near. In truth, he wanted to die, and be released from pain. For long hours he lay without thought or feeling or motion. Then his senses returned. The dawn of the third morning was breaking. Ah, the world seemed very beautiful to those worn eyes. Suddenly a great longing to live rose up in the lad's heart, and from his soul welled a deep and fervent prayer that Heaven would have mercy upon him and let him see his home and his friends once more. In that instant a soft, a faint, a far-off sound, but oh, how inexpressibly sweet to his waiting ear, came floating out of the distance, —

"Waw . . . . . . *he!* — waw . . . . . . *he!* — waw-he! — waw-he! — waw-he!"

"*That*, oh, *that* song is sweeter, a thousand times sweeter, than the voice of nightingale, thrush, or linnet, for it brings not mere hope, but *certainty* of succor; and now indeed am I saved! The sacred singer has chosen itself, as the oracle intended; the prophecy is fulfilled, and my life, my house, and my people are redeemed. The ass shall be sacred from this day!"

The divine music grew nearer and nearer, stronger and stronger, — and ever sweeter and sweeter to the perishing sufferer's ear. Down the declivity the docile little donkey wandered, cropping herbage and singing as he went; and when at last he saw the dead horse and the wounded king, he came and snuffed at them with simple and marvelling curiosity. The king petted him, and he knelt down as had been his wont when his little mistress desired to mount. With great labor and pain the lad drew himself upon the creature's back, and held himself there by aid of the generous ears. The ass went singing forth from the place and carried the king to the little peasant maid's hut. She gave him her pallet for a bed, refreshed him with goat's milk, and then flew to tell the great news to the first scouting party of searchers she might meet.

The king got well. His first act was to proclaim the sacredness and inviolability of the ass; his second was to add this particular ass to his cabinet and make him chief minister of the crown; his third was to have all the statues and effigies of nightingales throughout his kingdom destroyed, and replaced by statues and effigies of the sacred donkey; and his fourth was to announce that when the little peasant maid should reach her fifteenth year he would make her his queen, — and he kept his word.

Such is the legend. This explains why the mouldering image of the ass adorns all these old crumbling walls and arches; and it explains why, during many centuries, an ass was always the chief minister in that royal cabinet, just as is still the case in most cabinets to this day; and it also explains why, in that little kingdom, during many centuries, all great poems, all great speeches, all great books, all public solemnities, and all royal proclamations, always began with these stirring words, —

"Waw . . . . . . *he!* — waw . . . . . . *he!* — waw-he! — waw-he! — waw-he!"

SPEECH ON THE BABIES,

AT THE BANQUET, IN CHICAGO, GIVEN BY THE ARMY OF THE TENNESSEE TO THEIR FIRST COMMANDER, GENERAL U. S. GRANT, NOVEMBER, 1879.

[The fifteenth regular toast was "The Babies. — As they comfort us in our sorrows, let us not forget them in our festivities."]

I LIKE that. We have not all had the good fortune to be ladies. We have not all been generals, or poets, or statesmen; but when the toast works down to the babies, we stand on common ground. It is a shame that for a thousand years the world's banquets have utterly ignored the baby, as if he did n't amount to anything. If you will stop and think a minute, — if you will go back fifty or one hundred years to your early married life and recontemplate your first baby, — you will remember that he amounted to a good deal, and even something over. You soldiers all know that when that little fellow arrived at family headquarters you had to hand in your resignation. He took entire command. You became his lackey, his mere body-servant, and you had to stand around too. He was not a commander who made allowances for time, distance, weather, or anything else. You had to execute his order whether it was possible or not. And there was only one form of marching in his manual of tactics, and that was the double-quick. He treated you with every sort of insolence and disrespect, and the bravest of you did n't dare to say a word. You could face the death-storm at Donelson and Vicksburg, and give back blow for blow; but when he clawed your whiskers, and pulled your hair, and twisted your nose, you had to take it. When the thunders of war were sounding in your ears you set your faces toward the batteries, and advanced with steady tread; but when he turned on the terrors of his war-whoop you advanced in the other direction, and mighty glad of the chance too. When he called for soothing-syrup, did you venture to throw out any side remarks about certain services being unbecoming an officer and a gentleman? No. You got up and *got* it. When he ordered his pap bottle and it was not warm, did you talk back? Not you. You went to work and *warmed* it. You even descended so far in your menial office as to take a suck at that warm, insipid stuff yourself, to see if it was right, — three parts water to one of milk, a touch of sugar to modify the colic, and a drop of peppermint to kill those immortal hiccoughs. I can taste that stuff yet. And how many things you learned as you went along! Sentimental young folks still take stock in that beautiful old saying that when the baby smiles in his sleep, it is because the angels are whispering to him. Very pretty, but too thin, — simply wind on the stomach, my friends. If the baby proposed to take a walk at his usual hour, two o'clock in the morning, did n't you rise up promptly and remark, with a mental addition which would not improve a Sunday-school book *much*, that that was the very thing you were about to propose yourself? Oh! you were under good discipline, and as you went fluttering up and down the room in your undress uniform, you not only prattled undignified baby-talk, but even tuned up your martial voices and tried to *sing!*— "Rock-a-by baby in the tree-top," for instance. What a spectacle for an Army of the Tennessee! And what an affliction for the neighbors, too; for it is not everybody within a mile around that likes military music at three in the morning. And when you had been keeping this sort of thing up two or three hours, and your little velvet-head intimated that nothing suited him like exercise and noise, what did you do? [*"Go on"!*] You simply *went* on until you dropped in the last ditch. The idea that a *baby* does n't *amount* to anything! Why, *one* baby is just a house and a front yard full by itself. *One* baby can furnish more business than you and your whole Interior Department can attend to. He is enterprising, irrepressible, brimful of lawless activities. Do what you please, you can't make him stay on the reservation. Sufficient unto the day is one baby. As long as you are in your right mind don't you ever pray for twins. Twins amount to a permanent riot. And there ain't any real difference between triplets and an insurrection.

Yes, it was high time for a toast-master to recognize the importance of the babies. Think what is in store for the present crop! Fifty years from now we shall all be dead, I trust, and then this flag, if it still survive (and let us hope it may), will be floating over a Republic numbering 200,000,000 souls, according to the settled laws of our increase. Our present schooner of State will have grown into a political leviathan, — a Great Eastern. The cradled babies of to-day will be on deck. Let them be well trained, for we are going to leave a big contract on their hands. Among the three or four million cradles now rocking in the land are some which this nation would preserve for ages as sacred things, if we could know which ones they are. In one of these cradles the unconscious Farragut of the future is at this moment teething, — think of it! — and putting in a world of dead earnest, unarticulated, but perfectly justifiable profanity over it, too. In another the future renowned astronomer is blinking at the shining Milky Way with but a languid interest, — poor little chap! — and wondering what has become of that other one they call the wet-nurse. In another the future great historian is lying, — and doubtless will continue to lie until his earthly mission is ended. In another the future President is busying himself with no profounder problem of state than what the mischief has become of his hair so early; and in a mighty array of other cradles there are now some 60,000 future office-seekers, getting ready to furnish him occasion to grapple with that same old problem a second time. And in still one more cradle, somewhere under the flag, the future illustrious commander-in-chief of the American armies is so little burdened with his approaching grandeurs and responsibilities as to be giving his whole strategic mind at this moment to trying to find out some way to get his big toe into his mouth, — an achievement which, meaning no disrespect, the illustrious guest of this evening turned *his* entire attention to some fifty-six years ago; and if the child is but a prophecy of the man, there are mighty few who will doubt that he *succeeded.*

SPEECH ON THE WEATHER,

AT THE NEW ENGLAND SOCIETY'S SEVENTY-FIRST ANNUAL DINNER, NEW YORK CITY.

The next toast was: "The Oldest Inhabitant — The Weather of New England."

> Who can lose it and forget it?
> Who can have it and regret it?
>
> "Be interposer 'twixt us Twain."
>
> *Merchant of Venice.*

To this Samuel L. Clemens (Mark Twain) replied as follows: —

I REVERENTLY believe that the Maker who made us all makes everything in New England but the weather. I don't know who makes that, but I think it must be raw apprentices in the weather clerk's factory who experiment and learn how, in New England, for board and clothes, and then are promoted to make weather for countries that require a good article, and will take their custom elsewhere if they don't get it. There is a sumptuous variety about the New England weather that compels the stranger's admiration — and regret. The weather is always doing something there; always attending strictly to business; always getting up new designs and trying them on the people to see how they will go. But it gets through more business in spring than in any other season. In the spring I have counted one hundred and thirty-six different kinds of weather inside of four-and-twenty hours. It was I that made the fame and fortune of that man that had that marvellous collection of weather on exhibition at the Centennial, that so astounded the foreigners. He was going to travel all over the world and get specimens from all the climes. I said, "Don't you do it; you come to New England on a favorable spring day." I told him what we could do in the way of style, variety, and quantity. Well, he came and he made his collection in four days. As to variety, why, he confessed that he got hundreds of kinds of weather that he had never heard of before. And as to quantity — well, after he had picked out and discarded all that was blemished in any way, he not only had weather enough, but weather to spare; weather to hire out; weather to sell; to deposit; weather to invest; weather to give to the poor. The people of New England are by nature patient and forbearing, but there are some things which they will not stand. Every year they kill a lot of poets for writing about "Beautiful Spring." These are generally casual visitors, who bring their notions of spring from somewhere else, and cannot, of course, know how the natives feel about spring. And so the first thing they know the opportunity to inquire how they feel has permanently gone by. Old Probabilities has a mighty reputation for accurate prophecy, and thoroughly well deserves it. You take up the paper and observe how crisply and confidently he checks off what to-day's weather is going to be on the Pacific, down South, in the Middle States, in the Wisconsin region. See him sail along in the joy and pride of his power till he gets to New England, and then see his tail drop. *He* does n't know what the weather is going to be in New England. Well, he mulls over it, and by and by he gets out something about like this: Probable northeast to southwest winds, varying to the southward and westward and eastward, and points between, high and low barometer swapping around from place to place; probable areas of rain, snow, hail, and drought, succeeded or preceded by earthquakes, with thunder and lightning. Then he jots down this postscript from his wandering mind, to cover accidents. "But it is possible that the programme may be wholly changed in the mean time." Yes, one of the brightest gems in the New England weather is the dazzling uncertainty of it. There is only one thing certain about it: you are certain there is going to be plenty of it — a perfect grand review; but you never can tell which end of the procession is going to move first. You fix up for the drought; you leave your umbrella in the house and sally out, and two to one you get drowned. You make up your mind that the earthquake is due; you stand from under, and take hold of something to steady yourself, and the first thing you know you get struck by lightning. These are great disappointments; but they can't be helped. The lightning there is peculiar; it is so convincing, that when it strikes a thing it does n't leave enough of that thing behind for you to tell whether — Well, you 'd think it was something valuable, and a Congressman had been there. And the thunder. When the thunder begins to merely tune up and scrape and saw, and key up the instruments for the performance, strangers say, " Why, what awful thunder you have here ! " But when the baton is raised and the real concert begins, you 'll find that stranger down in the cellar with his head in the ash-barrel. Now as to the *size* of the weather in New England, — lengthways, I mean. It is utterly disproportioned to the size of that little country. Half the time, when it is packed as full as it can stick, you will see that New England weather sticking out beyond the edges and projecting around hundreds and hundreds of miles over the neighboring States. She can't hold a tenth part of her weather. You can see cracks all about where she has strained herself trying to do it. I could speak volumes about the inhuman perversity of the New England weather, but I will give but a single specimen. I like to hear rain on a tin roof. So I covered part of my roof with tin, with an eye to that luxury. Well, sir, do you think it ever rains on that tin ? No, sir : skips it every time. Mind, in this speech I have been trying merely to do honor to the New England weather, — no language could do it justice. But, after all, there is at least one or two things about that weather (or, if you please, effects produced by it) which we residents would not like to part with. If we had n't our bewitching autumn foliage, we should still have to credit the weather with one feature which compensates for all its bullying vagaries, — the ice-storm : when a leafless tree is clothed with ice from the bottom to the top, — ice that is as bright and clear as crystal ; when every bough and twig is strung with ice-beads, frozen dew-drops, and the whole tree sparkles cold and white, like the Shah of Persia's diamond plume. Then the wind waves the branches and the sun comes out and turns all those myriads of beads and drops to prisms that glow and burn and flash with all manner of colored fires, which change and change again with inconceivable rapidity from blue to red, from red to green, and green to gold, — the tree becomes a spraying fountain, a very explosion of dazzling jewels ; and it stands there the acme, the climax, the supremest possibility in art or nature, of bewildering, intoxicating, intolerable magnificence. One cannot make the words too strong.

CONCERNING THE AMERICAN LANGUAGE.[1]

THERE was an Englishman in our compartment, and he complimented me on — on what? But you would never guess. He complimented me on my English. He said Americans in general did not speak the English language as correctly as I did. I said I was obliged to him for his compliment, since I knew he meant it for one, but that I was not fairly entitled to it, for I did n't speak English at all, — I only spoke American.

He laughed, and said it was a distinction without a difference. I said no, the difference was not prodigious, but still it was considerable. We fell into a friendly dispute over the matter. I put my case as well as I could, and said, —

"The languages were identical several generations ago, but our changed conditions and the spread of our people far to the south and far to the west have made many alterations in our pronunciation, and have introduced new words among us and changed the

[1] Being part of a chapter which was crowded out of "A Tramp Abroad." — M. T.

meanings of many old ones. English people talk through their noses; we do not. We say *know*, English people say *näo;* we say *cow*, the Briton says *käow;* we — "

"Oh, come! that is pure Yankee; everybody knows that."

"Yes, it is pure Yankee; that is true. One cannot hear it in America outside of the little corner called New England, which is Yankee land. The English themselves planted it there, two hundred and fifty years ago, and there it remains; it has never spread. But England talks through her nose yet; the Londoner and the backwoods New-Englander pronounce 'know' and 'cow' alike, and then the Briton unconsciously satirizes himself by making fun of the Yankee's pronunciation."

We argued this point at some length; nobody won; but no matter, the fact remains, — Englishmen say *näo* and *käow* for "know" and "cow," and that is what the rustic inhabitant of a very small section of America does.

"You conferred your *a* upon New England, too, and there it remains; it has not travelled out of the narrow limits of those six little States in all these two hundred and fifty years. All England uses it, New England's small population — say four millions — use it, but we have forty-five millions who do not use it. You say 'glahs of wawtah,' so does New England; at least, New England says *glahs.* America at large flattens the *a*, and says 'glass of water.' These sounds are pleasanter than yours; you may think they are not right, — well, in English they are *not* right, but in 'American' they are. You say *flahsk*, and *bahsket*, and *jackahss;* we say 'flask,' 'basket,' 'jackass,'—sounding the *a* as it is in 'tallow,' 'fallow,' and so on. Up to as late as 1847 Mr. Webster's Dictionary had the impudence to still pronounce 'basket' *bahsket*, when he knew that outside of his little New England all America shortened the *a* and paid no attention to his English broadening of it. However, it called itself an English Dictionary, so it was proper enough that it should stick to English forms, perhaps. It still calls itself an English Dictionary to-day, but it has quietly ceased to pronounce 'basket' as if it were spelt *bahsket*. In the American language the *h* is respected; the *h* is not dropped or added improperly."

"The same is the case in England, — I mean among the educated classes, of course."

"Yes, that is true; but a nation's language is a very large matter. It is not simply a manner of speech obtaining among the educated handful; the manner obtaining among the vast uneducated multitude must be considered also. Your uneducated masses speak English, you will not deny that; our uneducated masses speak American, — it won't be fair for you to deny that, for you can see, yourself, that when your stable-boy says, 'It is n't the 'unting that 'urts the 'orse, but the 'ammer, 'ammer, 'ammer on the 'ard 'ighway,' and our stable-boy makes the same remark without suffocating a single *h*, these two people are manifestly talking two different languages. But if the signs are to be trusted, even your educated classes used to drop the *h*. They say *humble*, now, and *heroic*, and *historic*, etc., but I judge that they used to drop those *h*'s because your writers still keep up the fashion of putting *an* before those words, instead of *a*. This is what Mr. Darwin might call a 'rudimentary' sign that that *an* was justifiable once, and useful, — when your educated classes used to say *'umble*, and *'eroic*, and *'istorical*. Correct writers of the American language do not put *an* before those words."

The English gentleman had something to say upon this matter, but never mind what he said, — I 'm not arguing his case. I have him at a disadvantage, now. I proceeded : —

"In England you encourage an orator by exclaiming 'H'yaah! h'yaah!' We pronounce it *heer* in some sections, 'h'*yer*' in others, and so on; but our whites do not say 'h'yaah,' pronouncing the *a*'s like the *a* in *ah*. I have heard English ladies say 'don't you' — making two separate and distinct words of it; your Mr. Bernand has satirized it. But we always say 'dontchu.' This is much better. Your ladies say, 'Oh, it 's *o*ful nice!" Ours say, 'Oh, it 's *aw*ful nice!' We say, '*Four* hundred,' you say '*For*' — as in the word *or*. Your clergymen speak of 'the Lawd,' ours of 'the Lord'; yours speak of 'the gawds of the heathen,' ours of 'the gods of the heathen.' When you are exhausted, you say you are 'knocked up.' We don't. When you say you will do a thing 'directly,' you mean 'immediately'; in the American language — generally speaking — the word signifies 'after a little.' When you say 'clever,' you mean 'capable'; with us the word used to mean 'accommodating,' but I don't know what it means now. Your word 'stout' means 'fleshy'; our word 'stout' usually means 'strong.' Your words 'gentleman' and 'lady' have a very restricted meaning; with us they include the bar-maid, butcher, burglar, harlot, and horse-thief. You say, 'I have n't *got* any stockings on,' 'I have n't *got* any memory,' 'I have n't *got* any money in my purse'; we usually say, 'I have n't any stockings on,' 'I have n't any memory,' 'I have n't any money in my purse.' You say 'out of window'; we always put in a *the.* If one asks 'How old is that man?' the Briton answers, 'He will be about forty;' in the American language, we should say, 'He *is* about forty.' However, won't tire you, sir; but if I wanted to, I could pile up differences here until I not only convinced you that English and American are separate languages, but that when I speak my native tongue in its utmost purity an Englishman can't understand me at all."

"I don't wish to flatter you, but it is about all I can do to understand you *now.*"

That was a very pretty compliment, and it put us on the pleasantest terms directly, — I use the word in the English sense.

[*Later* — 1882. Æsthetes in many of our schools are now beginning to teach the pupils to broaden the *a*, and to say "don't you," in the elegant foreign way.]

ROGERS.

THIS man Rogers happened upon me and introduced himself at the town of ——, in the South of England, where I stayed awhile. His step-father had married a distant relative of mine who was afterwards hanged, and so he seemed to think a blood relationship existed between us. He came in every day and sat down and talked. Of all the bland, serene human curiosities I ever saw, I think he was the chiefest. He desired to look at my new chimney-pot hat. I was very willing, for I thought he would notice the name of the great Oxford Street hatter in it, and respect me accordingly. But he turned it about with a sort of grave compassion, pointed out two or three blemishes, and said that I, being so recently arrived, could not be expected to know where to supply myself. Said he would send me the address of *his* hatter. Then he said, "Pardon me," and proceeded to cut a neat circle of red tissue-paper; daintily notched the edges of it; took the mucilage and pasted it in my hat so as to cover the manufacturer's name. He said, "No one will know now where you got it. I will send you a hat-tip of my hatter, and you can paste it over this tissue circle." It was the calmest, coolest thing, — I never admired a man so much in my life. Mind, he did this while his own hat sat offensively near our noses, on the table, — an ancient extinguisher of the "slouch" pattern, limp and shapeless with age, discolored by vicissitudes of the weather, and banded by an equator of bear's grease that had stewed through.

Another time he examined my coat. I had no terrors, for over my tailor's door was the legend, "By Special Appointment Tailor to H. R. H. the Prince of Wales," etc. I did not know at the time that the most of the tailor shops had the same sign out, and that whereas it takes nine tailors to make an ordinary man, it takes a hundred and fifty to make a prince. He was full of compassion for my coat. Wrote down the address of his tailor for me. Did not tell me to mention my *nom de plume* and the tailor would put his best work on my garment, as complimentary people sometimes do, but said his tailor would hardly trouble himself for an unknown person (unknown person, when I thought I was so celebrated in England! — that was the cruelest cut), but cautioned me to mention *his* name, and it would be all right. Thinking to be facetious, I said, —

"But he might sit up all night and injure his health."

"Well, *let* him," said Rogers; "I 've done enough for him, for him to show some appreciation of it."

I might as well have tried to disconcert a mummy with my facetiousness. Said Rogers: "I get all my coats there, — they 're the only coats fit to be seen in."

I made one more attempt. I said, "I wish you had brought one with you — I would like to look at it."

"Bless your heart, have n't I got one on? — *this* article is Morgan's make."

I examined it. The coat had been bought ready-made, of a Chatham Street Jew, without any question — about 1848. It probably cost four dollars when it was new. It was ripped, it was frayed, it was napless and greasy. I could not resist showing him where it was ripped. It so affected him that I was almost sorry I had done it. First he seemed plunged into a bottomless abyss of grief. Then he roused himself, made a feint with his hands as if waving off the pity of a nation, and said, — with what seemed to me a manufactured emotion, — "No matter; no matter; don't mind me; do not bother about it. I can get another."

When he was thoroughly restored, so that he could examine the rip and command his feelings, he said, ah, *now* he understood it, — his servant must have done it while dressing him that morning.

His servant! There was something awe-inspiring in effrontery like this.

Nearly every day he interested himself in some article of my clothing. One would hardly have expected this sort of infatuation in a man who always wore the same suit, and it a suit that seemed coeval with the Conquest.

It was an unworthy ambition, perhaps, but I *did* wish I could make this man admire *something* about me or something I did, — you would have felt the same way. I saw my opportunity: I was about to return to London, and had "listed" my soiled linen for the wash. It made quite an imposing mountain in the corner of the room, — fifty-four pieces. I hoped he would fancy it was the accumulation of a single week. I took up the wash-list, as if to see that it was all right, and then tossed it on the table, with pretended forgetfulness. Sure enough, he took it up and ran his eye along down to the grand total. Then he said, "You get off easy," and laid it down again.

His gloves were the saddest ruin, but he told me where I could get some like them. His shoes would hardly hold walnuts without leaking, but he liked to put his feet up on the mantel-piece and contemplate them. He wore a dim glass breastpin, which he called a "morphylitic diamond," — whatever that may mean, — and said only two of them had ever been found, — the Emperor of China had the other one.

Afterward, in London, it was a pleasure to me to see this fantastic vagabond come marching into the lobby of the hotel in his grand-ducal way, for he always had some new imaginary grandeur to develop — there was nothing stale about him but his clothes. If he addressed me when strangers were about, he always raised his voice a little and called me "Sir Richard," or "General," or "Your Lordship," — and when people began to stare and look deferential, he would fall to inquiring in a casual way why I disappointed the Duke of Argyll the night before; and then remind me of our engagement at the Duke of Westminster's for the following day. I think that for the time being

18

these things were realities to him. He once came and invited me to go with him and spend the evening with the Earl of Warwick at his town house. I said I had received no formal invitation. He said that that was of no consequence, the Earl had no formalities for him or his friends. I asked if I might go just as I was. He said no, that would hardly do; evening dress was requisite at night in any gentleman's house. He said he would wait while I dressed, and then we would go to his apartments and I could take a bottle of champagne and a cigar while he dressed. I was very willing to see how this enterprise would turn out, so I dressed, and we started to his lodgings. He said if I did n't mind we would walk. So we tramped some four miles through the mud and fog, and finally found his "apartments:" they consisted of a single room over a barber's shop in a back street. Two chairs, a small table, an ancient valise, a wash-basin and pitcher (both on the floor in a corner), an unmade bed, a fragment of a looking-glass, and a flower-pot with a perishing little rose geranium in it, which he called a century plant, and said it had not bloomed now for upwards of two centuries — given to him by the late Lord Palmerston — been offered a prodigious sum for it) — these were the contents of the room. Also a brass candlestick and part of a candle. Rogers lit the candle, and told me to sit down and make myself at home. He said he hoped I was thirsty, because he would surprise my palate with an article of champagne that seldom got into a commoner's system; or would I prefer sherry, or port? Said he had port in bottles that were swathed in stratified cobwebs, every stratum representing a generation. And as for his cigars, — well, I should judge of them myself. Then he put his head out at the door and called, —

"Sackville!" No answer.

"Hi! — Sackville!" No answer.

"Now what the devil can have become of that butler? I *never* allow a servant to — Oh, confound that idiot, he's got the *keys*. Can't get into the other rooms without the keys."

(I was just wondering at his intrepidity in still keeping up the delusion of the champagne, and trying to imagine how he was going to get out of the difficulty.)

Now he stopped calling Sackville and began to call "Anglesy." But Anglesy did n't come. He said, "This is the *second* time that that equerry has been absent without leave. To-morrow I'll discharge him."

Now he began to whoop for "Thomas," but Thomas did n't answer. Then for "Theodore," but no Theodore replied.

"Well, I give it up," said Rogers. "The servants never expect me at this hour, and so they're all off on a lark. Might get along without the equerry and the page, but can't have any wine or cigars without the butler, and can't dress without my valet."

I offered to help him dress, but he would not hear of it; and besides, he said he would not feel comfortable unless dressed by a practised hand. However, he finally concluded that he was such old friends with the Earl that it would not make any difference how he was dressed. So we took a cab, he gave the driver some directions, and we started. By and by we stopped before a large house and got out. I never had seen this man with a collar on. He now stepped under a lamp and got a venerable paper collar out of his coat pocket, along with a hoary cravat, and put them on. He ascended the stoop, and entered. Presently he reappeared, descended rapidly, and said,—

"Come — quick!"

We hurried away, and turned the corner.

"Now we 're safe," he said, and took off his collar and cravat and returned them to his pocket.

"Made a mighty narrow escape," said he.

"How?" said I.

"B' George, the Countess was there!"

"Well, what of that? — don't she know you?"

"Know me? Absolutely worships me. I just did happen to catch a glimpse of her before she saw me — and out I shot. Have n't seen her for two months — to rush in on her without any warning might have been fatal. She could *not* have stood it. I did n't know *she* was in town — thought she was at the castle. Let me lean on you — just a moment — there; now I am better — thank you; thank you ever so much. Lord bless me, what an escape!"

So I never got to call on the Earl after all. But I marked his house for future reference. It proved to be an ordinary family hotel, with about a thousand plebeians roosting in it.

In most things Rogers was by no means a fool. In some things it was plain enough that he was a fool, but he certainly did not know it. He was in the "deadest" earnest in these matters. He died at sea, last summer, as the "Earl of Ramsgate."

THE LOVES OF ALONZO FITZ CLARENCE AND ROSANNAH ETHELTON.

IT was well along in the forenoon of a bitter winter's day. The town of Eastport, in the State of Maine, lay buried under a deep snow that was newly fallen. The customary bustle in the streets was wanting. One could look long distances down them and see nothing but a dead-white emptiness, with silence to match. Of course I do not mean that you could *see* the silence, — no, you could only hear it. The sidewalks were merely long, deep ditches, with steep snow walls on either side. Here and there you might hear the faint, far scrape of a wooden shovel, and if you were quick enough you might catch a glimpse of a distant black figure stooping and disappearing in one of those ditches, and reappearing the next moment with a motion which you would know meant the heaving out of a shovelful of snow. But you needed to be quick, for that black figure would not linger, but would soon drop that shovel and scud for the house, thrashing itself with its arms to warm them. Yes, it was too venomously cold for snow-shovellers or anybody else to stay out long.

Presently the sky darkened; then the wind rose and began to blow in fitful, vigorous gusts, which sent clouds of powdery snow aloft, and straight ahead, and everywhere. Under the impulse of one of these gusts, great white drifts banked themselves like graves across the streets; a moment later, another gust shifted them around the other way, driving a fine spray of snow from their sharp crests, as the gale drives the spume flakes from wave-crests at sea; a third gust swept that place as clean as your hand, if it saw fit. This was fooling, this was play; but each and all of the gusts dumped some snow into the sidewalk ditches, for that was business.

Alonzo Fitz Clarence was sitting in his snug and elegant little parlor, in a lovely blue silk dressing-gown, with cuffs and facings of crimson satin, elaborately quilted. The remains of his breakfast were before him, and the dainty and costly little table service added a harmonious charm to the grace, beauty, and richness of the fixed appointments of the room. A cheery fire was blazing on the hearth.

A furious gust of wind shook the windows, and a great wave of snow washed against them with a drenching sound, so to speak. The handsome young bachelor murmured, —

"That means, no going out to-day. Well, I am content. But what to do for company? Mother is well enough, Aunt Susan is well enough; but these, like the poor, I have with me always. On so grim a day as this, one needs a new interest, a fresh element, to whet the dull edge of captivity. That was very neatly said, but it does n't mean anything. One does n't *want* the edge of captivity sharpened up, you know, but just the reverse."

He glanced at his pretty French mantel-clock.

"That clock 's wrong again. That clock hardly ever knows what time it is; and when it does know, it lies about it, — which amounts to the same thing. Alfred!"

There was no answer.

"Alfred! . . . Good servant, but as uncertain as the clock."

Alonzo touched an electrical bell-button in the wall. He waited a moment, then touched it again; waited a few moments more, and said, —

"Battery out of order, no doubt. But now that I have started, I *will* find out what time it is." He stepped to a speaking-tube in the wall, blew its whistle, and called, "Mother!" and repeated it twice.

"Well, *that's* no use. Mother's battery is out of order, too. Can't raise anybody down-stairs, — that is plain."

He sat down at a rosewood desk, leaned his chin on the left-hand edge of it, and spoke, as if to the floor: "Aunt Susan!"

A low, pleasant voice answered, "Is that you, Alonzo?"

"Yes. I'm too lazy and comfortable to go down-stairs; I am in extremity, and I can't seem to scare up any help."

"Dear me, what is the matter?"

"Matter enough, I can tell you!"

"Oh, don't keep me in suspense, dear! What *is* it?"

"I want to know what time it is."

"You abominable boy, what a turn you did give me! Is that all?"

"All, — on my honor. Calm yourself. Tell me the time, and receive my blessing."

"Just five minutes after nine. No charge, — keep your blessing."

"Thanks. It would n't have impoverished me, aunty, nor so enriched you that you could live without other means." He got up, murmuring, "Just five minutes after nine," and faced his clock. "Ah," said he, "you are doing better than usual. You are only thirty-four minutes wrong. Let me see . . . let me see. . . . Thirty-three and twenty-one are fifty-four; four times fifty-four are two hundred and thirty-six. One off, leaves two hundred and thirty-five. That 's right."

He turned the hands of his clock forward till they marked twenty-five minutes to one, and said, "Now see if you can't keep right for a while . . . else I 'll raffle you!"

He sat down at the desk again, and said, "Aunt Susan!"

"Yes, dear."

"Had breakfast?"

"Yes indeed, an hour ago."

"Busy?"

"No, — except sewing. Why?"

"Got any company?"

"No, but I expect some at half past nine."

"I wish *I* did. I'm lonesome. I want to talk to somebody."

"Very well, talk to me."

"But this is very private."

"Don't be afraid, — talk right along; there 's nobody here but me."

"I hardly know whether to venture or not, but — "

"But what? Oh, don't stop there! You *know* you can trust me, Alonzo, — you know you can."

"I feel it, aunt, but this is very serious. It affects me deeply, — me, and all the family, — even the whole community."

"Oh, Alonzo, tell me! I will never breathe a word of it. What is it?"

"Aunt, if I might dare — "

"Oh, please go on! I love you, and can feel for you. Tell me all. Confide in me. What *is* it?"

"The weather!"

"Plague take the weather! I don't see how you can have the heart to serve me so, Lon."

"There, there, aunty dear, I 'm sorry; I am, on my honor. I won't do it again. Do you forgive me?"

"Yes, since you seem so sincere about it, though I know I ought n't to. You will fool me again as soon as I have forgotten this time."

"No, I won't, honor bright. But such weather, oh, such weather! You 've *got* to keep your spirits up artificially. It is snowy, and blowy, and gusty, and bitter cold! How is the weather with you?"

"Warm and rainy and melancholy. The mourners

go about the streets with their umbrellas running streams from the end of every whalebone. There 's an elevated double pavement of umbrellas stretching down the sides of the streets as far as I can see. I 've got a fire for cheerfulness, and the windows open to keep cool. But it is vain, it is useless: nothing comes in but the balmy breath of December, with its burden of mocking odors from the flowers that possess the realm outside, and rejoice in their lawless profusion whilst the spirit of man is low, and flaunt their gaudy splendors in his face whilst his soul is clothed in sack-cloth and ashes and his heart breaketh."

Alonzo opened his lips to say, "You ought to print that, and get it framed," but checked himself, for he heard his aunt speaking to some one else. He went and stood at the window and looked out upon the wintry prospect. The storm was driving the snow before it more furiously than ever; window shutters were slamming and banging; a forlorn dog, with bowed head and tail withdrawn from service, was pressing his quaking body against a windward wall for shelter and protection; a young girl was ploughing knee-deep through the drifts, with her face turned from the blast, and the cape of her water-proof blowing straight rear-ward over her head. Alonzo shuddered, and said with a sigh, "Better the slop, and the sultry rain, and even the insolent flowers, than this!"

He turned from the window, moved a step, and stopped in a listening attitude. The faint, sweet notes of a familiar song caught his ear. He remained there, with his head unconsciously bent forward, drinking in the melody, stirring neither hand nor foot, hardly breathing. There was a blemish in the execution of the song, but to Alonzo it seemed an added charm instead of a defect. This blemish consisted of a marked flatting of the third, fourth, fifth, sixth, and seventh notes of the refrain or chorus of the piece. When the music ended, Alonzo drew a deep breath, and said, "Ah, I never have heard 'In the Sweet By and By' sung like that before!"

He stepped quickly to the desk, listened a moment, and said in a guarded, confidential voice, "Aunty, who is this divine singer?"

"She is the company I was expecting. Stays with me a month or two. I will introduce you. Miss —"

"For goodness' sake, wait a moment, Aunt Susan! You never stop to think what you are about!"

He flew to his bed-chamber, and returned in a moment perceptibly changed in his outward appearance, and remarking, snappishly, —

"Hang it, she would have introduced me to this angel in that sky-blue dressing-gown with red-hot lapels! Women never think, when they get a-going."

He hastened and stood by the desk, and said eagerly, "Now, Aunty, I am ready," and fell to smiling and bowing with all the persuasiveness and elegance that were in him.

"Very well. Miss Rosannah Ethelton, let me introduce to you my favorite nephew, Mr. Alonzo Fitz Clarence. There! You are both good people, and I like you; so I am going to trust you together while I attend to a few household affairs. Sit down, Rosannah; sit down, Alonzo. Good-by; I shan't be gone long."

Alonzo had been bowing and smiling all the while, and motioning imaginary young ladies to sit down in imaginary chairs, but now he took a seat himself, mentally saying, "Oh, this is luck! Let the winds blow now, and the snow drive, and the heavens frown! Little I care!"

While these young people chat themselves into an acquaintanceship, let us take the liberty of inspecting the sweeter and fairer of the two. She sat alone, at her graceful ease, in a richly furnished apartment which was manifestly the private parlor of a refined and sensible lady, if signs and symbols may go for anything. For instance, by a low, comfortable chair stood a dainty, top-heavy work-stand, whose summit was a fancifully embroidered shallow basket, with vari-colored crewels, and other strings and odds and ends, protruding from under the gaping lid and hanging down in negligent profusion. On the floor lay bright shreds of Turkey red, Prussian blue, and kindred fabrics, bits of ribbon, a spool or two, a pair of scissors, and a roll or so of tinted silken stuffs. On a luxurious sofa, upholstered with some sort of soft Indian goods wrought in black and gold threads inter-webbed with other threads not so pronounced in color, lay a great square of coarse white stuff, upon whose surface a rich bouquet of flowers was growing, under the deft cultivation of the crochet needle. The household cat was asleep on this work of art. In a bay-window stood an easel with an unfinished picture on it, and a palette and brushes on a chair beside it. There were books everywhere: Robertson's Sermons, Tennyson, Moody and Sankey, Hawthorne, "Rab and his Friends," cook-books, prayer-books, pattern-books,—and books about all kinds of odious and exasperating pottery, of course. There was a piano, with a deck-load of music, and more in a tender. There was a great plenty of pictures on the walls, on the shelves of the mantel-piece, and around generally; where coigns of vantage offered were statuettes, and quaint and pretty gimcracks, and rare and costly specimens of peculiarly devilish china. The bay-window gave upon a garden that was ablaze with foreign and domestic flowers and flowering shrubs.

But the sweet young girl was the daintiest thing those premises, within or without, could offer for contemplation: delicately chiselled features, of Grecian cast; her complexion the pure snow of a japonica that is receiving a faint reflected enrichment from some scarlet neighbor of the garden; great, soft blue eyes fringed with long, curving lashes; an expression made up of the trustfulness of a child and the gentleness of a fawn; a beautiful head crowned with its own prodigal gold; a lithe and rounded figure, whose every attitude and movement were instinct with native grace.

Her dress and adornment were marked by that exquisite harmony that can come only of a fine natural taste perfected by culture. Her gown was of a simple magenta tulle, cut bias, traversed by three rows of light blue flounces, with the selvage edges turned up with ashes-of-roses chenille; overdress of dark bay tarlatan, with scarlet satin lambrequins; corn-colored polonaise, *en panier*, looped with mother-of-pearl buttons and silver cord, and hauled aft and made fast by buff-velvet lashings; basque of lavender reps, picked out with valenciennes; low neck, short sleeves; maroon-velvet necktie edged with delicate pink silk; inside handkerchief of some simple three-ply ingrain fabric of a soft saffron tint; coral bracelets and locket-chain; coiffure of forget-me-nots and lilies of the valley massed around a noble calla.

This was all; yet even in this subdued attire she was divinely beautiful. Then what must she have been when adorned for the festival or the ball?

All this time she has been busily chatting with Alonzo, unconscious of our inspection. The minutes still sped, and still she talked. But by and by she happened to look up, and saw the clock. A crimson blush sent its rich flood through her cheeks, and she exclaimed, —

"There, good-by, Mr. Fitz Clarence; I must go now!"

She sprang from her chair with such haste that she hardly heard the young man's answering good-by. She stood radiant, graceful, beautiful, and gazed, wondering, upon the accusing clock. Presently her pouting lips parted, and she said, —

"Five minutes after eleven! Nearly two hours, and it did not seem twenty minutes! Oh, dear, what will he think of me!"

At the self-same moment Alonzo was staring at *his* clock. And presently he said, —

"Twenty-five minutes to three! Nearly two hours, and I did n't believe it was two minutes! Is it possible that this clock is humbugging again? Miss Ethelton! Just one moment, please. Are you there yet?"

"Yes, but be quick; I 'm going right away."

"Would you be so kind as to tell me what time it is?"

The girl blushed again, murmured to herself, "It 's right down cruel of him to ask me!" and then spoke up and answered with admirably counterfeited unconcern, "Five minutes after eleven."

"Oh, thank you! You have to go, now, have you?"

"Yes."

"I 'm sorry."

No reply.

"Miss Ethelton!"

"Well?"

"You — you 're there yet, *ain't* you?"

"Yes; but please hurry. What did you want to say?"

"Well, I — well, nothing in particular. It 's very lonesome here. It 's asking a great deal, I know, but would you mind talking with me again by and by, — that is, if it will not trouble you too much?"

"I don't know — but I 'll think about it. I 'll try."

"Oh, thanks! Miss Ethelton? . . . Ah me, she 's gone, and here are the black clouds and the whirling snow and the raging winds come again! But she said *good-by!* She did n't say good-morning, she said good-by! . . . The clock was right, after all. What a lightning-winged two hours it was!"

He sat down, and gazed dreamily into his fire for a while, then heaved a sigh and said, —

"How wonderful it is! Two little hours ago I was a free man, and now my heart's in San Francisco!"

About that time Rosannah Ethelton, propped in the window-seat of her bed-chamber, book in hand, was gazing vacantly out over the rainy seas that washed the Golden Gate, and whispering to herself, "How different he is from poor Burley, with his empty head and his single little antic talent of mimicry!"

II.

FOUR weeks later Mr. Sidney Algernon Burley was entertaining a gay luncheon company, in a sumptuous drawing-room on Telegraph Hill, with some capital imitations of the voices and gestures of certain popular actors and San Franciscan literary people and Bonanza grandees. He was elegantly upholstered, and was a handsome fellow, barring a trifling cast in his eye. He seemed very jovial, but nevertheless he kept his eye on the door with an expectant and uneasy watchfulness. By and by a nobby lackey appeared, and delivered a message to the mistress, who nodded her head understandingly. That seemed to settle the thing for Mr. Burley; his vivacity decreased little by little, and a dejected look began to creep into one of his eyes and a sinister one into the other.

The rest of the company departed in due time, leaving him with the mistress, to whom he said, —

"There is no longer any question about it. She avoids me. She continually excuses herself. If I could see her, if I could speak to her only a moment, — but this suspense — "

"Perhaps her seeming avoidance is mere accident, Mr. Burley. Go to the small drawing-room up-stairs and amuse yourself a moment. I will despatch a household order that is on my mind, and then I will go to her room. Without doubt she will be persuaded to see you."

Mr. Burley went up-stairs, intending to go to the small drawing-room, but as he was passing "Aunt Susan's" private parlor, the door of which stood slightly ajar, he heard a joyous laugh which he recognized; so without knock or announcement he stepped confidently in. But before he could make his presence known he heard words that harrowed up his soul and chilled his young blood. He heard a voice say, —

"Darling, it has come!"

Then he heard Rosannah Ethelton, whose back was toward him, say, —

"So has yours, dearest!"

He saw her bowed form bend lower; he heard her kiss something, — not merely once, but again and again! His soul raged within him. The heart-breaking conversation went on, —

"Rosannah, I knew you must be beautiful, but this is dazzling, this is blinding, this is intoxicating!"

"Alonzo, it is such happiness to hear you say it. I

know it is not true, but I am *so* grateful to have you think it is, nevertheless! I knew you must have a noble face, but the grace and majesty of the reality beggar the poor creation of my fancy."

Burley heard that rattling shower of kisses again.

"Thank you, my Rosannah! The photograph flatters me, but you must not allow yourself to think of that. Sweetheart?"

"Yes, Alonzo."

"I am so happy, Rosannah."

"Oh, Alonzo, none that have gone before me knew what love was, none that come after me will ever know what happiness is. I float in a gorgeous cloudland, a boundless firmament of enchanted and bewildering ecstasy!"

"Oh, my Rosannah! — for you are mine, are you not?"

"Wholly, oh, wholly yours, Alonzo, now and forever! All the day long, and all through my nightly dreams, one song sings itself, and its sweet burden is, 'Alonzo Fitz Clarence, Alonzo Fitz Clarence, Eastport, State of Maine!'"

"Curse him, I 've got his address, any way!" roared Burley, inwardly, and rushed from the place.

Just behind the unconscious Alonzo stood his mother, a picture of astonishment. She was so muffled from head to heel in furs that nothing of herself was visible but her eyes and nose. She was a good allegory of winter, for she was powdered all over with snow.

Behind the unconscious Rosannah stood "Aunt Susan," another picture of astonishment. She was a good allegory of summer, for she was lightly clad, and was vigorously cooling the perspiration on her face with a fan.

Both of these women had tears of joy in their eyes.

"So ho!" exclaimed Mrs. Fitz Clarence, "this explains why nobody has been able to drag you out of your room for six weeks, Alonzo!"

"So ho!" exclaimed Aunt Susan, "this explains why you have been a hermit for the past six weeks, Rosannah!"

The young couple were on their feet in an instant, abashed, and standing like detected dealers in stolen goods awaiting Judge Lynch's doom.

"Bless you, my son! I am happy in your happiness. Come to your mother's arms, Alonzo!"

"Bless you, Rosannah, for my dear nephew's sake! Come to my arms!"

Then was there a mingling of hearts and of tears of rejoicing on Telegraph Hill and in Eastport Square.

Servants were called by the elders, in both places. Unto one was given the order, "Pile this fire high with hickory wood, and bring me a roasting-hot lemonade."

Unto the other was given the order, "Put out this fire, and bring me two palm-leaf fans and a pitcher of ice-water."

Then the young people were dismissed, and the elders sat down to talk the sweet surprise over and make the wedding plans.

Some minutes before this Mr. Burley rushed from the mansion on Telegraph Hill without meeting or taking formal leave of anybody. He hissed through his teeth, in unconscious imitation of a popular favorite in melodrama, "Him shall she never wed! I have sworn it! Ere great Nature shall have doffed her winter's ermine to don the emerald gauds of spring, she shall be mine!"

III.

Two weeks later. Every few hours, during some three or four days, a very prim and devout-looking Episcopal clergyman, with a cast in his eye, had visited Alonzo. According to his card, he was the Rev. Melton Hargrave, of Cincinnati. He said he had retired from the ministry on account of his health. If he had said on account of ill health, he would probably have erred, to judge by his wholesome looks and firm build. He was the inventor of an improvement in telephones, and hoped to make his bread by selling the privilege of using it. "At present," he continued, "a man may go and tap a telegraph wire which is conveying a song or a concert from one State to another, and he can attach his private telephone and steal a hearing of that music as it passes along. My invention will stop all that."

"Well," answered Alonzo, "if the owner of the music could not miss what was stolen, why should he care?"

"He should n't care," said the Reverend.

"Well?" said Alonzo inquiringly.

"Suppose," replied the Reverend, "suppose that, instead of music that was passing along and being stolen, the burden of the wire was loving endearments of the most private and sacred nature?"

Alonzo shuddered from head to heel. "Sir, it is a priceless invention," said he; "I must have it at any cost."

But the invention was delayed somewhere on the road from Cincinnati, most unaccountably. The impatient Alonzo could hardly wait. The thought of Rosannah's sweet words being shared with him by some ribald thief was galling to him. The Reverend came frequently and lamented the delay, and told of measures he had taken to hurry things up. This was some little comfort to Alonzo.

One forenoon the Reverend ascended the stairs and knocked at Alonzo's door. There was no response. He entered, glanced eagerly around, closed the door softly, then ran to the telephone. The exquisitely soft and remote strains of the "Sweet By and By" came floating through the instrument. The singer was flatting, as usual, the five notes that follow the first two in the chorus, when the Reverend interrupted her with this word, in a voice which was an exact imitation of Alonzo's, with just the faintest flavor of impatience added, —

"Sweetheart?"

"Yes, Alonzo?"

"Please don't sing that any more this week, — try something modern."

The agile step that goes with a happy heart was heard on the stairs, and the Reverend, smiling diabolically, sought sudden refuge behind the heavy folds of the velvet window curtains. Alonzo entered and flew to the telephone. Said he, —

"Rosannah, dear, shall we sing something together?"

"Something *modern?*" asked she, with sarcastic bitterness.

"Yes, if you prefer."

"Sing it yourself, if you like!"

This snappishness amazed and wounded the young man. He said, —

"Rosannah, that was not like you."

"I suppose it becomes me as much as your very polite speech became you, Mr. Fitz Clarence."

"*Mister* Fitz Clarence! Rosannah, there was nothing impolite about my speech."

"Oh, indeed! Of course, then, I misunderstood you, and I most humbly beg your pardon, ha-ha-ha! No doubt you said, 'Don't sing it any more *to-day.*'"

"Sing *what* any more to-day?"

"The song you mentioned, of course. How very obtuse we are, all of a sudden!"

"I never mentioned any song."

"Oh, you *did n't!*"

"No, I *did n't!*"

"I am compelled to remark that you *did.*"

"And I am obliged to reiterate that I *did n't.*"

"A second rudeness! That is sufficient, sir. I will never forgive you. All is over between us."

Then came a muffled sound of crying. Alonzo hastened to say, —

"Oh, Rosannah, unsay those words! There is some dreadful mystery here, some hideous mistake. I am utterly earnest and sincere when I say I never said anything about any song. I would not hurt you for the whole world . . . Rosannah, dear? . . . Oh, speak to me, won't you?"

There was a pause; then Alonzo heard the girl's sobbings retreating, and knew she had gone from the telephone. He rose with a heavy sigh and hastened from the room, saying to himself, "I will ransack the charity missions and the haunts of the poor for my mother. She will persuade her that I never meant to wound her."

A minute later, the Reverend was crouching over the telephone like a cat that knoweth the ways of the prey. He had not very many minutes to wait. A soft, repentant voice, tremulous with tears, said, —

"Alonzo, dear, I have been wrong. You *could* not have said so cruel a thing. It must have been some one who imitated your voice in malice or in jest."

The Reverend coldly answered, in Alonzo's tones, —

"You have said all was over between us. So let it be. I spurn your proffered repentance, and despise it!"

Then he departed, radiant with fiendish triumph, to return no more with his imaginary telephonic invention forever.

Four hours afterward, Alonzo arrived with his mother from her favorite haunts of poverty and vice. They summoned the San Francisco household; but there was no reply. They waited, and continued to wait, upon the voiceless telephone.

At length, when it was sunset in San Francisco, and three hours and a half after dark in Eastport, an answer came to the oft-repeated cry of "Rosannah!"

But, alas, it was Aunt Susan's voice that spake. She said, —

"I have been out all day; just got in. I will go and find her."

The watchers waited two minutes — five minutes — — ten minutes. Then came these fatal words, in a frightened tone, —

"She is gone, and her baggage with her. To visit another friend, she told the servants. But I found this note on the table in her room. Listen: 'I am gone; seek not to trace me out; my heart is broken; you will never see me more. Tell him I shall always think of him when I sing my poor "Sweet By and By," but never of the unkind words he said about it.' That is her note. Alonzo, Alonzo, what does it mean? What has happened?"

But Alonzo sat white and cold as the dead. His mother threw back the velvet curtains and opened a window. The cold air refreshed the sufferer, and he told his aunt his dismal story. Meantime his mother was inspecting a card which had disclosed itself upon the floor when she cast the curtains back. It read, "Mr. Sidney Algernon Burley, San Francisco."

"The miscreant!" shouted Alonzo, and rushed forth to seek the false Reverend and destroy him; for the card explained everything, since in the course of the lovers' mutual confessions they had told each other all about all the sweethearts they had ever had, and thrown no end of mud at their failings and foibles, — for lovers always do that. It has a fascination that ranks next after billing and cooing.

IV.

During the next two months many things happened. It had early transpired that Rosannah, poor suffering orphan, had neither returned to her grandmother in Portland, Oregon, nor sent any word to her save a duplicate of the woful note she had left in the mansion on Telegraph Hill. Whosoever was sheltering her — if she was still alive — had been persuaded not to betray her whereabouts, without doubt; for all efforts to find trace of her had failed.

Did Alonzo give her up? Not he. He said to himself, "She will sing that sweet song when she is sad; I shall find her." So he took his carpet sack and a portable telephone, and shook the snow of his native city from his arctics, and went forth into the world. He wandered far and wide and in many States. Time and again, strangers were astounded to see a wasted, pale, and woe-worn man laboriously climb a telegraph pole in wintry and lonely places, perch sadly there an hour, with his ear at a little box, then come sighing down, and wander wearily away. Sometimes they shot at him, as peasants do at aeronauts, thinking him mad and dangerous. Thus his clothes were much shredded by bullets and his person grievously lacerated. But he bore it all patiently.

In the beginning of his pilgrimage he used often to say, "Ah, if I could but hear the 'Sweet By and By'!" But toward the end of it he used to shed tears of anguish and say, "Ah, if I could but hear something else!"

Thus a month and three weeks drifted by, and at last some humane people seized him and confined him in a private mad-house in New York. He made no moan, for his strength was all gone, and with it all heart and all hope. The superintendent, in pity, gave up his own comfortable parlor and bed-chamber to him and nursed him with affectionate devotion.

At the end of a week the patient was able to leave his bed for the first time. He was lying, comfortably pillowed, on a sofa, listening to the plaintive Miserere of the bleak March winds, and the muffled sound of tramping feet in the street below, — for it was about six in the evening, and New York was going home from work. He had a bright fire and the added cheer of a couple of student lamps. So it was warm and snug within, though bleak and raw without; it was light and bright within, though outside it was as dark and dreary as if the world had been lit with Hartford gas. Alonzo smiled feebly to think how his loving vagaries had made him a maniac in the eyes of the world, and was proceeding to pursue his line of thought further, when a faint, sweet strain, the very ghost of sound, so remote and attenuated it seemed, struck upon his ear. His pulses stood still; he listened with parted lips and bated breath. The song flowed on, — he waiting, listening, rising slowly and unconsciously from his recumbent position. At last he exclaimed, —

"It is! it is she! Oh, the divine flatted notes!"

He dragged himself eagerly to the corner whence the sounds proceeded, tore aside a curtain, and discovered a telephone. He bent over, and as the last note died away he burst forth with the exclamation, —

"Oh, thank Heaven, found at last! Speak to me, Rosannah, dearest! The cruel mystery has been unravelled; it was the villain Burley who mimicked my voice and wounded you with insolent speech!"

There was a breathless pause, a waiting age to Alonzo; then a faint sound came, framing itself into language, —

"Oh, say those precious words again, Alonzo!"

"They are the truth, the veritable truth, my Rosannah, and you shall have the proof, ample and abundant proof!"

"Oh, Alonzo, stay by me! Leave me not for a moment! Let me feel that you are near me! Tell me we shall never be parted more! Oh, this happy hour, this blessed hour, this memorable hour!"

"We will make record of it, my Rosannah; every year, as this dear hour chimes from the clock, we will celebrate it with thanksgivings, all the years of our life."

"We will, we will, Alonzo!"

"Four minutes after six, in the evening, my Rosannah, shall henceforth —"

"Twenty-three minutes after twelve, afternoon, shall —"

"Why, Rosannah, darling, where are you?"

"In Honolulu, Sandwich Islands. And where are

you? Stay by me; do not leave me for a moment. I cannot bear it. Are you at home?"

"No, dear, I am in New York, — a patient in the doctor's hands."

An agonizing shriek came buzzing to Alonzo's ear, like the sharp buzzing of a hurt gnat; it lost power in travelling five thousand miles. Alonzo hastened to say, —

"Calm yourself, my child. It is nothing. Already I am getting well under the sweet healing of your presence. Rosannah?"

"Yes, Alonzo? Oh, how you terrified me! Say on."

"Name the happy day, Rosannah!"

There was a little pause. Then a diffident small voice replied, "I blush — but it is with pleasure, it is with happiness. Would — would you like to have it soon?"

"This very night, Rosannah! Oh, let us risk no more delays. Let it be now! — this very night, this very moment!"

"Oh, you impatient creature! I have nobody here but my good old uncle, a missionary for a generation, and now retired from service, — nobody but him and his wife. I would so dearly like it if your mother and your aunt Susan — "

"*Our* mother and *our* aunt Susan, my Rosannah."

"Yes, *our* mother and *our* aunt Susan, — I am content to word it so if it pleases you; I would so like to have them present."

"So would I. Suppose you telegraph Aunt Susan. How long would it take her to come?"

"The steamer leaves San Francisco day after to-morrow. The passage is eight days. She would be here the 31st of March."

"Then name the 1st of April: do, Rosannah, dear."

"Mercy, it would make us April fools, Alonzo!"

"So we be the happiest ones that that day's sun looks down upon in the whole broad expanse of the globe, why need we care? Call it the 1st of April, dear."

"Then the 1st of April it shall be, with all my heart!"

"Oh, happiness! Name the hour, too, Rosannah."

"I like the morning, it is so blithe. Will eight in the morning do, Alonzo?"

"The loveliest hour in the day, — since it will make you mine."

There was a feeble but frantic sound for some little time, as if wool-lipped, disembodied spirits were exchanging kisses; then Rosannah said, "Excuse me just a moment, dear; I have an appointment, and am called to meet it."

The young girl sought a large parlor and took her place at a window which looked out upon a beautiful scene. To the left one could view the charming Nuuana Valley, fringed with its ruddy flush of tropical flowers and its plumed and graceful cocoa palms; its rising foot-hills clothed in the shining green of lemon, citron, and orange groves; its storied precipice beyond, where the first Kamehameha drove his defeated foes over to their destruction, — a spot that had forgotten its grim history, no doubt, for now it was smiling, as almost always at noonday, under the glowing arches of a succession of rainbows. In front of the window one could see the quaint town, and here and there a picturesque group of dusky natives, enjoying the blistering weather; and far to the right lay the restless ocean, tossing its white mane in the sunshine.

Rosannah stood there, in her filmy white raiment, fanning her flushed and heated face, waiting. A Kanaka boy, clothed in a damaged blue neck-tie and part of a silk hat, thrust his head in at the door, and announced, "'Frisco *haole!*"

"Show him in," said the girl, straightening herself up and assuming a meaning dignity. Mr. Sidney Algernon Burley entered, clad from head to heel in dazzling snow, — that is to say, in the lightest and whitest of Irish linen. He moved eagerly forward, but the girl made a gesture and gave him a look which checked him suddenly. She said coldly, "I am here, as I promised. I believed your assertions, I yielded to your importunities, and said I would name the day. I name the 1st of April, — eight in the morning. Now go!"

"Oh, my dearest, if the gratitude of a life-time — "

"Not a word. Spare me all sight of you, all communication with you, until that hour. No, — no supplications; I will have it so."

When he was gone, she sank exhausted in a chair, for the long siege of troubles she had undergone had wasted her strength. Presently she said, "What a narrow escape! If the hour appointed had been an hour earlier — Oh, horror, what an escape I have made! And to think I had come to imagine I was loving this beguiling, this truthless, this treacherous monster! Oh, he shall repent his villany!"

Let us now draw this history to a close, for little more needs to be told. On the 2d of the ensuing April, the Honolulu "Advertiser" contained this notice:—

MARRIED. — In this city, by telephone, yesterday morning, at eight o'clock, by Rev. Nathan Hays, assisted by Rev. Nathaniel Davis, of New York, Mr. Alonzo Fitz Clarence, of Eastport, Maine, U. S., and Miss Rosannah Ethelton, of Portland, Oregon, U. S. Mrs. Susan Howland, of San Francisco, a friend of the bride, was present, she being the guest of the Rev. Mr. Hays and wife, uncle and aunt of the bride. Mr. Sidney Algernon Burley, of San Francisco, was also present, but did not remain till the conclusion of the marriage service. Captain Hawthorne's beautiful yacht, tastefully decorated, was in waiting, and the happy bride and her friends immediately departed on a bridal trip to Lahaina and Haleakala.

The New York papers of the same date contained this notice:—

MARRIED. — In this city, yesterday, by telephone, at half past two in the morning, by Rev. Nathaniel Davis, assisted by Rev. Nathan Hays, of Honolulu, Mr. Alonzo Fitz Clarence, of Eastport, Maine, and Miss Rosannah Ethelton, of Portland, Oregon. The parents and several friends of the bridegroom were present, and enjoyed a sumptuous breakfast and much festivity until nearly sunrise, and then departed on a bridal trip to the Aquarium, the bridegroom's state of health not admitting of a more extended journey.

Toward the close of that memorable day, Mr. and Mrs. Alonzo Fitz Clarence were buried in sweet converse concerning the pleasures of their several bridal tours, when suddenly the young wife exclaimed: "Oh, Lonny, I forgot! I did what I said I would."

"Did you, dear?"

"Indeed I did. I made *him* the April fool! And I told him so, too! Ah, it was a charming surprise! There he stood, sweltering in a black dress suit, with the mercury leaking out of the top of the thermometer, waiting to be married. You should have seen the look he gave when I whispered it in his ear! Ah, his wickedness cost me many a heartache and many a tear, but the score was all squared up, then. So the vengeful feeling went right out of my heart, and I begged him to stay, and said I forgave him everything. But he would n't. He said he would live to be avenged; said he would make our lives a curse to us. But he can't, *can* he, dear?"

"Never in this world, my Rosannah!"

Aunt Susan, the Oregonian grandmother, and the young couple and their Eastport parents, are all happy at this writing, and likely to remain so. Aunt Susan brought the bride from the Islands, accompanied her across our continent, and had the happiness of witnessing the rapturous meeting between an adoring husband and wife who had never seen each other until that moment.

A word about the wretched Burley, whose wicked machinations came so near wrecking the hearts and lives of our poor young friends, will be sufficient. In a murderous attempt to seize a crippled and helpless artisan who he fancied had done him some small offence, he fell into a caldron of boiling oil and expired before he could be extinguished.

University Press: John Wilson and Son, Cambridge.

211 TREMONT STREET, BOSTON,
Spring, 1882.

A LIST OF BOOKS

PUBLISHED BY

Messrs. James R. Osgood & Co.

☞ *Any book on this list sent* POSTPAID *on receipt of the advertised price.*

AMERICAN-ACTOR SERIES (THE). Edited by Laurence Hutton. A series of volumes by the best writers, embracing the lives of the most famous American Actors. Illustrated with Portraits, Views, etc. Each in one vol. 12mo. $1.25. Now ready: *Forrest, the Booths, the Jeffersons, Charlotte Cushman.*

ANGELL'S (DR. HENRY C.) Records of the Late William M. Hunt. With Illustrations. 1 vol. Small 4to. $1.50.

ARNOLD'S (GEORGE) Poems. Edited, with a Biographical Sketch of the Poet, by WILLIAM WINTER. With Portrait and Illustrations. 1 vol. 16mo. Cloth, $1.50. Half-calf, $3.00. Morocco antique or tree-calf, $4.00.

BARRETT'S (LAWRENCE) Edwin Forrest. Vol. I. of the American-Actor Series. 1 vol. 12mo. $1.25.

BLACKBURN'S (HENRY) Breton Folk: An Artistic Tour in Normandy. With 170 Illustrations by RANDOLPH CALDECOTT. 1 vol. Small 4to. $1.50.

BLAINE'S (HON. JAMES G.) Eulogy on James Abram Garfield. 1 vol. 16mo. With Portrait. 50 cents.

BOSTON, The Memorial History of. Including the Present County of Suffolk. 1630-1880. Seventy eminent Collaborators. 4 vols. 4to. Copiously illustrated. *Send for Prospectus.*

BUDGE'S (ERNEST A.) The History of Esarhaddon, King of Assyria, B.C. 681–668. 8vo. Gilt top. $4 00.

BURNETT'S (MRS. FRANCES HODGSON) A Fair Barbarian. 1 vol. 16mo. $1.00.

CESNOLA'S (GEN. L. P. DI) The Cesnola Collection of Cyprus Antiquities. A Descriptive and Pictorial Atlas. Large folio. 500 Plates. *Sold by subscription only.*

CHAMBERLAIN'S (BASIL HALL) The Classical Poetry of the Japanese. 1 vol. 8vo. Gilt top. $3.00.

CHENOWETH'S (MRS. C. VAN D.) Stories of the Saints. Beautifully illustrated. 1 vol. 12mo. $2.00.

CLARKE'S (MRS. ASIA BOOTH) The Elder and the Younger Booth. Vol. III. American-Actor Series. Illustrated. $1.25.

CLARKE'S (REV. JAMES FREEMAN) Self-Culture. 1 vol. 12mo. Cloth, $1.50. Half-calf, $3.00.

——— Events and Epochs in Religious History: 1 vol. Crown octavo. With many Illustrations. $3.00.

CLEMENT'S (CLARA ERSKINE) A Handbook of Legendary and Mythological Art. With a profusion of Descriptive Illustrations. Fourteenth Edition, with Revisions and New Illustrations. 1 vol. Crown 8vo. $3.00. Half-calf, $5.00. Tree-calf, $7.00.

——— Painters, Sculptors, Architects, Engravers, and their Works. Sixth Edition, with Revisions and New Illustrations. 1 vol. Crown 8vo. $3.00. Half-calf, $5.00. Tree-calf, $7.00.

——— Eleanor Maitland. A Novel. 16mo. $1.25.

——— Charlotte Cushman. Vol. IV. of the American-Actor Series. Illustrated. 1 vol. 12mo. $1.25.

COLLING'S (J. K.) Art-Foliage. Entirely new Plates from the latest enlarged London Edition. Cloth. Folio. $10.00.

CONGDON'S (CHARLES T.) Reminiscences of a Journalist. 1 vol. 12mo. With Portrait. Cloth, $1.50. Half-calf, $3.00.

COOKE'S (GEORGE WILLIS) Ralph Waldo Emerson: His Life, Writings, and Philosophy. 1 vol. Crown octavo. $2.00.

COOKE'S (MRS. LAURA S. H.) Dimple Dopp. A beautiful illustrated juvenile. Small quarto, elegantly bound. $1.50.

COOKE'S (ROSE TERRY) Somebody's Neighbors. 1 vol. 12mo. Cloth, $1.50. Half-calf, $3.00.

CRANE'S (WALTER) The First of May. A Fairy Masque. With 57 designs by Walter Crane. 1 vol. Oblong folio. $2.50.

DAHLGREN'S (MRS. MADELEINE VINTON) South-Sea Sketches. 1 vol. 12mo. $1.50.

DAMEN'S GHOST. Vol. VI. of the Round-Robin Series of anonymous novels. 16mo. $1.00.

DAVIDSON'S (J. MORRISON) Eminent English Liberals. 1 vol. 16mo. $1.25.

DICKENS'S (CHARLES) Works. *University Edition*. 15 vols. 12mo. 200 Illustrations. Half-calf, $50.00. *Sold only in sets.*

——— Child's History of England. 24 Illustrations. New plates and large type. 12mo. $1.00.

——— *The Same.* 100 fine Illustrations. 12mo. Bevelled and gilt extra. $2.50.

DODGE'S (THEODORE A., U.S.A.) The Campaign of Chancellorsville. 1 vol. 8vo. With 4 colored maps. $3.00.

DOROTHEA. Vol. X. of the Round-Robin Series of anonymous novels. 16mo. $1.00.

DU MAURIER'S (GEORGE) Pictures from Society. 50 full-page Pictures from *Punch*. 1 vol. 4to. Full gilt. $5.00.

EASTWICK'S (EDWARD B.) The Gulistan; or, Rose Garden of Shekh Mushlin'ddin Sâdi of Shiraz. 8vo. Gilt top. $3.50.

EWING'S (MRS. EMMA P.) Cooking and Castle-Building. 1 vol. 16mo. $1.00.

FAVORITE-AUTHORS SERIES:

FAVORITE AUTHORS. A Companion-Book of Prose and Poetry. With Steel Portraits. 1 vol. 12mo. Full gilt. Cloth, $2.00. Half-calf, $4.00. Morocco antique, $5.00.

HOUSEHOLD FRIENDS for Every Season. 10 Steel Portraits. 1 vol. 12mo. Full gilt. Cloth, $2.00. Half-calf, $4.00. Morocco antique, $5.00.

GOOD COMPANY FOR EVERY DAY IN THE YEAR. With Steel Portraits. 1 vol. 12mo. Full gilt. Cloth, $2.00. Half-calf, $4.00. Morocco antique, $5.00.

FEATHERMAN'S (A.) Aramæans: Social History of the Races of Mankind. 8vo. Uncut edges. Gilt top. 664 pages. $5.00.

FORBES'S (ARCHIBALD) Glimpses through the Cannon-Smoke. 1 vol. 12mo. $1.00.

FULLER'S (ALBERT W.) Artistic Homes in City and Country. Oblong folio. 44 full-page Illustrations. $3.50.

GARDNER'S (E. C.) Homes, and How to Make Them. In Letters between an Architect and a Family Man seeking a House. 30 Illustrations. 1 vol. Square 12mo. $1.50.

——— Home Interiors. Illustrated with 62 Plates designed by the Author. 1 vol. Square 12mo. $1.50.

——— Illustrated Homes. Illustrated with 51 Plates designed by the Author. 1 vol. Square 12mo. $1.50.

GEORGIANS (THE). Vol. III. of the Round-Robin Series of anonymous novels. 16mo. $1.00.

GERALDINE: A Souvenir of the St. Lawrence. A poetical romance. 1 vol. 16mo. $1.25.

GOETHE'S Faust. Translated into English Prose by A. HAYWARD. 1 vol. 16mo. $1.25.

GORDON'S (GEN. GEORGE H.) A War Diary of Events in the War of the Great Rebellion, 1863-1865. With three Maps and three Illustrations. 1 vol. 12mo. $3.00.

GRANT'S (ROBERT) Confessions of a Frivolous Girl. 1 vol. 16mo. $1.25.

GREENOUGH'S (MRS. RICHARD) Mary Magdalene: a Poem. In unique London binding. 1 vol. 12mo. $1.50.

GUSTAFSON'S (MRS. ZADEL BARNES) Genevieve Ward. A Biographical Sketch. With Illustrations. 1 vol. $1.25.

HALE'S (LUCRETIA P.) The Peterkin Papers. 8 Illustrations. 1 vol. 16mo. $1.00.

HALL'S (G. STANLEY, PH.D.) Aspects of German Culture. A Volume of Essays and Criticisms. 1 vol. 12mo. $1.50.

HARTING'S (JAMES EDMUND, F.L.S., F.Z.S.) British Animals Extinct within Historic Times. With Some Account of British Wild White Cattle. Illustrated. 1 vol. 8vo. Gilt top. $4.50.

HARTT'S (PROFESSOR C. F.) Geology and Physical Geography of Brazil. With 72 Illustrations. *In preparation.*

HASSARD'S (JOHN R. G.) A Pickwickian Pilgrimage. 1 vol. Small 16mo. $1.00.

HAYWARD'S (ALMIRA L.) The Illustrated Birthday Book of American Poets. Revised and enlarged Edition, with Index for names, and portraits of thirteen great American Poets. 1 vol. 18mo. $1.00. Half-calf, $2.25. Flexible morocco, $3.50.

HELIOTYPE GALLERIES. Elegant quarto volumes, richly stamped in gold and colors, with Descriptive Text and full-page Heliotype Engravings.

GEMS OF THE DRESDEN GALLERY. 20 Heliotypes, with Descriptions. 4to. Full gilt. $7.50.

THE GOETHE GALLERY. 21 Heliotypes, from the Original Drawings by WILHELM VON KAULBACH. With Explanatory Text. 4to. Full gilt. $7.50.

ENGRAVINGS FROM LANDSEER. 20 full-page Heliotypes. With a Biography of Landseer. 4to. Full gilt. $7.50.

STUDIES FROM RAPHAEL. 24 choice Heliotypes from Raphael's celebrated Paintings. With Text by M. T. B. EMERIC-DAVID, of the Institute of France. 4to. Full gilt. $7.50.

THE TITIAN GALLERY. 20 large Heliotypes of Titian's *chef-d'œuvres.* With a Biography. 4to. Full gilt. $7.50.

TOSCHI'S ENGRAVINGS, from Frescos by Correggio and Parmegiano. 24 Heliotypes. 4to. Full gilt. $7.50.

THE GALLERY OF GREAT COMPOSERS. Fine Portraits of Bach, Handel, Gluck, Haydn, Mozart, Beethoven, Schubert, Von Weber, Mendelssohn, Schuman, Meyerbeer, and Wagner. Biographies by RIMBAULT. 4to. Full gilt. $7.50.

HINSDALE'S (PRESIDENT B. A.) President Garfield and Education. 1 vol. 12mo. With Steel Portraits of President Garfield, Mrs. Garfield, etc. $1.50.

HOMOSELLE. Vol. V. of the Round-Robin Series of anonymous novels. 16mo. $1.00.

HOUSE'S (EDWARD H.) Japanese Episodes. 1 vol. 16mo. $1.00.

HOWARD'S (BLANCHE WILLIS) Aunt Serena. 1 vol. 16mo. $1.25.

HOWELLS'S (WILLIAM D.) A Fearful Responsibility, and other Stories. 1 vol. 12mo. $1.50.

——— Dr. Breen's Practice. 1 vol. 12mo. $1.50.

HOWITT'S (MARY) Mabel on Midsummer Day: a Story of the Olden Time. With 12 Silhouettes by HELEN M. HINDS. 1 vol. 4to. $1.50.

HUNNEWELL'S (J. F.) Bibliography of Charlestown, Mass., and Bunker Hill. 1 vol. 8vo. Illustrated. $2.00.

HUTCHINSON'S (ELLEN M.) Songs and Lyrics. 1 vol. 16mo. $1.25.

JOHNSTON'S (ELIZABETH BRYANT) Original Portraits of Washington. About seventy Heliotype Portraits, with Descriptive History of each. 1 vol. Quarto. $10.00.

KEENE'S (CHARLES) Our People. 400 Pictures from *Punch.* 1 vol. 4to. Full gilt. $5.00.

KENDRICK'S (PROFESSOR A. C.) Our Poetical Favorites. First, Second, and Third Series. 3 vols. 12mo. In cloth, per vol., $2.00. Half-calf, per vol., $4.00. Morocco or tree-calf, $5.00.

KING'S (CLARENCE) Mountaineering in the Sierra Nevada. 1 vol. 12mo. With Maps. Revised and Enlarged Edition. $2.00.

LATHROP'S (GEORGE PARSONS) In the Distance. A novel. 1 vol. 16mo. $1.25.

LEOPARDI (GIACOMO), The Essays and Dialogues of. Translated from the Italian, with Biographical Sketch, by Charles Edwardes. 1 vol.

LESSON IN LOVE (A). Vol. II. of the Round-Robin Series of anonymous novels. 16mo. $1.00.

LONGFELLOW'S (H. W.) Poems. Illustrated Family Edition. Full gilt. Elegantly stamped. 1 vol. 8vo. Cloth, $2.50. Half-calf, $5.00. Morocco antique or tree-calf, $7.50.

MADAME LUCAS. Vol. VIII. of the Round-Robin Series of anonymous novels. 16mo. $1.00.

MADDEN'S (FREDERIC W., M.R.A.S., M. Num. Soc., etc.) The Coins of the Jews. Illustrated with 270 Woodcuts (chiefly by the eminent artist-antiquary, F. W. Fairholt, F.S.A.) and a plate of Alphabets. $12.00.

MEREDITH'S (OWEN) Lucile. Entirely new edition, from new plates, with 160 Illustrations. Elegantly bound, with full gilt edges, in box. 1 vol. 8vo. $6.00. Morocco or tree-calf, $10.

——— With 24 Illustrations by GEO. DU MAURIER. 1 vol. 8vo. Cloth, $5.00. Morocco antique or tree-calf, $9.00.

NAMELESS NOBLEMAN (A). Vol. I. of the Round-Robin Series of novels. 16mo. $1.00.

NARJOUX'S (FELIX) Journey of an Architect in the North-west of Europe Fully illustrated. 8vo. $2.00.

NORMAN'S (HENRY) An Account of the Harvard Greek Play. With 15 Heliotypes from life. 1 vol. Small quarto. $2.50.

NORTON'S (C. B.) American Inventions and Improvements in Breech-Loading Small Arms, Heavy Ordnance, Machine Guns, Magazine Arms, Fixed Ammunition, Pistols, Projectiles, Explosives, and other Munitions of War. Second edition. Quarto With 75 engravings, steel plates, and plates in color. $10.00.

OSGOOD'S AMERICAN GUIDE-BOOKS:

NEW ENGLAND. With 17 Maps and Plans. 1 vol. 16mo. Flexible cloth. $1.50.

THE MIDDLE STATES. With 22 Maps and Plans. 1 vol. 16mo. Flexible cloth. $1.50.

THE MARITIME PROVINCES. With 9 Maps and Plans. 1 vol. 16mo. Flexible cloth. $1.50.

THE WHITE MOUNTAINS. With 6 Maps and 6 Panoramas. 1 vol. 16mo. Flexible cloth. $1.50.

PALMER'S (MRS. HENRIETTA LEE) Home-Life in the Bible. Edited by John W. Palmer. With 220 Illustrations. Full octavo. $5.00. *By subscription only.*

PATTY'S PERVERSITIES. Vol. IV. of the Round-Robin Series of anonymous novels. 16mo. $1.00.

PENINSULAR CAMPAIGN (THE) of General McClellan in 1862. 1 vol. 8vo. With maps. $3.00.

PERKINS'S (F. B.) Congressional District Vote Map of the United States. In cloth case. 50 cents.

PERRY'S (NORA) A Book of Love Stories. 1 vol. 16mo. $1.00.

PLYMPTON'S (MISS A. G.) The Glad Year Round. A new juvenile, beautifully printed in colors with original and entertaining poetry. Square octavo, with illuminated covers. $2.50.

POETS AND ETCHERS. A sumptuous volume of twenty full-page etchings, by James D. Smillie, Samuel Colman, A. F. Bellows, H. Farrer, R. Swain Gifford, illustrating poems by Longfellow, Whittier, Bryant, Aldrich, etc. Quarto. Elegantly bound. $10.00. *Also limited editions on China and Japan paper.*

PREBLE'S (REAR-ADMIRAL GEORGE HENRY) History of the Flag of the United States of America. Third revised edition. Illustrated with 10 colored plates, 206 engravings on wood, 6 maps, and 18 autographs. 1 vol. Full royal octavo. 815 pages. $7.50.

PRESTON'S (MISS HARRIET W.) The Georgics of Vergil. 1 vol. 18mo. $1.00.

———— *The Same.* With red rules and initials. Illustrated. 1 vol. Small quarto. $2.00.

PUTNAM'S (J. PICKERING) The Open Fireplace in all Ages. With 300 cuts (53 full-page) and 20 new plates of Interior Decorations by American Architects. *Second thousand, revised and enlarged.* 1 vol. 12mo. $4.00.

RENAN'S (ERNEST) English Conferences. Rome and Christianity: Marcus Aurelius. Translated by CLARA ERSKINE CLEMENT. 1 vol. 16mo. 75 cents.

ROSEMARY AND RUE. Vol. VII. of the Round-Robin Series of anonymous novels. 16mo. $1.00.

ROUND-ROBIN SERIES (THE). A new series of anonymous novels by the best writers. 16mo. $1.00. Now ready: *A Nameless Nobleman, A Lesson in Love, The Georgians, Patty's Perversities, Homoselle, Damen's Ghost, Rosemary and Rue, Madame Lucas, A Tallahassee Girl, Dorothea.*

SANBORN'S (KATE) Purple and Gold. Choice Poems on the Golden Rod and Aster. Illustrated by Rosina Emmet. Printed on leaflets, bound with purple ribbon. $1.25.

SARGENT'S (MRS. JOHN T.) Sketches and Reminiscences of the Radical Club. Illustrated. 1 vol. 12mo. Cloth, $2.00. Full gilt, $2.50. Half-calf, $4.00.

SENSIER'S (ALFRED) Jean-Francois Millet: Peasant and Painter. Translated by HELENA DE KAY. With Illustrations. 1 vol. Square 8vo. $3.00.

SHAKESPEARE'S WORKS. Handy-Volume Edition. 13 vols. Illustrated. 32mo. In neat box. Cloth, $7.50.

SHALER (PROFESSOR N. S) and *DAVIS'S* (WM. M.) Illustrations of the Earth's Surface. Part I. Glaciers. Copiously illustrated with Heliotypes. Large folio. $10.00.

SHEDD'S (MRS. JULIA A.) Famous Painters and Paintings. Revised and enlarged edition. With 13 full-page Heliotypes. 1 vol. 12mo. Cloth, $3.00. Half-calf, $5.00. Tree-calf, $7.00.

———— Famous Sculptors and Sculpture. With 13 full-page Heliotypes. 1 vol. 12mo. Cloth, $3.00. Half-calf, $5.00. Tree-calf, $7.00.

SHERRATT'S (R. J) The Elements of Hand-Railing. 38 Plates. Small folio. $2.00.

SIKES'S (WIRT) British Goblins, Welsh Folk-Lore, Fairy Mythology, Legends, and Traditions. Illustrated. 8vo. $4.00.

SIMCOX'S (EDITH) Episodes in the Lives of Men, Women, and Lovers. 1 vol. Crown 8vo.

SPOONER (SAMUEL) and *CLEMENT'S* (MRS. CLARA ERSKINE) A Biographical History of the Fine Arts. *In Press.*

STILLMANN'S (J. D. B., A.M., M.D.) The Horse in Motion, as shown in a series of views by instantaneous photography, with a study on animal mechanics, founded on the revelations of the camera, in which the theory of quadrupedal locomotion is demonstrated. With anatomical illustrations in chromo, after drawings by Hahn. With a preface by Leland Stanford. Royal quarto. $10.00.

SWEETSER'S (M. F.) Artist-Biographies. Illustrated with 12 Heliotypes in each volume. 16mo. Cloth. $1.50 each.

Vol. I. Raphael, Leonardo da Vinci, Michael Angelo.
Vol. II. Titian, Guido Reni, Claude Lorraine.
Vol. III. Sir Joshua Reynolds, Turner, Landseer.
Vol. IV. Dürer, Rembrandt, Van Dyck.
Vol. V. Fra Angelico, Murillo, Allston.

The set, in cloth, 5 vols., $7.50. Half-calf, $15.00. Tree-calf, $25.00. The same are also published in smaller volumes, one biography in each. 15 vols. 18mo. Per vol., 50 cents.

SYMONDS'S (JOHN ADDINGTON) New and Old: a Volume of Verse. 1 vol. 12mo. $1.50.

TALLAHASSEE GIRL (A). Vol. IX. of the Round-Robin Series of anonymous novels. 16mo. $1.00.

TENNYSON'S (ALFRED) A Dream of Fair Women. 40 Illustrations. Arabesque binding. 1 vol. 4to. $5.00. In Morocco antique or tree-calf, $9.00.

———— Ballads and other Poems. Author's Edition, with Portrait. 1 vol. 16mo. 50 cents.

———— Poems. Illustrated Family Edition. Full gilt. Elegantly stamped. 1 vol. 8vo. $2.50. Half-calf, $5.00. Morocco antique or tree-calf, $7.50.

THACKERAY (WILLIAM M.), The Ballads of. Complete illustrated edition. Small quarto. Elegantly bound. $3.00.

THOMAS À KEMPIS'S The Imitation of Christ. With 300 Mediæval Vignettes. 16mo. Red edges. $2.00.

TOWNSEND'S (MARY ASHLEY) Down the Bayou, and other Poems. 1 vol. 12mo. $1.50.

TOWNSEND'S (S. NUGENT) Our Indian Summer in the Far West. Illustrated with full-page Photographs of Scenes in Kansas, Colorado, New Mexico, Texas, etc. 1 vol. 4to. $25.00.

TWAIN'S (MARK) The Prince and the Pauper. With 200 Illustrations by the best artists. Elegantly bound. 1 vol. Square 8vo. *Sold by subscription only.*

UNDERWOOD'S (FRANCIS H.) James Russell Lowell. A Biographical Sketch. 1 vol. 8vo. $1.50.

UPTON'S (GEORGE P.) Woman in Music. With Heliotype Illustrations. 1 vol. 12mo. $2.00. Half-calf, $4.00.

VIOLLET-LE-DUC'S (E.-E.) Discourses on Architecture. With 18 large Plates and 110 Woodcuts. Vol. I. 8vo. $5.00.

——— Discourses on Architecture. Vol. II. With Steel Plates, Chromos, and Woodcuts. 8vo. $5.00.

——— The Story of a House. Illustrated with 62 Plates and Cuts. 1 vol. 12mo. $2.00.

——— The Habitations of Man in all Ages. With 103 Illustrations. 1 vol. 12mo. $2.00.

——— Annals of a Fortress. With 85 Illustrations. 1 vol. 12mo. $2.00.

WARNER'S (CHARLES DUDLEY) The American Newspaper: an Essay. 1 vol. 32mo. 25 cents.

WARREN'S (JOSEPH H., M.D.) A Practical Treatise on Hernia. New Edition, enlarged and revised. 8vo. $5.00.

WHEELER'S (WILLIAM A.) Familiar Allusions. A Handbook of Miscellaneous Information, including the names of celebrated statues, paintings, palaces, country-seats, ruins, churches, ships, streets, clubs, natural curiosities, etc. Completed and edited by Charles G. Wheeler. 1 vol. Royal 8vo. $3.00. Half-calf, $5.50.

WHIST, American or Standard. By G. W. P. Second Edition revised. 1 vol. 16mo. $1.00.

WHITMAN'S (WALT) Leaves of Grass. Containing the matter comprised in his former volumes, with his latest Poems. With Portrait. 1 vol. 12mo. $2.50.

WHITTIER'S (JOHN G.) Poems. Illustrated Family Edition. Full gilt. 8vo. $2.50. Half-calf, $5. Morocco or tree-calf, $7.50.

WILLIAMS'S (ALFRED M.) The Poets and Poetry of Ireland. With Historical and Critical Notes. 12mo. $2.00.

WINCKELMANN'S (JOHN) The History of Ancient Art. Translated by Dr. G. H. LODGE. 78 Copper-plate Engravings. 2 vols. 8vo. Cloth, $9.00. Half-calf, $18.00. Morocco antique or tree-calf, $25.00.

——— *The Same*. Large-paper Edition. Large 4to. $30.00. *Only* 100 *copies printed.*

WINTER'S (WILLIAM) Poems. Revised Edition. 1 vol. 16mo. Cloth, $1.50. Half-calf, $3.00. Morocco or tree-calf, $4.00.

——— The Trip to England. With full-page Illustrations by JOSEPH JEFFERSON. 1 vol. 16mo. $2.00. Half-calf, $4.00. Morocco antique or tree-calf, $5.00.

——— The Life, Stories, and Poems of John Brougham. Edited by W. WINTER. 1 vol. 12mo. Illustrated. $2.00.

——— Fitz-James O'Brien's Tales, Sketches, and Poems. Edited by W. WINTER. 12mo. $2.00.

——— The Jeffersons. Vol. II. of the American-Actor Series. 1 vol. 12mo. $1.25.

In Preparation.

BACON'S (HENRY) Parisian Art and Artists. Copiously illustrated. 1 vol. Square 8vo. $3.00.

BARTLETT'S (T. H.) The Life of the Late Dr. William Rimmer. With illustrations from his Paintings, Drawings, and Sculpture. 1 vol. Quarto. Full gilt. $10.00.

DAHLGREN'S (MRS. MADELEINE VINTON) South-Mountain Magic. 1 vol. 12mo.

ESSAYS FROM THE CRITIC. By John Burroughs, E. C. Stedman, Walt Whitman, R. H. Stoddard, F. B. Sanborn, E. W. Gosse, etc. 1 vol. 16mo.

GONSE'S (LOUIS) Eugène Fromentin, Painter and Writer: translated from the French. Copiously illustrated. 1 vol. Square 8vo. $3.00.

IRELAND'S (JOSEPH N.) Mrs. Duff. Vol. V. of the American-Actor Series. 1 vol. 12mo. Illustrated. $1.25.

NORTON'S (C. B.) Heavy Ordnance, Siege and Naval Guns, Light Artillery, Machine Guns, Life-Saving Ordnance and Projectiles, as manufactured by the South Boston Iron Company. 1 vol. Quarto.

PERCY'S (TOWNSEND) A Dictionary of the Stage. 1 vol. 12mo. $2.00.

POCKET GUIDE TO EUROPE (A). 1 vol. 32mo. With 6 Maps and Plans.

SHALER'S (PROFESSOR N. S.) and *DAVIS'S* (WILLIAM M.) Illustrations of the Earth's Surface. Volume II. Quarto, with many Heliotypes. $10.00.

UNDERWOOD'S (FRANCIS H.) Henry Wadsworth Longfellow. A Biographical Sketch. 1 vol. Small quarto, fully illustrated. Fine Steel Portrait, and numerous wood-cuts.

WALKER'S (REV. J. B. R.) A New and Enlarged Concordance to the Holy Scriptures. The most perfect Concordance of the Bible in the English language. It contains over *forty thousand*, or *one-fifth*, more references and quotations than Cruden's Unabridged, which has been the standard for a century. It contains *three times* as many names of persons and places as Cruden's, each one accentuated, so as to show its exact pronunciation, and having also copious and exhaustive references and quotations. 1 vol. 8vo.

WARE'S (PROFESSOR WILLIAM R.) Modern Perspective. For Architects, Artists, and Draughtsmen. 1 vol. 12mo. With Atlas of Plates in oblong folio.

WEEKS'S (LYMAN H.) Among the Azores. Copiously illustrated. 1 vol. 16mo.

TOM SAWYER,

DETECTIVE

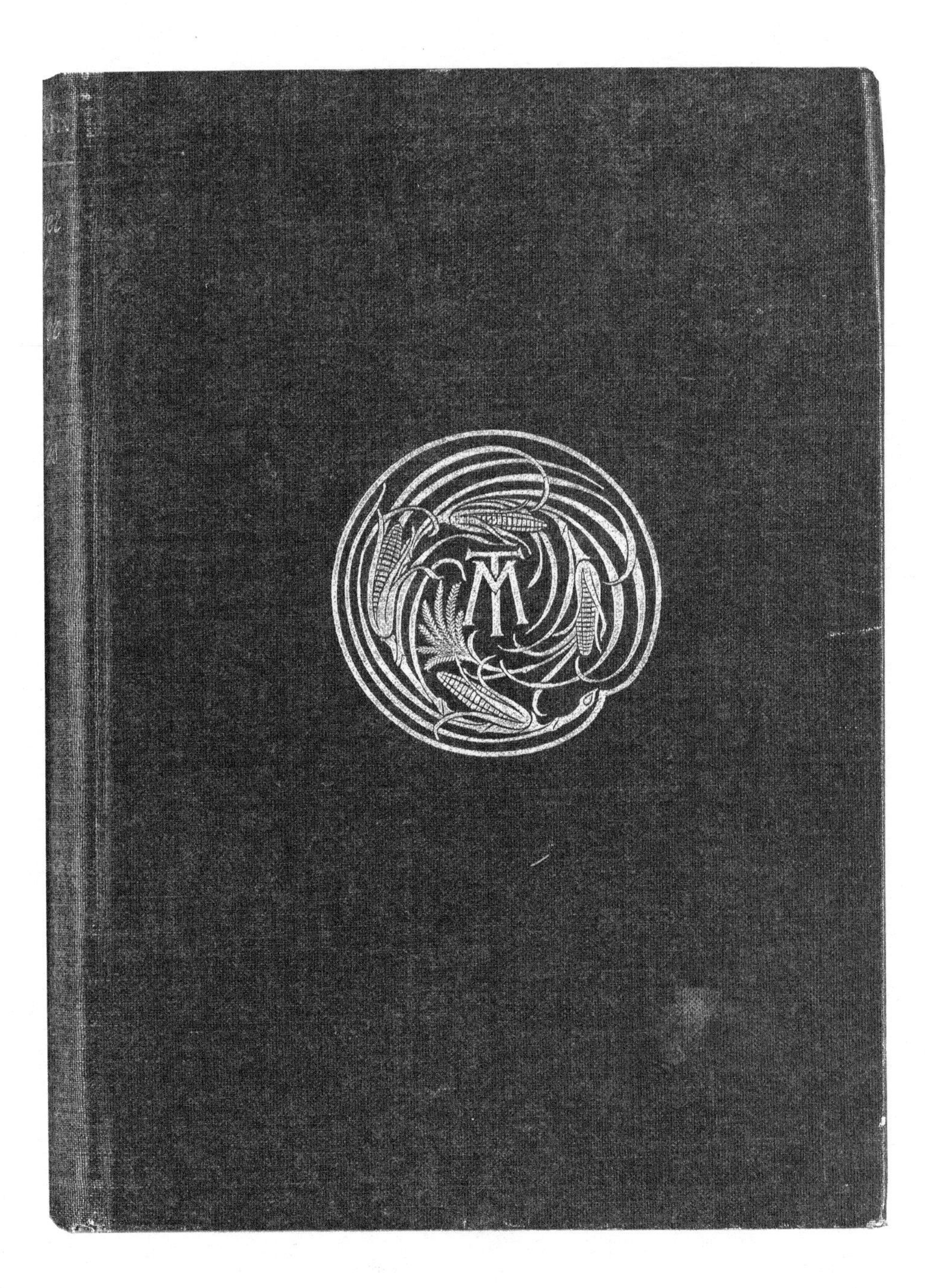

TOM SAWYER ABROAD

TOM SAWYER, DETECTIVE

AND OTHER STORIES

Etc., Etc.

By MARK TWAIN

ILLUSTRATED

NEW YORK
HARPER & BROTHERS PUBLISHERS
1896

TOM SAWYER, DETECTIVE

TOM SAWYER, DETECTIVE *

CHAPTER I

WELL, it was the next spring after me and Tom Sawyer set our old nigger Jim free, the time he was chained up for a runaway slave down there on Tom's uncle Silas's farm in Arkansaw. The frost was working out of the ground, and out of the air too, and it was getting closer and closer onto barefoot time every day; and next it would be marble time, and next mumbletypeg, and next tops and hoops, and next kites, and then right away it would be summer and going in a-swimming. It just makes a boy homesick to look ahead like that and see how far off summer is. Yes, and it sets him to sighing and saddening around, and there's something the matter with him, he don't know what. But anyway, he gets out by himself and mopes and thinks; and mostly he hunts for a lonesome place high up on the hill in the edge of the woods, and sets there and looks away off on the big Mississippi down there a-reaching miles and miles around the points where the timber looks smoky and dim it's so far

* Strange as the incidents of this story are, they are not inventions, but facts—even to the public confession of the accused. I take them from an old-time Swedish criminal trial, change the actors, and transfer the scene to America. I have added some details, but only a couple of them are important ones.—M. T.

off and still, and everything's so solemn it seems like everybody you've loved is dead and gone, and you 'most wish you was dead and gone too, and done with it all.

Don't you know what that is? It's spring fever. That is what the name of it is. And when you've got it, you want —oh, you don't quite know what it is you *do* want, but it just fairly makes your heart ache, you want it so! It seems to you that mainly what you want is to get away; get away from the same old tedious things you're so used to seeing and so tired of, and see something new. That is the idea; you want to go and be a wanderer; you want to go wandering far away to strange countries where everything is mysterious and wonderful and romantic. And if you can't do that, you'll put up with considerable less; you'll go anywhere you *can* go, just so as to get away, and be thankful of the chance, too.

Well, me and Tom Sawyer had the spring fever, and had it bad, too; but it warn't any use to think about Tom trying to get away, because, as he said, his aunt Polly wouldn't let him quit school and go traipsing off somers wasting time; so we was pretty blue. We was setting on the front steps one day about sundown talking this way, when out comes his aunt Polly with a letter in her hand and says—

"Tom, I reckon you've got to pack up and go down to Arkansaw—your aunt Sally wants you."

I 'most jumped out of my skin for joy. I reckoned Tom would fly at his aunt and hug her head off; but if you believe me he set there like a rock, and never said a word. It made me fit to cry to see him act so foolish, with such a noble chance as this opening up. Why, we might lose it if he didn't speak up and show he was thankful and grateful. But he set there and studied and studied till I was that distressed I didn't know what to do; then he says, very ca'm, and I could a shot him for it:

"Well," he says, "I'm right down sorry, Aunt Polly, but I reckon I got to be excused—for the present."

His aunt Polly was knocked so stupid and so mad at the cold impudence of it that she couldn't say a word for as much as a half a minute, and this give me a chance to nudge Tom and whisper:

"Ain't you got any sense? Sp'iling such a noble chance as this and throwing it away?"

But he warn't disturbed. He mumbled back:

"Huck Finn, do you want me to let her *see* how bad I want to go? Why, she'd begin to doubt, right away, and imagine a lot of sicknesses and dangers and objections, and first you know she'd take it all back. You lemme alone; I reckon I know how to work her."

Now I never would 'a' thought of that. But he was right. Tom Sawyer was always right—the levelest head I ever see, and always *at* himself and ready for anything you might spring on him. By this time his aunt Polly was all straight again, and she left fly. She says:

"You'll be excused! *You* will! Well, I never heard the like of it in all my days! The idea of you talking like that to *me!* Now take yourself off and pack your traps; and if I hear another word out of you about what you'll be excused from and what you won't, I lay *I'll* excuse you—with a hickory!"

She hit his head a thump with her thimble as we dodged by, and he let on to be whimpering as we struck for the stairs. Up in his room he hugged me, he was so out of his head for gladness because he was going travelling. And he says:

"Before we get away she'll wish she hadn't let me go, but she won't know any way to get around it now. After what she's said, her pride won't let her take it back."

Tom was packed in ten minutes, all except what his aunt and Mary would finish up for him; then we waited ten more for her to get cooled down and sweet and gentle again; for Tom said it took her ten minutes to unruffle in times when half of her feathers was up, but twenty when they was all up, and this was one of the times when they was all up. Then we went down, being in a sweat to know what the letter said.

She was setting there in a brown study, with it laying in her lap. We set down, and she says:

"They're in considerable trouble down there, and they think you and Huck 'll be a kind of a diversion for them—'comfort,' they say. Much of that they'll get out of you and Huck Finn, I reckon. There's a neighbor named Brace Dunlap that's been wanting to marry their Benny for three months, and at last they told him pine blank and once for all, he *couldn't;* so he has soured on them, and they're worried about it. I reckon he's somebody they think they better be on the good side of, for they've tried to please him by hiring his no-account brother to help on the farm when they can't hardly afford it, and don't want him around anyhow. Who are the Dunlaps?"

"They live about a mile from Uncle Silas's place, Aunt Polly—all the farmers live about a mile apart down there—and Brace Dunlap is a long sight richer than any of the others, and owns a whole grist of niggers. He's a widower, thirty-six years old, without any children, and is proud of his money and overbearing, and everybody is a little afraid of him. I judge he thought he could have any girl he wanted, just for the asking, and it must have set him back a good deal when he found he couldn't get Benny. Why, Benny's only half as old as he is, and just as sweet and lovely as—well, you've seen her. Poor old Uncle Silas—why, it's pitiful, him trying to curry favor that way—so

"'I RECKON I GOT TO BE EXCUSED'"

hard pushed and poor, and yet hiring that useless Jubiter Dunlap to please his ornery brother."

"What a name—Jubiter! Where'd he get it?"

"It's only just a nickname. I reckon they've forgot his real name long before this. He's twenty-seven, now, and has had it ever since the first time he ever went in swimming. The school-teacher seen a round brown mole the size of a dime on his left leg above his knee, and four little bits of moles around it, when he was naked, and he said it minded him of Jubiter and his moons; and the children thought it was funny, and so they got to calling him Jubiter, and he's Jubiter yet. He's tall, and lazy, and sly, and sneaky, and ruther cowardly, too, but kind of good-natured, and wears long brown hair and no beard, and hasn't got a cent, and Brace boards him for nothing, and gives him his old clothes to wear, and despises him. Jubiter is a twin."

"What's t'other twin like?"

"Just exactly like Jubiter—so they say; used to was, anyway, but he hain't been seen for seven years. He got to robbing when he was nineteen or twenty, and they jailed him; but he broke jail and got away—up North here, somers. They used to hear about him robbing and burglaring now and then, but that was years ago. He's dead, now. At least that's what they say. They don't hear about him any more."

"What was his name?"

"Jake."

There wasn't anything more said for a considerable while; the old lady was thinking. At last she says:

"The thing that is mostly worrying your aunt Sally is the tempers that that man Jubiter gets your uncle into."

Tom was astonished, and so was I. Tom says:

"Tempers? Uncle Silas? Land, you must be joking! I didn't know he *had* any temper."

"Works him up into perfect rages, your aunt Sally says; says he acts as if he would really hit the man, sometimes."

"Aunt Polly, it beats anything I ever heard of. Why, he's just as gentle as mush."

"Well, she's worried, anyway. Says your uncle Silas is like a changed man, on account of all this quarrelling. And the neighbors talk about it, and lay all the blame on your uncle, of course, because he's a preacher and hain't got any business to quarrel. Your aunt Sally says he hates to go into the pulpit he's so ashamed; and the people have begun to cool toward him, and he ain't as popular now as he used to was."

"Well, ain't it strange? Why, Aunt Polly, he was always so good and kind and moony and absent-minded and chuckle-headed and lovable—why, he was just an angel! What *can* be the matter of him, do you reckon?"

CHAPTER II

WE had powerful good luck; because we got a chance in a stern-wheeler from away North which was bound for one of them bayous or one-horse rivers away down Louisiana way, and so we could go all the way down the Upper Mississippi and all the way down the Lower Mississippi to that farm in Arkansaw without having to change steamboats at St. Louis: not so very much short of a thousand miles at one pull.

A pretty lonesome boat; there warn't but few passengers, and all old folks, that set around, wide apart, dozing, and was very quiet. We was four days getting out of the "upper river," because we got aground so much. But it warn't dull—couldn't be for boys that was travelling, of course.

From the very start me and Tom allowed that there was somebody sick in the state-room next to ourn, because the meals was always toted in there by the waiters. By-and-by we asked about it—Tom did—and the waiter said it was a man, but he didn't look sick.

"Well, but *ain't* he sick?"

"I don't know; maybe he is, but 'pears to me he's just letting on."

"What makes you think that?"

"Because if he was sick he would pull his clothes off *some* time or other—don't you reckon he would? Well, this one don't. At least he don't ever pull off his boots, anyway."

"The mischief he don't! Not even when he goes to bed?"

"No."

It was always nuts for Tom Sawyer—a mystery was. If you'd lay out a mystery and a pie before me and him, you wouldn't have to say take your choice; it was a thing that would regulate itself. Because in my nature I have always run to pie, whilst in his nature he has always run to mystery. People are made different. And it is the best way. Tom says to the waiter:

"What's the man's name?"

"Phillips."

"Where'd he come aboard?"

"I think he got aboard at Elexandria, up on the Iowa line."

"What do you reckon he's a-playing?"

"I hain't any notion—I never thought of it."

I says to myself, here's another one that runs to pie.

"Anything peculiar about him?—the way he acts or talks?"

"No—nothing, except he seems so scary, and keeps his doors locked night and day both, and when you knock he won't let you in till he opens the door a crack and sees who it is."

"By jimminy, it's int'resting! I'd like to get a look at him. Say—the next time you're going in there, don't you reckon you could spread the door and—"

"No, indeedy! He's always behind it. He would block that game."

Tom studied over it, and then he says:

"Looky here. You lend me your apern and let me take him his breakfast in the morning. I'll give you a quarter."

The boy was plenty willing enough, if the head steward wouldn't mind. Tom says that's all right, he reckoned he could fix it with the head steward; and he done it. He fixed it so as we could both go in with aperns on and toting vittles.

He didn't sleep much, he was in such a sweat to get in there and find out the mystery about Phillips; and moreover he done a lot of guessing about it all night, which warn't no use, for if you are going to find out the facts of a thing, what's the sense in guessing out what ain't the facts and wasting ammunition? I didn't lose no sleep. I wouldn't give a dern to know what's the matter of Phillips, I says to myself.

Well, in the morning we put on the aperns and got a couple of trays of truck, and Tom he knocked on the door. The man opened it a crack, and then he let us in and shut it quick. By Jackson, when we got a sight of him, we 'most dropped the trays! and Tom says:

"Why, Jubiter Dunlap, where'd *you* come from?"

Well, the man was astonished, of course; and first off he looked like he didn't know whether to be scared, or glad, or both, or which, but finally he settled down to being glad; and then his color come back, though at first his face had turned pretty white. So we got to talking together while he et his breakfast. And he says:

"But I ain't Jubiter Dunlap. I'd just as soon tell you who I am, though, if you'll swear to keep mum, for I ain't no Phillips, either."

Tom says:

"We'll keep mum, but there ain't any need to tell who you are if you ain't Jubiter Dunlap."

"Why?"

"Because if you ain't him you're t'other twin, Jake. You're the spit'n image of Jubiter."

"Well, I *am* Jake. But looky here, how do you come to know us Dunlaps?"

Tom told about the adventures we'd had down there at

his uncle Silas's last summer, and when he see that there warn't anything about his folks—or him either, for that matter—that we didn't know, he opened out and talked perfectly free and candid. He never made any bones about his own case; said he'd been a hard lot, was a hard lot yet, and reckoned he'd *be* a hard lot plumb to the end. He said of course it was a dangerous life, and—

He give a kind of gasp, and set his head like a person that's listening. We didn't say anything, and so it was very still for a second or so, and there warn't no sounds but the screaking of the wood-work and the chug-chugging of the machinery down below.

Then we got him comfortable again, telling him about his people, and how Brace's wife had been dead three years, and Brace wanted to marry Benny and she shook him, and Jubiter was working for Uncle Silas, and him and Uncle Silas quarrelling all the time—and then he let go and laughed.

"Land!" he says, "it's like old times to hear all this tittle-tattle, and does me good. It's been seven years and more since I heard any. How do they talk about me these days?"

"Who?"

"The farmers—and the family."

"Why, they don't talk about you at all—at least only just a mention, once in a long time."

"The nation!" he says, surprised; "why is that?"

"Because they think you are dead long ago."

"No! Are you speaking true?—honor bright, now." He jumped up, excited.

"Honor bright. There ain't anybody thinks you are alive."

"Then I'm saved, I'm saved, sure! I'll go home. They'll hide me and save my life. You keep mum. Swear

"'SWEAR YOU'LL BE GOOD TO ME AND HELP ME SAVE MY LIFE'"

you'll keep mum — swear you'll never, never tell on me. Oh, boys, be good to a poor devil that's being hunted day and night, and dasn't show his face! I've never done you any harm; I'll never do you any, as God is in the heavens; swear you'll be good to me and help me save my life."

We'd a swore it if he'd been a dog; and so we done it. Well, he couldn't love us enough for it or be grateful enough, poor cuss; it was all he could do to keep from hugging us.

We talked along, and he got out a little hand-bag and begun to open it, and told us to turn our backs. We done it, and when he told us to turn again he was perfectly different to what he was before. He had on blue goggles and the naturalest-looking long brown whiskers and mustashes you ever see. His own mother wouldn't 'a' knowed him. He asked us if he looked like his brother Jubiter, now.

"No," Tom said; "there ain't anything left that's like him except the long hair."

"All right, I'll get that cropped close to my head before I get there; then him and Brace will keep my secret, and I'll live with them as being a stranger, and the neighbors won't ever guess me out. What do you think?"

Tom he studied a while, then he says:

"Well, of course me and Huck are going to keep mum there, but if you don't keep mum yourself there's going to be a little bit of a risk—it ain't much, maybe, but it's a little. I mean, if you talk, won't people notice that your voice is just like Jubiter's; and mightn't it make them think of the twin they reckoned was dead, but maybe after all was hid all this time under another name?"

"By George," he says, "you're a sharp one! You're perfectly right. I've got to play deef and dumb when there's a neighbor around. If I'd a struck for home and forgot that little detail— However, I wasn't striking for home. I was breaking for any place where I could get away from these fellows that are after me; then I was going to put on this disguise and get some different clothes, and—"

He jumped for the outside door and laid his ear against it and listened, pale and kind of panting. Presently he whispers:

"Sounded like cocking a gun! Lord, what a life to lead!"

Then he sunk down in a chair all limp and sick like, and wiped the sweat off of his face.

"'SOUNDED LIKE COCKING A GUN!'"

CHAPTER III

FROM that time out, we was with him 'most all the time, and one or t'other of us slept in his upper berth. He said he had been so lonesome, and it was such a comfort to him to have company, and somebody to talk to in his troubles. We was in a sweat to find out what his secret was, but Tom said the best way was not to seem anxious, then likely he would drop into it himself in one of his talks, but if we got to asking questions he would get suspicious and shet up his shell. It turned out just so. It warn't no trouble to see that he *wanted* to talk about it, but always along at first he would scare away from it when he got on the very edge of it, and go to talking about something else. The way it come about was this: He got to asking us, kind of indifferent like, about the passengers down on deck. We told him about them. But he warn't satisfied; we warn't particular enough. He told us to describe them better. Tom done it. At last, when Tom was describing one of the roughest and raggedest ones, he gave a shiver and a gasp and says:

"Oh, lordy, that's one of them! They're aboard sure—I just knowed it. I sort of hoped I had got away, but I never believed it. Go on."

Presently when Tom was describing another mangy, rough deck passenger, he give that shiver again and says—

"That's him!—that's the other one. If it would only come a good black stormy night and I could get ashore. You see, they've got spies on me. They've got a right to come up and buy drinks at the bar yonder forrard, and they take that chance to bribe somebody to keep watch on me—porter or boots or somebody. If I was to slip ashore without anybody seeing me, they would know it inside of an hour."

So then he got to wandering along, and pretty soon, sure enough, he was telling! He was poking along through his ups and downs, and when he come to that place he went right along. He says:

"It was a confidence game. We played it on a julery-shop in St. Louis. What we was after was a couple of noble big di'monds as big as a hazel-nuts, which everybody was running to see. We was dressed up fine, and we played it on them in broad daylight. We ordered the di'monds sent to the hotel for us to see if we wanted to buy, and when we was examining them we had paste counterfeits all ready, and *them* was the things that went back to the shop when we said the water wasn't quite fine enough for twelve thousand dollars."

"Twelve—thousand—dollars!" Tom says. "Was they really worth all that money, do you reckon?"

"Every cent of it."

"And you fellows got away with them?"

"As easy as nothing. I don't reckon the julery people know they've been robbed yet. But it wouldn't be good sense to stay around St. Louis, of course, so we considered where we'd go. One was for going one way, one another, so we throwed up, heads or tails, and the Upper Mississippi won. We done up the di'monds in a paper and put our names on it and put it in the keep of the hotel clerk, and told him not to ever let either of us have it again without the others was on hand to see it done; then we went down town, each by his own self—because I reckon maybe we all had the same notion. I don't know for certain, but I reckon maybe we had."

"What notion?" Tom says.

"To rob the others."

"What—one take everything, after all of you had helped to get it?"

"Cert'nly."

It disgusted Tom Sawyer, and he said it was the orneriest, low-downest thing he ever heard of. But Jake Dunlap said it warn't unusual in the profession. Said when a person was in that line of business he'd got to look out for his own intrust, there warn't nobody else going to do it for him. And then he went on. He says:

"You see, the trouble was, you couldn't divide up two di'monds amongst three. If there'd been three— But never mind about that, there *warn't* three. I loafed along the back streets studying and studying. And I says to myself, I'll hog them di'monds the first chance I get, and I'll have a disguise all ready, and I'll give the boys the slip, and when I'm safe away I'll put it on, and then let them find me if they can. So I got the false whiskers and the goggles and this countrified suit of clothes, and fetched them along back in a hand-bag; and when I was passing a shop where they sell all sorts of things, I got a glimpse of one of my pals through the window. It was Bud Dixon. I was glad, you bet. I says to myself, I'll see what he buys. So I kept shady, and watched. Now what do you reckon it was he bought?"

"Whiskers?" said I.

"No."

"Goggles?"

"No."

"Oh, keep still, Huck Finn, can't you, you're only just hendering all you can. What *was* it he bought, Jake?"

"You'd never guess in the world. It was only just a screw-driver—just a wee little bit of a screw-driver."

"Well, I declare! What did he want with that?"

"That's what *I* thought. It was curious. It clean stumped me. I says to myself, what can he want with that thing? Well, when he come out I stood back out of sight, and then tracked him to a second-hand slop-shop and see him buy a red flannel shirt and some old ragged clothes—just the ones he's got on now, as you've described. Then I went down to the wharf and hid my things aboard the up-river boat that we had picked out, and then started back and had another streak of luck. I seen our other pal lay in *his* stock of old rusty second-handers. We got the di'monds and went aboard the boat.

"But now we was up a stump, for we couldn't go to bed. We had to set up and watch one another. Pity, that was; pity to put that kind of a strain on us, because there was bad blood between us from a couple of weeks back, and we was only friends in the way of business. Bad anyway, seeing there was only two di'monds betwixt three men. First we had supper, and then tramped up and down the deck together smoking till most midnight; then we went and set down in my state-room and locked the doors and looked in the piece of paper to see if the di'monds was all right, then laid it on the lower berth right in full sight; and there we set, and set, and by-and-by it got to be dreadful hard to keep awake. At last Bud Dixon he dropped off. As soon as he was snoring a good regular gait that was likely to last, and had his chin on his breast and looked permanent, Hal Clayton nodded towards the di'monds and then towards the outside door, and I understood. I reached and got the paper, and then we stood up and waited perfectly still; Bud never stirred; I turned the key of the outside door very soft and slow, then turned the knob the same way, and we went tiptoeing out onto the guard, and shut the door very soft and gentle.

"'WE STOOD UP AND WAITED, PERFECTLY STILL'"

"There warn't nobody stirring anywhere, and the boat was slipping along, swift and steady, through the big water in the smoky moonlight. We never said a word, but went straight up onto the hurricane-deck and plumb back aft, and set down on the end of the skylight. Both of us knowed what that meant, without having to explain to one another. Bud Dixon would wake up and miss the swag, and would come straight for us, for he ain't afeard of anything or anybody, that man ain't. He would come, and we would heave him overboard, or get killed trying. It made me shiver, because I ain't as brave as some people, but if I showed the white feather—well, I knowed better than do that. I kind of hoped the boat would land somers, and we could skip ashore and not have to run the risk of this row, I was so scared of Bud Dixon, but she was an upper-river tub and there warn't no real chance of that.

"Well, the time strung along and along, and that fellow never come! Why, it strung along till dawn begun to break, and still he never come. 'Thunder,' I says, 'what do you make out of this?—ain't it suspicious?' 'Land!' Hal says, 'do you reckon he's playing us?—open the paper!' I done it, and by gracious there warn't anything in it but a couple of little pieces of loaf-sugar! *That's* the reason he could set there and snooze all night so comfortable. Smart? Well, I reckon! He had had them two papers all fixed and ready, and he had put one of them in place of t'other right under our noses.

"We felt pretty cheap. But the thing to do, straight off, was to make a plan; and we done it. We would do up the paper again, just as it was, and slip in, very elaborate and soft, and lay it on the bunk again, and let on *we* didn't know about any trick, and hadn't any idea he was a-laughing at us behind them bogus snores of his'n; and we would stick by him, and the first night we was ashore we would get him drunk and search him, and get the di'monds; and *do* for him, too, if it warn't too risky. If we got the swag, we'd *got* to do for him, or he would hunt us down and do for us, sure. But I didn't have no real hope. I knowed we could get him drunk—he was always ready for that—but what's the good of it? You might search him a year and never find—

"Well, right there I catched my breath and broke off my thought! For an idea went ripping through my head that tore my brains to rags—and land, but I felt gay and good! You see, I had had my boots off, to unswell my feet, and just then I took up one of them to put it on, and I catched a glimpse of the heel-bottom, and it just took my breath away. You remember about that puzzlesome little screw-driver?"

"You bet I do," says Tom, all excited.

"Well, when I catched that glimpse of that boot heel, the idea that went smashing through my head was, *I* know where he's hid the di'monds! You look at this boot heel, now. See, it's bottomed with a steel plate, and the plate is fastened on with little screws. Now there wasn't a screw about that feller anywhere but in his boot heels; so, if he needed a screw-driver, I reckoned I knowed why."

"Huck, ain't it bully!" says Tom.

"Well, I got my boots on, and we went down and slipped in and laid the paper of sugar on the berth, and sat down soft and sheepish and went to listening to Bud Dixon snore. Hal Clayton dropped off pretty soon, but I didn't; I wasn't ever so wide-awake in my life. I was spying out from under the shade of my hat brim, searching the floor for leather. It took me a long time, and I begun to think maybe my guess was wrong, but at last I struck it. It laid over by the bulkhead, and was nearly the color of the carpet. It was a little round plug about as thick as the end of your little finger, and I says to myself there's a di'mond in the nest you've come from. Before long I spied out the plug's mate.

"Think of the smartness and coolness of that blatherskite! He put up that scheme on us and reasoned out what we would do, and we went ahead and done it perfectly exact, like a couple of pudd'n heads. He set there and took his own time to unscrew his heel-plates and cut out his plugs and stick in the di'monds and screw on his plates again. He allowed we would steal the bogus swag and wait all night for him to come up and get drownded, and by George it's just what we done! *I* think it was powerful smart."

"You bet your life it was!" says Tom, just full of admiration.

CHAPTER IV

"WELL, all day we went through the humbug of watching one another, and it was pretty sickly business for two of us and hard to act out, I can tell you. About night we landed at one of them little Missouri towns high up toward Iowa, and had supper at the tavern, and got a room upstairs with a cot and a double bed in it, but I dumped my bag under a deal table in the dark hall whilst we was moving along it to bed, single file, me last, and the landlord in the lead with a tallow candle. We had up a lot of whiskey, and went to playing high-low-jack for dimes, and as soon as the whiskey begun to take hold of Bud we stopped drinking, but we didn't let him stop. We loaded him till he fell out of his chair and laid there snoring.

"We was ready for business now. I said we better pull our boots off, and his'n too, and not make any noise, then we could pull him and haul him around and ransack him without any trouble. So we done it. I set my boots and Bud's side by side, where they'd be handy. Then we stripped him and searched his seams and his pockets and his socks and the inside of his boots, and everything, and searched his bundle. Never found any di'monds. We found the screw-driver, and Hal says, 'What do you reckon he wanted with that?' I said I didn't know; but when he wasn't looking I hooked it. At last Hal he looked beat and discouraged, and said we'd got to give it up. That was what I was waiting for. I says:

"'There's one place we hain't searched.'

"SEARCHED HIS SEAMS AND HIS POCKETS AND HIS SOCKS"

"'What place is that?' he says.

"'His stomach.'

"'By gracious, I never thought of that! *Now* we're on the homestretch, to a dead moral certainty. How'll we manage?'

"'Well,' I says, 'just stay by him till I turn out and hunt up a drug-store, and I reckon I'll fetch something that'll make them di'monds tired of the company they're keeping.'

"He said that's the ticket, and with him looking straight at me I slid myself into Bud's boots instead of my own, and he never noticed. They was just a shade large for me, but that was considerable better than being too small. I got my bag as I went a-groping through the hall, and in about a minute I was out the back way and stretching up the river road at a five-mile gait.

"And not feeling so very bad, neither—walking on di'monds don't have no such effect. When I had gone fifteen minutes I says to myself, there's more'n a mile behind me, and everything quiet. Another five minutes and I says there's considerable more land behind me now, and there's a man back there that's begun to wonder what's the trouble. Another five and I says to myself he's getting real uneasy—he's walking the floor now. Another five, and I says to myself, there's two mile and a half behind me, and he's *awful* uneasy—beginning to cuss, I reckon. Pretty soon I says to myself, forty minutes gone—he *knows* there's something up! Fifty minutes—the truth's a-busting on him now! he is reckoning I found the di'monds whilst we was searching, and shoved them in my pocket and never let on—yes, and he's starting out to hunt for me. He'll hunt for new tracks in the dust, and they'll as likely send him down the river as up.

"Just then I see a man coming down on a mule, and before I thought I jumped into the bush. It was stupid! When he got abreast he stopped and waited a little for me to come out; then he rode on again. But I didn't feel gay any more. I says to myself I've botched my chances by that; I surely have, if he meets up with Hal Clayton.

"Well, about three in the morning I fetched Elexandria and see this stern-wheeler laying there, and was very glad, because I felt perfectly safe, now, you know. It was just daybreak. I went aboard and got this state-room and put on these clothes and went up in the pilot-house—to watch, though I didn't reckon there was any need of it. I set there and played with my di'monds and waited and waited for the boat to start, but she didn't. You see, they was mending her machinery, but I didn't know anything about it, not being very much used to steamboats.

"Well, to cut the tale short, we never left there till plumb noon; and long before that I was hid in this state-room; for before breakfast I see a man coming, away off, that had a gait like Hal Clayton's, and it made me just sick. I says to myself, if he finds out I'm aboard this boat, he's got me like a rat in a trap. All he's got to do is to have me watched, and wait—wait till I slip ashore, thinking he is a thousand miles away, then slip after me and dog me to a good place and make me give up the di'monds, and then he'll—oh, *I* know what he'll do! Ain't it awful—awful! And now to think the *other* one's aboard, too! Oh, ain't it hard luck, boys—ain't it hard! But you'll help save me, *won't* you?—oh, boys, be good to a poor devil that's being hunted to death, and save me—I'll worship the very ground you walk on!"

We turned in and soothed him down and told him we would plan for him and help him, and he needn't be so afeard; and so by-and-by he got to feeling kind of comfortable again, and unscrewed his heel-plates and held up his

WALKED ASHORE

di'monds this way and that, admiring them and loving them; and when the light struck into them they *was* beautiful, sure; why, they seemed to kind of bust, and snap fire out all around. But all the same I judged he was a fool. If I had been him I would a handed the di'monds to them pals and got them to go ashore and leave me alone. But he was made different. He said it was a whole fortune and he couldn't bear the idea.

Twice we stopped to fix the machinery and laid a good while, once in the night; but it wasn't dark enough, and he was afeard to skip. But the third time we had to fix it there was a better chance. We laid up at a country wood-yard about forty mile above Uncle Silas's place a little after one at night, and it was thickening up and going to storm. So Jake he laid for a chance to slide. We begun to take in wood. Pretty soon the rain come a-drenching down, and the wind blowed hard. Of course every boat-hand fixed a gunny sack and put it on like a bonnet, the way they do when they are toting wood, and we got one for Jake, and he slipped down aft with his hand-bag and come tramping forrard just like the rest, and walked ashore with them, and when we see him pass out of the light of the torch-basket and get swallowed up in the dark, we got our breath again and just felt grateful and splendid. But it wasn't for long. Somebody told, I reckon; for in about eight or ten minutes them two pals come tearing forrard as tight as they could jump and darted ashore and was gone. We waited plumb till dawn for them to come back, and kept hoping they would, but they never did. We was awful sorry and low-spirited. All the hope we had was that Jake had got such a start that they couldn't get on his track, and he would get to his brother's and hide there and be safe.

He was going to take the river road, and told us to find

out if Brace and Jubiter was to home and no strangers there, and then slip out about sundown and tell him. Said he would wait for us in a little bunch of sycamores right back of Tom's uncle Silas's tobacker-field on the river road, a lonesome place.

We set and talked a long time about his chances, and Tom said he was all right if the pals struck up the river instead of down, but it wasn't likely, because maybe they knowed where he was from; more likely they would go right, and dog him all day, him not suspecting, and kill him when it come dark, and take the boots. So we was pretty sorrowful.

CHAPTER V

We didn't get done tinkering the machinery till away late in the afternoon, and so it was so close to sundown when we got home that we never stopped on our road, but made a break for the sycamores as tight as we could go, to tell Jake what the delay was, and have him wait till we could go to Brace's and find out how things was there. It was getting pretty dim by the time we turned the corner of the woods, sweating and panting with that long run, and see the sycamores thirty yards ahead of us; and just then we see a couple of men run into the bunch and heard two or three terrible screams for help. "Poor Jake is killed, sure," we says. We was scared through and through, and broke for the tobacker-field and hid there, trembling so our clothes would hardly stay on; and just as we skipped in there, a couple of men went tearing by, and into the bunch they went, and in a second out jumps four men and took out up the road as tight as they could go, two chasing two.

We laid down, kind of weak and sick, and listened for more sounds, but didn't hear none for a good while but just our hearts. We was thinking of that awful thing laying yonder in the sycamores, and it seemed like being that close to a ghost, and it give me the cold shudders. The moon come a-swelling up out of the ground, now, powerful big and round and bright, behind a comb of trees, like a face looking through prison bars, and the black shadders and white places begun to creep around, and it was miserable quiet and still and night-breezy and graveyardy and scary. All of a sudden Tom whispers:

"Look!—what's that?"

"Don't!" I says. "Don't take a person by surprise that way. I'm 'most ready to die, anyway, without you doing that."

"Look, I tell you. It's something coming out of the sycamores."

"Don't, Tom!"

"It's terrible tall!"

"Oh, lordy-lordy! let's—"

"Keep still—it's a-coming this way."

He was so excited he could hardly get breath enough to whisper. I had to look. I couldn't help it. So now we was both on our knees with our chins on a fence-rail and gazing—yes, and gasping, too. It was coming down the road—coming in the shadder of the trees, and you couldn't see it good; not till it was pretty close to us; then it stepped into a bright splotch of moonlight and we sunk right down in our tracks—it was Jake Dunlap's ghost! That was what we said to ourselves.

We couldn't stir for a minute or two; then it was gone. We talked about it in low voices. Tom says:

"They're mostly dim and smoky, or like they're made out of fog, but this one wasn't."

"No," I says; "I seen the goggles and the whiskers perfectly plain."

"Yes, and the very colors in them loud countrified Sunday clothes—plaid breeches, green and black—"

"Cotton-velvet westcot, fire-red and yaller squares—"

"Leather straps to the bottoms of the breeches legs and one of them hanging unbuttoned—"

"Yes, and that hat—"

"What a hat for a ghost to wear!"

"IT WAS JAKE DUNLOP'S GHOST"

You see it was the first season anybody wore that kind—a black stiff-brim stove-pipe, very high, and not smooth, with a round top—just like a sugar-loaf.

"Did you notice if its hair was the same, Huck?"

"No—seems to me I did, then again it seems to me I didn't."

"I didn't either; but it had its bag along, I noticed that."

"So did I. How can there be a ghost-bag, Tom?"

"Sho! I wouldn't be as ignorant as that if I was you, Huck Finn. Whatever a ghost has, turns to ghost-stuff. They've got to have their things, like anybody else. You see, yourself, that its clothes was turned to ghost-stuff. Well, then, what's to hender its bag from turning, too? Of course it done it."

That was reasonable. I couldn't find no fault with it. Bill Withers and his brother Jack come along by, talking, and Jack says:

"What do you reckon he was toting?"

"I dunno; but it was pretty heavy."

"Yes, all he could lug. Nigger stealing corn from old Parson Silas, I judged."

"So did I. And so I allowed I wouldn't let on to see him."

"That's me, too."

Then they both laughed, and went on out of hearing. It showed how unpopular old Uncle Silas had got to be, now. They wouldn't 'a' let a nigger steal anybody else's corn and never done anything to him.

We heard some more voices mumbling along towards us and getting louder, and sometimes a cackle of a laugh. It was Lem Beebe and Jim Lane. Jim Lane says:

"Who?—Jubiter Dunlap?"

"Yes."

"Oh, I don't know. I reckon so. I seen him spading up some ground along about an hour ago, just before sun-down—him and the parson. Said he guessed he wouldn't go to-night, but we could have his dog if we wanted him."

"Too tired, I reckon."

"Yes—works so hard!"

"Oh, you bet!"

They cackled at that, and went on by. Tom said we better jump out and tag along after them, because they was going our way and it wouldn't be comfortable to run across the ghost all by ourselves. So we done it, and got home all right.

That night was the second of September—a Saturday. I sha'n't ever forget it. You'll see why, pretty soon.

CHAPTER VI

We tramped along behind Jim and Lem till we come to the back stile where old Jim's cabin was that he was captivated in, the time we set him free, and here come the dogs piling around us to say howdy, and there was the lights of the house, too; so we warn't afeard any more, and was going to climb over, but Tom says:

"Hold on; set down here a minute. By George!"

"What's the matter?" says I.

"Matter enough!" he says. "Wasn't you expecting we would be the first to tell the family who it is that's been killed yonder in the sycamores, and all about them rapscallions that done it, and about the di'monds they've smouched off of the corpse, and paint it up fine, and have the glory of being the ones that knows a lot more about it than anybody else?"

"Why, of course. It wouldn't be you, Tom Sawyer, if you was to let such a chance go by. I reckon it ain't going to suffer none for lack of paint," I says, "when you start in to scollop the facts."

"Well, now," he says, perfectly ca'm, "what would you say if I was to tell you I ain't going to start in at all?"

I was astonished to hear him talk so. I says:

"I'd say it's a lie. You ain't in earnest, Tom Sawyer."

"You'll soon see. Was the ghost barefooted?"

"No, it wasn't. What of it?"

"You wait—I'll show you what. Did it have its boots on?"

"Yes. I seen them plain."

"Swear it?"

"Yes, I swear it."

"So do I. Now do you know what that means?"

"No. What does it mean?"

"Means that them thieves *didn't get the di'monds.*"

"Jimminy! What makes you think that?"

"I don't only think it, I know it. Didn't the breeches and goggles and whiskers and hand-bag and every blessed thing turn to ghost-stuff? Everything it had on turned, didn't it? It shows that the reason its boots turned too was because it still had them on after it started to go ha'nting around, and if that ain't proof that them blatherskites didn't get the boots, I'd like to know what you'd *call* proof."

Think of that now. I never see such a head as that boy had. Why, *I* had eyes and I could see things, but they never meant nothing to me. But Tom Sawyer was different. When Tom Sawyer seen a thing it just got up on its hind legs and *talked* to him—told him everything it knowed. *I* never see such a head.

"Tom Sawyer," I says, "I'll say it again as I've said it a many a time before: I ain't fitten to black your boots. But that's all right—that's neither here nor there. God Almighty made us all, and some He gives eyes that's blind, and some He gives eyes that can see, and I reckon it ain't none of our lookout what He done it for; it's all right, or He'd 'a' fixed it some other way. Go on—I see plenty plain enough, now, that them thieves didn't get way with the di'monds. Why didn't they, do you reckon?"

"Because they got chased away by them other two men before they could pull the boots off of the corpse."

"That's so! I see it now. But looky here, Tom, why ain't we to go and tell about it?"

"'WAS THE GHOST BAREFOOTED?'"

"Oh, shucks, Huck Finn, can't you see? Look at it. What's a-going to happen? There's going to be an inquest in the morning. Them two men will tell how they heard the yells and rushed there just in time to not save the stranger. Then the jury 'll twaddle and twaddle and twaddle, and finally they'll fetch in a verdict that he got shot or stuck or busted over the head with something, and come to his death by the inspiration of God. And after they've buried him they'll auction off his things for to pay the expenses, and then's *our* chance."

"How, Tom?"

"Buy the boots for two dollars!"

Well, it 'most took my breath.

"My land! Why, Tom, *we'll* get the di'monds!"

"You bet. Some day there'll be a big reward offered for them—a thousand dollars, sure. That's our money! Now we'll trot in and see the folks. And mind you we don't know anything about any murder, or any di'monds, or any thieves—don't you forget that."

I had to sigh a little over the way he had got it fixed. *I*'d 'a' *sold* them di'monds—yes, sir—for twelve thousand dollars; but I didn't say anything. It wouldn't done any good. I says:

"But what are we going to tell your aunt Sally has made us so long getting down here from the village, Tom?"

"Oh, I'll leave that to you," he says. "I reckon you can explain it somehow."

He was always just that strict and delicate. He never would tell a lie himself.

We struck across the big yard, noticing this, that, and t'other thing that was so familiar, and we so glad to see it again, and when we got to the roofed big passageway betwixt the double log house and the kitchen part, there was everything hanging on the wall just as it used to was, even to Uncle Silas's old faded green baize working-gown with the hood to it, and raggedy white patch between the shoulders that always looked like somebody had hit him with a snowball; and then we lifted the latch and walked in. Aunt Sally she was just a-ripping and a-tearing around, and the children was huddled in one corner, and the old man he was huddled in the other and praying for help in time of need. She jumped for us with joy and tears running down her face and give us a whacking box on the ear, and then hugged us and kissed us and boxed us again, and just couldn't seem to get enough of it, she was so glad to see us; and she says

"Where *have* you been a-loafing to, you good-for-nothing trash! I've been that worried about you I didn't know what to do. Your traps has been here *ever* so long, and I've had supper cooked fresh about four times so as to have it hot and good when you come, till at last my patience is just plumb wore out, and I declare I—I—why I could skin you alive! You must be starving, poor things!—set down, set down, everybody; don't lose no more time."

It was good to be there again behind all that noble corn-pone and spareribs, and everything that you could ever want in this world. Old Uncle Silas he peeled off one of his bulliest old-time blessings, with as many layers to it as an onion, and whilst the angels was hauling in the slack of it I was trying to study up what to say about what kept us so long. When our plates was all loadened and we'd got a-going, she asked me, and I says:

"Well, you see,—er—Mizzes—"

"Huck Finn! Since when am I Mizzes to you? Have I ever been stingy of cuffs or kisses for you since the day you stood in this room and I took you for Tom Sawyer and blessed God for sending you to me, though you told me four thousand lies and I believed every one of them like a simpleton? Call me Aunt Sally—like you always done."

So I done it. And I says:

"Well, me and Tom allowed we would come along afoot and take a smell of the woods, and we run across Lem Beebe and Jim Lane, and they asked us to go with them blackberrying to-night, and said they could borrow Jubiter Dunlap's dog, because he had told them just that minute—"

"Where did they see him?" says the old man; and when I looked up to see how *he* come to take an intrust in a little thing like that, his eyes was just burning into me, he was that eager. It surprised me so it kind of throwed me off, but I pulled myself together again and says:

"It was when he was spading up some ground along with you, towards sundown or along there."

He only said, "Um," in a kind of a disappointed way, and didn't take no more intrust. So I went on. I says:

"Well, then, as I was a-saying—"

"That 'll do, you needn't go no furder." It was Aunt Sally. She was boring right into me with her eyes, and very indignant. "Huck Finn," she says, "how'd them men come to talk about going a-blackberrying in September—in *this* region?"

I see I had slipped up, and I couldn't say a word. She waited, still a-gazing at me, then she says:

"And how'd they come to strike that idiot idea of going a-blackberrying in the night?"

"Well, m'm, they—er—they told us they had a lantern, and—"

"Oh, *shet* up—do! Looky here; what was they going to do with a dog?—hunt blackberries with it?"

"I think, m'm, they—"

"Now, Tom Sawyer, what kind of a lie are you fixing *your* mouth to contribit to this mess of rubbage? Speak out—and I warn you before you begin, that I don't believe a word of it. You and Huck's been up to something you no business to—*I* know it perfectly well; *I* know you, *both* of you. Now you explain that dog, and them blackberries, and the lantern, and the rest of that rot—and mind you talk as straight as a string—do you hear?"

Tom he looked considerable hurt, and says, very dignified:

"It is a pity if Huck is to be talked to that away, just for making a little bit of a mistake that anybody could make."

"What mistake has he made?"

"Why, only the mistake of saying blackberries when of course he meant strawberries."

"Tom Sawyer, I lay if you aggravate me a little more, I'll—"

"Aunt Sally, without knowing it—and of course without intending it—you are in the wrong. If you'd 'a' studied natural history the way you ought, you would know that all over the world except just here in Arkansaw they *always* hunt strawberries with a dog—and a lantern—"

But she busted in on him there and just piled into him and snowed him under. She was so mad she couldn't get the words out fast enough, and she gushed them out in one everlasting freshet. That was what Tom Sawyer was after. He allowed to work her up and get her started and then leave her alone and let her burn herself out. Then she would be so aggravated with that subject that she wouldn't say another word about it, nor let anybody else. Well, it happened just so. When she was tuckered out and had to hold up, he says, quite ca'm:

"And yet, all the same, Aunt Sally—"

"Shet up!" she says, "I don't want to hear another word out of you."

So we was perfectly safe, then, and didn't have no more trouble about that delay. Tom done it elegant.

CHAPTER VII

Benny she was looking pretty sober, and she sighed some, now and then; but pretty soon she got to asking about Mary, and Sid, and Tom's aunt Polly, and then Aunt Sally's clouds cleared off and she got in a good humor and joined in on the questions and was her lovingest best self, and so the rest of the supper went along gay and pleasant. But the old man he didn't take any hand hardly, and was absent-minded and restless, and done a considerable amount of sighing; and it was kind of heart-breaking to see him so sad and troubled and worried.

By-and-by, a spell after supper, come a nigger and knocked on the door and put his head in with his old straw hat in his hand bowing and scraping, and said his Marse Brace was out at the stile and wanted his brother, and was getting tired waiting supper for him, and would Marse Silas please tell him where he was? I never see Uncle Silas speak up so sharp and fractious before. He says:

"Am *I* his brother's keeper?" And then he kind of wilted together, and looked like he wished he hadn't spoken so, and then he says, very gentle: "But you needn't say that, Billy; I was took sudden and irritable, and I ain't very well these days, and not hardly responsible. Tell him he ain't here."

And when the nigger was gone he got up and walked the floor, backwards and forwards, mumbling and muttering to himself and ploughing his hands through his hair. It was real pitiful to see him. Aunt Sally she whispered to us and told us not to take notice of him, it embarrassed him. She said he was always thinking and thinking, since these troubles come on, and she allowed he didn't more'n about half know what he was about when the thinking spells was on him; and she said he walked in his sleep considerable more now than he used to, and sometimes wandered around over the house and even outdoors in his sleep, and if we catched him at it we must let him alone and not disturb him. She said she reckoned it didn't do him no harm, and mav be it done him good. She said Benny was the only one that was much help to him these days. Said Benny appeared to know just when to try to soothe him and when to leave him alone.

So he kept on tramping up and down the floor and mut tering, till by-and-by he begun to look pretty tired; then Benny she went and snuggled up to his side and put one hand in his and one arm around his waist and walked with him; and he smiled down on her, and reached down and kissed her; and so, little by little the trouble went out of his face and she persuaded him off to his room. They had very petting ways together, and it was uncommon pretty to see.

Aunt Sally she was busy getting the children ready for bed; so by-and-by it got dull and tedious, and me and Tom took a turn in the moonlight, and fetched up in the watermelon-patch and et one, and had a good deal of talk. And Tom said he'd bet the quarrelling was all Jubiter's fault, and he was going to be on hand the first time he got a chance, and see; and if it was so, he was going to do his level best to get Uncle Silas to turn him off.

And so we talked and smoked and stuffed watermelon as much as two hours, and then it was pretty late, and when we got back the house was quiet and dark, and everybody gone to bed.

"SMOKED AND STUFFED WATERMELON"

Tom he always seen everything, and now he see that the old green baize work-gown was gone, and said it wasn't gone when he went out; and so we allowed it was curious, and then we went up to bed.

We could hear Benny stirring around in her room, which was next to ourn, and judged she was worried a good deal about her father and couldn't sleep. We found we couldn't, neither. So we set up a long time, and smoked and talked in a low voice, and felt pretty dull and down-hearted. We talked the murder and the ghost over and over again, and got so creepy and crawly we couldn't get sleepy nohow and noway.

By-and-by, when it was away late in the night and all the sounds was late sounds and solemn, Tom nudged me and whispers to me to look, and I done it, and there we see a man poking around in the yard like he didn't know just what he wanted to do, but it was pretty dim and we couldn't see him good. Then he started for the stile, and as he went over it the moon came out strong, and he had a long-handled shovel over his shoulder, and we see the white patch on the old work-gown. So Tom says:

"He's a-walking in his sleep. I wish we was allowed to follow him and see where he's going to. There, he's turned down by the tobacker-field. Out of sight now. It's a dreadful pity he can't rest no better."

We waited a long time, but he didn't come back any more, or if he did he come around the other way; so at last we was tuckered out and went to sleep and had nightmares, a million of them. But before dawn we was awake again, because meantime a storm had come up and been raging, and the thunder and lightning was awful, and the wind was a-thrashing the trees around, and the rain was driving down in slanting sheets, and the gullies was running rivers. Tom says:

"Looky here, Huck, I'll tell you one thing that's mighty curious. Up to the time we went out, last night, the family hadn't heard about Jake Dunlap being murdered. Now the men that chased Hal Clayton and Bud Dixon away would spread the thing around in a half an hour, and every neighbor that heard it would shin out and fly around from one farm to t'other and try to be the first to tell the news. Land, they don't have such a big thing as that to tell twice in thirty year! Huck, it's mighty strange; I don't understand it."

So then he was in a fidget for the rain to let up, so we could turn out and run across some of the people and see if they would say anything about it to us. And he said if they did we must be horribly surprised and shocked.

We was out and gone the minute the rain stopped. It was just broad day, then. We loafed along up the road, and now and then met a person and stopped and said howdy, and told them when we come, and how we left the folks at home, and how long we was going to stay, and all that, but none of them said a word about that thing; which was just astonishing, and no mistake. Tom said he believed if we went to the sycamores we would find that body laying there solitary and alone, and not a soul around. Said he believed the men chased the thieves so far into the woods that the thieves prob'ly seen a good chance and turned on them at last, and maybe they all killed each other, and so there wasn't anybody left to tell.

First we knowed, gabbling along that away, we was right at the sycamores. The cold chills trickled down my back and I wouldn't budge another step, for all Tom's persuading. But he couldn't hold in; he'd *got* to see if the boots was safe on that body yet. So he crope in—and the next minute out he come again with his eyes bulging he was so excited, and says:

"'HUCK, IT'S GONE!'"

"Huck, it's gone!"

I *was* astonished! I says:

"Tom, you don't mean it."

"It's gone, sure. There ain't a sign of it. The ground is trampled some, but if there was any blood it's all washed away by the storm, for it's all puddles and slush in there."

At last I give in, and went and took a look myself; and it was just as Tom said—there wasn't a sign of a corpse.

"Dern it," I says, "the di'monds is gone. Don't you reckon the thieves slunk back and lugged him off, Tom?"

"Looks like it. It just does. Now where'd they hide him, do you reckon?"

"I don't know," I says, disgusted, "and what's more I don't care. They've got the boots, and that's all *I* cared about. He'll lay around these woods a long time before *I* hunt him up."

Tom didn't feel no more intrust in him neither, only curiosity to know what come of him; but he said we'd lay low and keep dark and it wouldn't be long till the dogs or somebody rousted him out.

We went back home to breakfast ever so bothered and put out and disappointed and swindled. I warn't ever so down on a corpse before.

CHAPTER VIII

It warn't very cheerful at breakfast. Aunt Sally she looked old and tired and let the children snarl and fuss at one another and didn't seem to notice it was going on, which wasn't her usual style; me and Tom had a plenty to think about without talking; Benny she looked like she hadn't had much sleep, and whenever she'd lift her head a little and steal a look towards her father you could see there was tears in her eyes; and as for the old man, his things stayed on his plate and got cold without him knowing they was there, I reckon, for he was thinking and thinking all the time, and never said a word and never et a bite.

By-and-by when it was stillest, that nigger's head was poked in at the door again, and he said his Marse Brace was getting powerful uneasy about Marse Jubiter, which hadn't come home yet, and would Marse Silas please—

He was looking at Uncle Silas, and he stopped there, like the rest of his words was froze; for Uncle Silas he rose up shaky and steadied himself leaning his fingers on the table, and he was panting, and his eyes was set on the nigger, and he kept swallowing, and put his other hand up to his throat a couple of times, and at last he got his words started, and says.

"Does he—does he—think—*what* does he think! Tell him—tell him—" Then he sunk down in his chair limp and weak, and says, so as you could hardly hear him: "Go away—go away!"

The nigger looked scared, and cleared out, and we all

"'WHAT DOES HE THINK?'"

felt—well, I don't know how we felt, but it was awful, with the old man panting there, and his eyes set and looking like a person that was dying. None of us could budge; but Benny she slid around soft, with her tears running down, and stood by his side, and nestled his old gray head up against her and begun to stroke it and pet it with her hands, and nodded to us to go away, and we done it, going out very quiet, like the dead was there.

Me and Tom struck out for the woods mighty solemn, and saying how different it was now to what it was last summer when we was here and everything was so peaceful and happy and everybody thought so much of Uncle Silas, and he was so cheerful and simple-hearted and pudd'n-headed and good—and now look at him. If he hadn't lost his mind he wasn't much short of it. That was what we allowed.

It was a most lovely day, now, and bright and sunshiny; and the further and further we went over the hill towards the prairie the lovelier and lovelier the trees and flowers got to be and the more it seemed strange and somehow wrong that there had to be trouble in such a world as this. And then all of a sudden I catched my breath and grabbed Tom's arm, and all my livers and lungs and things fell down into my legs.

"There it is!" I says. We jumped back behind a bush, shivering, and Tom says:

"'Sh!—don't make a noise."

It was setting on a log right in the edge of the little prairie, thinking. I tried to get Tom to come away, but he wouldn't, and I dasn't budge by myself. He said we mightn't ever get another chance to see one, and he was going to look his fill at this one if he died for it. So I looked too, though it give me the fan-tods to do it. Tom he *had* to talk, but he talked low. He says:

"Poor Jakey, it's got all its things on, just as he said he would. *Now* you see what we wasn't certain about—its hair. It's not long, now, the way it was; it's got it cropped close to its head, the way he said he would. Huck, I never see anything look any more naturaler than what It does."

"Nor I neither," I says; "I'd recognize it anywheres."

"So would I. It looks perfectly solid and genuwyne, just the way it done before it died."

So we kept a-gazing. Pretty soon Tom says:

"Huck, there's something mighty curious about this one, don't you know? *It* oughtn't to be going around in the daytime."

"That's so, Tom—I never heard the like of it before."

"No, sir, they don't ever come out only at night—and then not till after twelve. There's something wrong about this one, now you mark my words. I don't believe it's got any right to be around in the daytime. But don't it look natural! Jake was going to play deef and dumb here, so the neighbors wouldn't know his voice. Do you reckon it would do that if we was to holler at it?"

"Lordy, Tom, don't talk so! If you was to holler at it I'd die in my tracks."

"Don't you worry, I ain't going to holler at it. Look, Huck, it's a-scratching its head—don't you see?"

"Well, what of it?"

"Why, this. What's the sense of it scratching its head? There ain't anything there to itch; its head is made out of fog or something like that, and *can't* itch. A fog can't itch; any fool knows that."

"Well, then, if it don't itch and can't itch, what in the nation is it scratching it for? Ain't it just habit, don't you reckon?"

"No, sir, I don't. I ain't a bit satisfied about the way this one acts. I've a blame good notion it's a bogus one—

I have, as sure as I'm a-sitting here. Because, if it—Huck!"

"Well, what's the matter now?"

"*You can't see the bushes through it!*"

"Why, Tom, it's so, sure! It's as solid as a cow. I sort of begin to think—"

"Huck, it's biting off a chaw of tobacker! By George, *they* don't chaw—they hain't got anything to chaw *with*. Huck!"

"I'm a-listening."

"It ain't a ghost at all. It's Jake Dunlap his own self!"

"Oh, your granny!" I says.

"Huck Finn, did we find any corpse in the sycamores?"

"No."

"Or any sign of one?"

"No."

"Mighty good reason. Hadn't ever been any corpse there."

"Why, Tom, you know we heard—"

"Yes, we did—heard a howl or two. Does that prove anybody was killed? Course it don't. And we seen four men run, then this one come walking out and we took it for a ghost. No more ghost than you are. It was Jake Dunlap his own self, and it's Jake Dunlap now. He's been and got his hair cropped, the way he said he would, and he's playing himself for a stranger, just the same as he said he would. Ghost? Hum!—he's as sound as a nut."

Then I see it all, and how we had took too much for granted. I was powerful glad he didn't get killed, and so was Tom, and we wondered which he would like the best—for us to never let on to know him, or how? Tom reckoned the best way would be to go and ask him. So he started; but I kept a little behind, because I didn't know but it might be a ghost, after all. When Tom got to where he was, he says:

"Me and Huck's mighty glad to see you again, and you needn't be afeard we'll tell. And if you think it 'll be safer for you if we don't let on to know you when we run across you, say the word and you'll see you can depend on us, and would ruther cut our hands off than get you into the least little bit of danger."

First off he looked surprised to see us, and not very glad, either; but as Tom went on he looked pleasanter, and when he was done he smiled, and nodded his head several times, and made signs with his hands, and says:

"Goo-goo—goo-goo," the way deef and dummies does.

Just then we see some of Steve Nickerson's people coming that lived t'other side of the prairie, so Tom says:

"You do it elegant; I never see anybody do it better. You're right; play it on us, too; play it on us same as the others; it 'll keep you in practice and prevent you making blunders. We'll keep away from you and let on we don't know you, but any time we can be any help, you just let us know."

Then we loafed along past the Nickersons, and of course they asked if that was the new stranger yonder, and where'd he come from, and what was his name, and which communion was he, Babtis' or Methodis', and which politics, Whig or Democrat, and how long is he staying, and all them other questions that humans always asks when a stranger comes, and animals does too. But Tom said he warn't able to make anything out of deef and dumb signs, and the same with goo-gooing. Then we watched them go and bullyrag Jake; because we was pretty uneasy for him. Tom said it would take him days to get so he wouldn't forget he was a deef and dummy sometimes, and speak out before he thought. When we had watched long enough to

"'GOO-GOO—GOO-GOO'"

see that Jake was getting along all right and working his signs very good, we loafed along again, allowing to strike the school-house about recess time, which was a three-mile tramp.

I was so disappointed not to hear Jake tell about the row in the sycamores, and how near he come to getting killed, that I couldn't seem to get over it, and Tom he felt the same, but said if we was in Jake's fix we would want to go careful and keep still and not take any chances.

The boys and girls was all glad to see us again, and we had a real good time all through recess. Coming to school the Henderson boys had come across the new deef and dummy and told the rest; so all the scholars was chuck full of him and couldn't talk about anything else, and was in a sweat to get a sight of him because they hadn't ever seen a deef and dummy in their lives, and it made a powerful excitement.

Tom said it was tough to have to keep mum now; said we would be heroes if we could come out and tell all we knowed; but after all, it was still more heroic to keep mum, there warn't two boys in a million could do it. That was Tom Sawyer's idea about it, and I reckoned there warn't anybody could better it.

CHAPTER IX

In the next two or three days Dummy he got to be powerful popular. He went associating around with the neighbors, and they made much of him, and was proud to have such a rattling curiosity amongst them. They had him to breakfast, they had him to dinner, they had him to supper; they kept him loaded up with hog and hominy, and warn't ever tired staring at him and wondering over him, and wishing they knowed more about him, he was so uncommon and romantic. His signs warn't no good; people couldn't understand them and he prob'ly couldn't himself, but he done a sight of goo-gooing, and so everybody was satisfied, and admired to hear him go it. He toted a piece of slate around, and a pencil; and people wrote questions on it and he wrote answers; but there warn't anybody could read his writing but Brace Dunlap. Brace said he couldn't read it very good, but he could manage to dig out the meaning most of the time. He said Dummy said he belonged away off somers and used to be well off, but got busted by swindlers which he had trusted, and was poor now, and hadn't any way to make a living.

Everybody praised Brace Dunlap for being so good to that stranger. He let him have a little log-cabin all to himself, and had his niggers take care of it, and fetch him all the vittles he wanted.

Dummy was at our house some, because old Uncle Silas was so afflicted himself, these days, that anybody else that was afflicted was a comfort to him. Me and Tom didn't let on that we had knowed him before, and he didn't let on that he had knowed us before. The family talked their troubles out before him the same as if he wasn't there, but we reckoned it wasn't any harm for him to hear what they said. Generly he didn't seem to notice, but sometimes he did.

Well, two or three days went along, and everybody got to getting uneasy about Jubiter Dunlap. Everybody was asking everybody if they had any idea what had become of him. No, they hadn't, they said; and they shook their heads and said there was something powerful strange about it. Another and another day went by; then there was a report got around that praps he was murdered. You bet it made a big stir! Everybody's tongue was clacking away after that. Saturday two or three gangs turned out and hunted the woods to see if they could run across his remainders. Me and Tom helped, and it was noble good times and exciting. Tom he was so brim full of it he couldn't eat nor rest. He said if we could find that corpse we would be celebrated, and more talked about than if we got drownded.

The others got tired and give it up; but not Tom Sawyer—that warn't his style. Saturday night he didn't sleep any, hardly, trying to think up a plan; and towards daylight in the morning he struck it. He snaked me out of bed and was all excited, and says—

"Quick, Huck, snatch on your clothes—I've got it! Blood-hound!"

In two minutes we was tearing up the river road in the dark towards the village. Old Jeff Hooker had a blood-hound, and Tom was going to borrow him. I says—

"The trail's too old, Tom—and, besides, it's rained, you know."

"It don't make any difference, Huck. If the body's hid

in the woods anywhere around the hound will find it. If he's been murdered and buried, they wouldn't bury him deep, it ain't likely, and if the dog goes over the spot he'll scent him, sure. Huck, we're going to be celebrated, sure as you're born!"

He was just a-blazing; and whenever he got afire he was most likely to get afire all over. That was the way this time. In two minutes he had got it all ciphered out, and wasn't only just going to find the corpse—no, he was going to get on the track of that murderer and hunt *him* down, too; and not only that, but he was going to stick to him till—

"Well," I says, "you better find the corpse first; I reckon that's a-plenty for to-day. For all we know, there *ain't* any corpse and nobody hain't been murdered. That cuss could 'a' gone off somers and not been killed at all."

That gravelled him, and he says—

"Huck Finn, I never see such a person as you to want to spoil everything. As long as *you* can't see anything hopeful in a thing, you won't let anybody else. What good can it do you to throw cold water on that corpse and get up that selfish theory that there ain't been any murder? None in the world. I don't see how you can act so. I wouldn't treat you like that, and you know it. Here we've got a noble good opportunity to make a ruputation, and—"

"Oh, go ahead," I says; "I'm sorry, and I take it all back. I didn't mean nothing. Fix it any way you want it. *He* ain't any consequence to me. If he's killed, I'm as glad of it as you are; and if he—"

"I never said anything about being glad; I only—"

"Well, then, I'm as *sorry* as you are. Any way you druther have it, that is the way *I* druther have it. He—"

"There ain't any druthers *about* it, Huck Finn; nobody said anything about druthers. And as for—"

He forgot he was talking, and went tramping along, studying. He begun to get excited again, and pretty soon he says—

"Huck, it 'll be the bulliest thing that ever happened if we find the body after everybody else has quit looking, and then go ahead and hunt up the murderer. It won't only be an honor to us, but it 'll be an honor to Uncle Silas because it was us that done it. It 'll set him up again, you see if it don't."

But old Jeff Hooker he throwed cold water on the whole business when we got to his blacksmith-shop and told him what we come for.

"You can take the dog," he says, "but you ain't a-going to find any corpse, because there ain't any corpse to find. Everybody's quit looking, and they're right. Soon as they come to think, they knowed there warn't no corpse. And I'll tell you for why. What does a person kill another person *for*, Tom Sawyer?—answer me that."

"Why, he—er—"

"Answer up! You ain't no fool. What does he kill him *for?*"

"Well, sometimes it's for revenge, and—"

"Wait. One thing at a time. Revenge, says you; and right you are. Now who ever had anything agin that poor trifling no-account? Who do you reckon would want to kill *him?*—that rabbit!"

Tom was stuck. I reckon he hadn't thought of a person having to have a *reason* for killing a person before, and now he sees it warn't likely anybody would have that much of a grudge against a lamb like Jubiter Dunlap. The blacksmith says, by-and-by—

"The revenge idea won't work, you see. Well, then, what's next? Robbery? B'gosh, that must 'a' been it, Tom! Yes, sirree, I reckon we've struck it this time. Some feller wanted his gallus-buckles, and so he—"

But it was so funny he busted out laughing, and just went *on* laughing and laughing and laughing till he was 'most dead, and Tom looked so put out and cheap that I knowed he was ashamed he had come, and he wished he hadn't. But old Hooker never let up on him. He raked up everything a person ever could want to kill another person about, and any fool could see they didn't any of them fit this case, and he just made no end of fun of the whole business and of the people that had been hunting the body; and he said—

"If they'd had any sense they'd 'a' knowed the lazy cuss slid out because he wanted a loafing spell after all this work. He'll come pottering back in a couple of weeks, and then how 'll you fellers feel? But, laws bless you, take the dog, and go and hunt his remainders. Do, Tom."

Then he busted out, and had another of them forty-rod laughs of hisn. Tom couldn't back down after all this, so he said, "All right, unchain him;" and the blacksmith done it, and we started home and left that old man laughing yet.

It was a lovely dog. There ain't any dog that's got a lovelier disposition than a blood-hound, and this one knowed us and liked us. He capered and raced around ever so friendly, and powerful glad to be free and have a holiday; but Tom was so cut up he couldn't take any intrust in him, and said he wished he'd stopped and thought a minute before he ever started on such a fool errand. He said old Jeff Hooker would tell everybody, and we'd never hear the last of it.

So we loafed along home down the back lanes, feeling pretty glum and not talking. When we was passing the far corner of our tobacker-field we heard the dog set up a long howl in there, and we went to the place and he was scratching the ground with all his might, and every now and then canting up his head sideways and fetching another howl.

It was a long square, the shape of a grave; the rain had

"FETCHING ANOTHER HOWL"

made it sink down and show the shape. The minute we come and stood there we looked at one another and never said a word. When the dog had dug down only a few inches he grabbed something and pulled it up, and it was an arm and a sleeve. Tom kind of gasped out, and says—

"Come away, Huck—it's found."

I just felt awful. We struck for the road and fetched the first men that come along. They got a spade at the crib and dug out the body, and you never see such an excitement. You couldn't make anything out of the face, but you didn't need to. Everybody said—

"Poor Jubiter; it's his clothes, to the last rag!"

Some rushed off to spread the news and tell the justice of the peace and have an inquest, and me and Tom lit out for the house. Tom was all afire and 'most out of breath when we come tearing in where Uncle Silas and Aunt Sally and Benny was. Tom sung out—

"Me and Huck's found Jubiter Dunlap's corpse all by ourselves with a blood-hound, after everybody else had quit hunting and given it up; and if it hadn't a been for us it never *would* 'a' been found; and he *was* murdered too—they done it with a club or something like that; and I'm going to start in and find the murderer, next, and I bet I'll do it!"

Aunt Sally and Benny sprung up pale and astonished, but Uncle Silas fell right forward out of his chair onto the floor and groans out—

"Oh, my God, you've found him *now!*"

CHAPTER X

THEM awful words froze us solid. We couldn't move hand or foot for as much as half a minute. Then we kind of come to, and lifted the old man up and got him into his chair, and Benny petted him and kissed him and tried to comfort him, and poor old Aunt Sally she done the same; but, poor things, they was so broke up and scared and knocked out of their right minds that they didn't hardly know what they was about. With Tom it was awful; it 'most petrified him to think maybe he had got his uncle into a thousand times more trouble than ever, and maybe it wouldn't ever happened if he hadn't been so ambitious to get celebrated, and let the corpse alone the way the others done. But pretty soon he sort of come to himself again and says—

"Uncle Silas, don't you say another word like that. It's dangerous, and there ain't a shadder of truth in it."

Aunt Sally and Benny was thankful to hear him say that, and they said the same; but the old man he wagged his head sorrowful and hopeless, and the tears run down his face, and he says—

"No—I done it; poor Jubiter, I done it!"

It was dreadful to hear him say it. Then he went on and told about it, and said it happened the day me and Tom come—along about sundown. He said Jubiter pestered him and aggravated him till he was so mad he just sort of lost his mind and grabbed up a stick and hit him over the head with all his might, and Jubiter dropped in his tracks. Then he was scared and sorry, and got down on his knees and lifted his head up, and begged him to speak and say he wasn't dead; and before long he come to, and when he see who it was holding his head, he jumped like he was 'most scared to death, and cleared the fence and tore into the woods, and was gone. So he hoped he wasn't hurt bad.

"But laws," he says, "it was only just fear that gave him that last little spurt of strength, and of course it soon played out and he laid down in the bush, and there wasn't anybody to help him, and he died."

Then the old man cried and grieved, and said he was a murderer and the mark of Cain was on him, and he had disgraced his family and was going to be found out and hung. But Tom said—

"No, you ain't going to be found out. You *didn't* kill him. *One* lick wouldn't kill him. Somebody else done it."

"Oh yes," he says, "I done it—nobody else. Who else had anything against him? Who else *could* have anything against him?"

He looked up kind of like he hoped some of us could mention somebody that could have a grudge against that harmless no-account, but of course it warn't no use—he *had* us; we couldn't say a word. He noticed that, and he saddened down again, and I never see a face so miserable and so pitiful to see. Tom had a sudden idea, and says—

"But hold on!—somebody *buried* him. Now who—"

He shut off sudden. I knowed the reason. It give me the cold shudders when he said them words, because right away I remembered about us seeing Uncle Silas prowling around with a long-handled shovel away in the night that night. And I knowed Benny seen him, too, because she was talking about it one day. The minute Tom shut off he changed the subject and went to begging Uncle Silas to keep mum, and the rest of us done the same, and said he *must*, and said it wasn't his business to tell on himself, and if he kept mum nobody would ever know; but if it was found out and any harm come to him it would break the family's hearts and kill them, and yet never do anybody any good. So at last he promised. We was all of us more comfortable, then, and went to work to cheer up the old man. We told him all he'd got to do was to keep still, and it wouldn't be long till the whole thing would blow over and be forgot. We all said there wouldn't anybody ever suspect Uncle Silas, nor ever dream of such a thing, he being so good and kind, and having such a good character; and Tom says, cordial and hearty, he says—

"Why, just look at it a minute; just consider. Here is Uncle Silas, all these years a preacher — at his own expense; all these years doing good with all his might and every way he can think of — at his own expense, all the time; always been loved by everybody, and respected; always been peaceable and minding his own business, the very last man in this whole deestrict to touch a person, and everybody knows it. Suspect *him?* Why, it ain't any more possible than—"

"By authority of the State of Arkansaw, I arrest you for the murder of Jubiter Dunlap!" shouts the sheriff at the door.

It was awful. Aunt Sally and Benny flung themselves at Uncle Silas, screaming and crying, and hugged him and hung to him, and Aunt Sally said go away, she wouldn't ever give him up, they shouldn't have him, and the niggers they come crowding and crying to the door and — well, I couldn't stand it; it was enough to break a person's heart; so I got out.

They took him up to the little one-horse jail in the village, and we all went along to tell him good-by; and Tom was feeling elegant, and says to me, "We'll have a most noble good time and heaps of danger some dark night getting him out of there, Huck, and it 'll be talked about everywheres and we will be celebrated;" but the old man busted that scheme up the minute he whispered to him about it. He said no, it was his duty to stand whatever the law done to him, and he would stick to the jail plumb through to the end, even if there warn't no door to it. It disappointed Tom and gravelled him a good deal, but he had to put up with it.

But he felt responsible and bound to get his uncle Silas free; and he told Aunt Sally, the last thing, not to worry, because he was going to turn in and work night and day and beat this game and fetch Uncle Silas out innocent; and she was very loving to him and thanked him and said she knowed he would do his very best. And she told us to help Benny take care of the house and the children, and then we had a good-by cry all around and went back to the farm, and left her there to live with the jailer's wife a month till the trial in October.

CHAPTER XI

WELL, that was a hard month on us all. Poor Benny, she kept up the best she could, and me and Tom tried to keep things cheerful there at the house, but it kind of went for nothing, as you may say. It was the same up at the jail. We went up every day to see the old people, but it was awful dreary, because the old man warn't sleeping much, and was walking in his sleep considerable, and so he got to looking fagged and miserable, and his mind got shaky, and we all got afraid his troubles would break him down and kill him. And whenever we tried to persuade him to feel cheerfuler, he only shook his head and said if we only knowed what it was to carry around a murderer's load on your heart we wouldn't talk that way. Tom and all of us kept telling him it *wasn't* murder, but just accidental killing, but it never made any difference—it was murder, and he wouldn't have it any other way. He actu'ly begun to come out plain and square towards trial-time and acknowledge that he *tried* to kill the man. Why, that was awful, you know. It made things seem fifty times as dreadful, and there warn't no more comfort for Aunt Sally and Benny. But he promised he wouldn't say a word about his murder when others was around, and we was glad of that.

Tom Sawyer racked the head off of himself all that month trying to plan some way out for Uncle Silas, and many's the night he kept me up 'most all night with this kind of tiresome work, but he couldn't seem to get on the

"'KEPT ME UP 'MOST ALL NIGHT'"

right track no way. As for me, I reckoned a body might as well give it up, it all looked so blue and I was so downhearted; but he wouldn't. He stuck to the business right along, and went on planning and thinking and ransacking his head.

So at last the trial come on, towards the middle of October, and we was all in the court. The place was jammed of course. Poor old Uncle Silas, he looked more like a dead person than a live one, his eyes was so hollow and he looked so thin and so mournful. Benny she set on one side of him and Aunt Sally on the other, and they had veils on, and was full of trouble. But Tom he set by our lawyer, and had his finger in everywheres, of course. The lawyer let him, and the judge let him. He 'most took the business out of the lawyer's hands sometimes; which was well enough, because that was only a mud-turtle of a back-settlement lawyer and didn't know enough to come in when it rains, as the saying is.

They swore in the jury, and then the lawyer for the prostitution got up and begun. He made a terrible speech against the old man, that made him moan and groan, and made Benny and Aunt Sally cry. The way *he* told about the murder kind of knocked us all stupid it was so different from the old man's tale. He said he was going to prove that Uncle Silas was *seen* to kill Jubiter Dunlap by two good witnesses, and done it deliberate, and *said* he was going to kill him the very minute he hit him with the club; and they seen him hide Jubiter in the bushes, and they seen that Jubiter was stone-dead. And said Uncle Silas come later and lugged Jubiter down into the tobacker-field, and two men seen him do it. And said Uncle Silas turned out, away in the night, and buried Jubiter, and a man seen him at it.

I says to myself, poor old Uncle Silas has been lying

about it because he reckoned nobody seen him and he couldn't bear to break Aunt Sally's heart and Benny's; and right he was: as for me, I would 'a' lied the same way, and so would anybody that had any feeling, to save them such misery and sorrow which *they* warn't no ways responsible for. Well, it made our lawyer look pretty sick; and it knocked Tom silly, too, for a little spell, but then he braced up and let on that he warn't worried—but I knowed he *was*, all the same. And the people—my, but it made a stir amongst them!

And when that lawyer was done telling the jury what he was going to prove, he set down and begun to work his witnesses.

First, he called a lot of them to show that there was bad blood betwixt Uncle Silas and the diseased; and they told how they had heard Uncle Silas threaten the diseased, at one time and another, and how it got worse and worse and everybody was talking about it, and how diseased got afraid of his life, and told two or three of them he was certain Uncle Silas would up and kill him some time or another.

Tom and our lawyer asked them some questions; but it warn't no use, they stuck to what they said.

Next, they called up Lem Beebe, and he took the stand. It come into my mind, then, how Lem and Jim Lane had come along talking, that time, about borrowing a dog or something from Jubiter Dunlap; and that brought up the blackberries and the lantern; and that brought up Bill and Jack Withers, and how *they* passed by, talking about a nigger stealing Uncle Silas's corn; and that fetched up our old ghost that come along about the same time and scared us so—and here *he* was too, and a privileged character, on accounts of his being deef and dumb and a stranger, and they had fixed him a chair inside the railing, where he

OUR LAWYER

could cross his legs and be comfortable, whilst the other people was all in a jam so they couldn't hardly breathe. So it all come back to me just the way it was that day; and it made me mournful to think how pleasant it was up to then, and how miserable ever since.

Lem Beebe, sworn, said: "I was a-coming along, that day, second of September, and Jim Lane was with me, and it was towards sundown, and we heard loud talk, like quarrelling, and we was very close, only the hazel bushes between (that's along the fence); and we heard a voice say, 'I've told you more'n once I'd kill you,' and knowed it was this prisoner's voice; and then we see a club come up above the bushes and down out of sight again, and heard a smashing thump and then a groan or two; and then we crope soft to where we could see, and there laid Jubiter Dunlap dead, and this prisoner standing over him with the club; and the next he hauled the dead man into a clump of bushes and hid him, and then we stooped low, to be out of sight, and got away."

Well, it was awful. It kind of froze everybody's blood to hear it, and the house was 'most as still whilst he was telling it as if there warn't nobody in it. And when he was done, you could hear them gasp and sigh, all over the house, and look at one another the same as to say, "Ain't it perfectly terrible—ain't it awful!"

Now happened a thing that astonished me. All the time the first witnesses was proving the bad blood and the threats and all that, Tom Sawyer was alive and laying for them; and the minute they was through, he went for them, and done his level best to catch them in lies and spile their testimony. But now, how different. When Lem first begun to talk, and never said anything about speaking to Jubiter or trying to borrow a dog off of him, he was all alive and laying for Lem, and you could see he was getting ready to cross-question him to death pretty soon, and then I judged him and me would go on the stand by-and-by and tell what we heard him and Jim Lane say. But the next time I looked at Tom I got the cold shivers. Why, he was in the brownest study you ever see—miles and miles away. He warn't hearing a word Lem Beebe was saying; and when he got through he was still in that brown-study, just the same. Our lawyer joggled him, and then he looked up startled, and says, "Take the witness if you want him. Lemme alone—I want to think."

Well, that beat me. I couldn't understand it. And Benny and her mother—oh, they looked sick, they was so troubled. They shoved their veils to one side and tried to get his eye, but it warn't any use, and I couldn't get his eye either. So the mud-turtle he tackled the witness, but it didn't amount to nothing; and he made a mess of it.

Then they called up Jim Lane, and he told the very same story over again, exact. Tom never listened to this one at all, but set there thinking and thinking, miles and miles away. So the mud-turtle went in alone again and come out just as flat as he done before. The lawyer for the prostitution looked very comfortable, but the judge looked disgusted. You see, Tom was just the same as a regular lawyer, nearly, because it was Arkansaw law for a prisoner to choose anybody he wanted to help his lawyer, and Tom had had Uncle Silas shove him into the case, and now he was botching it and you could see the judge didn't like it much.

All that the mud-turtle got out of Lem and Jim was this: he asked them—

"Why didn't you go and tell what you saw?"

"We was afraid we would get mixed up in it ourselves. And we was just starting down the river a-hunting for all the week besides; but as soon as we come back we found out they'd been searching for the body, so then we went and told Brace Dunlap all about it."

"When was that?"

"Saturday night, September 9th."

The judge he spoke up and says—

"Mr. Sheriff, arrest these two witnesses on suspicions of being accessionary after the fact to the murder."

The lawyer for the prostitution jumps up all excited, and says—

"Your honor! I protest against this extraordi—"

"Set down!" says the judge, pulling his bowie and laying it on his pulpit. "I beg you to respect the Court."

So he done it. Then he called Bill Withers.

Bill Withers, sworn, said: "I was coming along about sundown, Saturday, September 2d, by the prisoner's field, and my brother Jack was with me, and we seen a man toting off something heavy on his back and allowed it was a nigger stealing corn; we couldn't see distinct; next we made out that it was one man carrying another; and the way it hung, so kind of limp, we judged it was somebody that was drunk; and by the man's walk we said it was Parson Silas, and we judged he had found Sam Cooper drunk in the road, which he was always trying to reform him, and was toting him out of danger."

It made the people shiver to think of poor old Uncle Silas toting off the diseased down to the place in his tobacker-field where the dog dug up the body, but there warn't much sympathy around amongst the faces, and I heard one cuss say, "'Tis the coldest-blooded work I ever struck, lugging a murdered man around like that, and going to bury him like a animal, and him a preacher at that."

Tom he went on thinking, and never took no notice; so our lawyer took the witness and done the best he could, and it was plenty poor enough.

Then Jack Withers he come on the stand and told the same tale, just like Bill done.

And after him comes Brace Dunlap, and he was looking very mournful, and most crying; and there was a rustle and a stir all around, and everybody got ready to listen, and lots of the women folks said, "Poor cretur, poor cretur," and you could see a many of them wiping their eyes.

Brace Dunlap, sworn, said: "I was in considerable trouble a long time about my poor brother, but I reckoned things warn't near so bad as he made out, and I couldn't make myself believe anybody would have the heart to hurt a poor harmless cretur like that"—[by jings, I was sure I seen Tom give a kind of a faint little start, and then look disappointed again]—"and you know I *couldn't* think a preacher would hurt him—it warn't natural to think such an onlikely thing—so I never paid much attention, and now I sha'n't ever, ever forgive myself; for if I had a done different, my poor brother would be with me this day, and not laying yonder murdered, and him so harmless." He kind of broke down there and choked up, and waited to get his voice; and people all around said the most pitiful things, and women cried; and it was very still in there, and solemn, and old Uncle Silas, poor thing, he give a groan right out so everybody heard him. Then Brace he went on, "Saturday, September 2d, he didn't come home to supper. By-and-by I got a little uneasy, and one of my niggers went over to this prisoner's place, but come back and said he warn't there. So I got uneasier and uneasier, and couldn't rest. I went to bed, but I couldn't sleep; and turned out, away late in the night, and went wandering over to this prisoner's place and all around about there a good while, hoping I would run across my poor brother, and never knowing he was out of his troubles and gone to a better shore—" So he broke down and choked up again, and most all the women was crying now. Pretty soon he got another start and says: "But it warn't no use; so at last I went home and tried to get some sleep, but couldn't. Well, in a day or two everybody was uneasy, and they got to talking about this prisoner's threats, and took to the idea, which I didn't take no stock in, that my brother was murdered; so they hunted around and tried to find his body, but couldn't and give it up. And so I reckoned he was gone off somers to have a little peace, and would come back to us when his troubles was kind of healed. But late Saturday night, the 9th, Lem Beebe and Jim Lane come to my house and told me all—told me the whole awful 'sassination, and my heart was broke. And *then* I remembered something that hadn't took no hold of me at the time, because reports said this prisoner had took to walking in his sleep and doing all kind of things of

"'SET DOWN!' SAYS THE JUDGE"

no consequence, not knowing what he was about. I will tell you what that thing was that come back into my memory. Away late that awful Saturday night when I was wandering around about this prisoner's place, grieving and troubled, I was down by the corner of the tobacker-field and I heard a sound like digging in a gritty soil; and I crope nearer and peeped through the vines that hung on the rail fence and seen this prisoner *shovelling*—shovelling with a long-handled shovel—heaving earth into a big hole that was most filled up; his back was to me, but it was bright moonlight and I knowed him by his old green baize work-gown with a splattery white patch in the middle of the back like somebody had hit him with a snowball. *He was burying the man he'd murdered!*"

And he slumped down in his chair crying and sobbing, and 'most everybody in the house busted out wailing, and crying, and saying, "Oh, it's awful—awful—horrible!" and there was a most tremenduous excitement, and you couldn't hear yourself think; and right in the midst of it up jumps old Uncle Silas, white as a sheet, and sings out—

"*It's true, every word—I murdered him in cold blood!*"

By Jackson, it petrified them! People rose up wild all over the house, straining and staring for a better look at him, and the judge was hammering with his mallet and the sheriff yelling "Order—order in the court—order!"

And all the while the old man stood there a-quaking and his eyes a-burning, and not looking at his wife and daughter, which was clinging to him and begging him to keep still, but pawing them off with his hands and saying he *would* clear his black soul from crime, he *would* heave off this load that was more than he could bear, and he *wouldn't* bear it another hour! And then he raged right along with his awful tale, everybody a-staring and gasping, judge, jury, lawyers, and everybody, and Benny and Aunt Sally crying their hearts out. And by George, Tom Sawyer never looked at him once! Never once—just set there gazing with all his eyes at something else, I couldn't tell what. And so

12

the old man raged right along, pouring his words out like a stream of fire:

"I killed him! I am guilty! But I never had the notion in my life to hurt him or harm him, spite of all them lies about my threatening him, till the very minute I raised the club—then my heart went cold!—then the pity all went out of it, and I struck to kill! In that one moment all my wrongs come into my mind; all the insults that that man and the scoundrel his brother, there, had put upon me, and how they laid in together to ruin me with the people, and take away my good name, and *drive* me to some deed that would destroy me and my family that hadn't ever done *them* no harm, so help me God! And they done it in a mean revenge—for why? Because my innocent pure girl here at my side wouldn't marry that rich, insolent, ignorant coward, Brace Dunlap, who's been snivelling here over a brother he never cared a brass farthing for"—[I see Tom give a jump and look glad *this* time, to a dead certainty]—"and in that moment I've told you about, I forgot my God and remembered only my heart's bitterness, God forgive me, and I struck to kill. In one second I was miserably sorry—oh, filled with remorse; but I thought of my poor family, and I *must* hide what I'd done for their sakes; and I did hide that corpse in the bushes; and presently I carried it to the tobacker-field; and in the deep night I went with my shovel and buried it where—"

Up jumps Tom and shouts—

"*Now*, I've got it!" and waves his hand, oh, ever so fine and starchy, towards the old man, and says—

"Set down! A murder *was* done, but you never had no hand in it!"

Well, sir, you could a heard a pin drop. And the old man he sunk down kind of bewildered in his seat and Aunt Sally and Benny didn't know it, because they was so astonished

"'A MURDER WAS DONE'"

and staring at Tom with their mouths open and not knowing what they was about. And the whole house the same. *I* never seen people look so helpless and tangled up, and I hain't ever seen eyes bug out and gaze without a blink the way theirn did. Tom says, perfectly ca'm—

"Your honor, may I speak?"

"For God's sake, yes—go on!" says the judge, so astonished and mixed up he didn't know what he was about hardly.

Then Tom he stood there and waited a second or two—that was for to work up an "effect," as he calls it—then he started in just as ca'm as ever, and says:

"For about two weeks, now, there's been a little bill sticking on the front of this court-house offering two thousand dollars reward for a couple of big di'monds — stole at St. Louis. Them di'monds is worth twelve thousand dollars. But never mind about that till I get to it. Now about this murder. I will tell you all about it—how it happened—who done it—every *de*tail."

You could see everybody nestle, now, and begin to listen for all they was worth.

"This man here, Brace Dunlap, that's been snivelling so about his dead brother that *you* know he never cared a straw for, wanted to marry that young girl there, and she wouldn't have him. So he told Uncle Silas he would make him sorry. Uncle Silas knowed how powerful he was, and how little chance he had against such a man, and he was scared and worried, and done everything he could think of to smooth him over and get him to be good to him: he even took his no-account brother Jubiter on the farm and give him wages and stinted his own family to pay them; and Jubiter done everything his brother could contrive to insult Uncle Silas, and fret and worry him, and try to drive Uncle Silas into doing him a hurt, so as to injure Uncle Silas with the people.

And it done it. Everybody turned against him and said the meanest kind of things about him, and it graduly broke his heart — yes, and he was so worried and distressed that often he warn't hardly in his right mind.

"Well, on that Saturday that we've had so much trouble about, two of these witnesses here, Lem Beebe and Jim Lane, come along by where Uncle Silas and Jubiter Dunlap was at work—and that much of what they've said is true, the rest is lies. They didn't hear Uncle Silas say he would kill Jubiter; they didn't hear no blow struck; they didn't see no dead man, and they didn't see Uncle Silas hide anything in the bushes. Look at them now—how they set there, wishing they hadn't been so handy with their tongues; anyway, they'll wish it before I get done.

"That same Saturday evening Bill and Jack Withers *did* see one man lugging off another one. That much of what they said is true, and the rest is lies. First off they thought it was a nigger stealing Uncle Silas's corn — you notice it makes them look silly, now, to find out somebody overheard them say that. That's because they found out by-and-by who it was that was doing the lugging, and *they* know best why they swore here that they took it for Uncle Silas by the gait — which it *wasn't*, and they knowed it when they swore to that lie.

"A man out in the moonlight *did* see a murdered person put under ground in the tobacker-field—but it wasn't Uncle Silas that done the burying. He was in his bed at that very time.

"Now, then, before I go on, I want to ask you if you've ever noticed this: that people, when they're thinking deep, or when they're worried, are most always doing something with their hands, and they don't know it, and don't notice what it is their hands are doing. Some stroke their chins; some stroke their noses; some stroke up *under* their chin

"'I STRUCK TO KILL'"

with their hand; some twirl a chain, some fumble a button, then there's some that draws a figure or a letter with their finger on their cheek, or under their chin or on their under lip. That's *my* way. When I'm restless, or worried, or thinking hard, I draw capital V's on my cheek or on my under lip or under my chin, and never anything *but* capital V's—and half the time I don't notice it and don't know I'm doing it."

That was odd. That is just what I do; only I make an O. And I could see people nodding to one another, same as they do when they mean "*that's* so."

"Now then, I'll go on. That same Saturday—no, it was the night before—there was a steamboat laying at Flagler's Landing, forty miles above here, and it was raining and storming like the nation. And there was a thief aboard, and he had them two big di'monds that's advertised out here on this court-house door; and he slipped ashore with his hand-bag and struck out into the dark and the storm, and he was a-hoping he could get to this town all right and be safe. But he had two pals aboard the boat, hiding, and he knowed they was going to kill him the first chance they got and take the di'monds; because all three stole them, and then this fellow he got hold of them and skipped.

"Well, he hadn't been gone more'n ten minutes before his pals found it out, and they jumped ashore and lit out after him. Prob'ly they burnt matches and found his tracks. Anyway, they dogged along after him all day Saturday and kept out of his sight; and towards sundown he come to the bunch of sycamores down by Uncle Silas's field, and he went in there to get a disguise out of his hand-bag and put it on before he showed himself here in the town — and mind you he done that just a little after the time that Uncle Silas was hitting Jubiter Dunlap over the head with a club—for he *did* hit him.

"But the minute the pals see that thief slide into the bunch of sycamores, they jumped out of the bushes and slid in after him.

"They fell on him and clubbed him to death.

"Yes, for all he screamed and howled so, they never had no mercy on him, but clubbed him to death. And two men that was running along the road heard him yelling that way, and they made a rush into the sycamore bunch—which was where they was bound for, anyway—and when the pals saw them they lit out and the two new men after them a-chasing them as tight as they could go. But only a minute or two—then these two new men slipped back very quiet into the sycamores.

"*Then* what did they do? I will tell you what they done. They found where the thief had got his disguise out of his carpet-sack to put on; so one of them strips and puts on that disguise."

Tom waited a litttle here, for some more "effect"—then he says, very deliberate—

"The man that put on that dead man's disguise was—*Jubiter Dunlap!*"

"Great Scott!" everybody shouted, all over the house, and old Uncle Silas he looked perfectly astonished.

"Yes, it was Jubiter Dunlap. Not dead, you see. Then they pulled off the dead man's boots and put Jubiter Dunlap's old ragged shoes on the corpse and put the corpse's boots on Jubiter Dunlap. Then Jubiter Dunlap stayed where he was, and the other man lugged the dead body off in the twilight; and after midnight he went to Uncle Silas's house, and took his old green work-robe off of the peg where it always hangs in the passage betwixt the house and the kitchen and put it on, and stole the long-handled shovel and went off down into the tobacker-field and buried the murdered man."

"AND THERE WAS THE MURDERED MAN"

He stopped, and stood a half a minute. Then—

"And who do you reckon the murdered man *was?* It was—*Jake* Dunlap, the long-lost burglar!"

"Great Scott!"

"And the man that buried him was—*Brace* Dunlap, his brother!"

"Great Scott!"

"And who do you reckon is this mowing idiot here that's letting on all these weeks to be a deef and dumb stranger? It's—*Jubiter* Dunlap!"

My land, they all busted out in a howl, and you never see the like of that excitement since the day you was born. And Tom he made a jump for Jupiter and snaked off his goggles and his false whiskers, and there was the murdered man, sure enough, just as alive as anybody! And Aunt Sally and Benny they went to hugging and crying and kissing and smothering old Uncle Silas to that degree he was more muddled and confused and mushed up in his mind than he ever was before, and that is saying considerable. And next, people begun to yell—

"Tom Sawyer! Tom Sawyer! Shut up everybody, and let him go on! Go on, Tom Sawyer!"

Which made him feel uncommon bully, for it was nuts for Tom Sawyer to be a public character thataway, and a hero, as he calls it. So when it was all quiet, he says—

"There ain't much left, only this. When that man there, Brace Dunlap, had most worried the life and sense out of Uncle Silas till at last he plum lost his mind and hit this other blatherskite his brother with a club, I reckon he seen his chance. Jubiter broke for the woods to hide, and I reckon the game was for him to slide out, in the night, and leave the country. Then Brace would make everybody believe Uncle Silas killed him and hid his body somers; and that would ruin Uncle Silas and drive *him* out of the country—hang him, maybe; I dunno. But when they found their dead brother in the sycamores without knowing him, because he was so battered up, they see they had a better thing; disguise *both* and bury Jake and dig him up presently all dressed up in Jubiter's clothes, and hire Jim Lane and Bill Withers and the others to swear to some handy lies—which they done. And there they set, now, and I told them they would be looking sick before I got done, and that is the way they're looking now.

"Well, me and Huck Finn here, we come down on the boat with the thieves, and the dead one told us all about the di'monds, and said the others would murder him if they got the chance; and we was going to help him all we could. We was bound for the sycamores when we heard them killing him in there; but we was in there in the early morning after the storm and allowed nobody hadn't been killed, after all. And when we see Jubiter Dunlap here spreading around in the very same disguise Jake told us *he* was going to wear, we thought it was Jake his own self — and he was goo-gooing deef and dumb, and *that* was according to agreement.

"Well, me and Huck went on hunting for the corpse after the others quit, and we found it. And was proud, too; but Uncle Silas he knocked us crazy by telling us *he* killed the man. So we was mighty sorry we found the body, and was bound to save Uncle Silas's neck if we could; and it was going to be tough work, too, because he wouldn't let us break him out of prison the way we done with our old nigger Jim.

"I done everything I could the whole month to think up some way to save Uncle Silas, but I couldn't strike a thing. So when we come into court to-day I come empty, and couldn't see no chance anywheres. But by-and-by I had a glimpse of something that set me thinking—just a little wee

"WHICH MADE HIM FEEL UNCOMMON BULLY"

glimpse—only that, and not enough to make sure; but it set me thinking hard—and *watching*, when I was only letting on to think; and by-and-by, sure enough, when Uncle Silas was piling out that stuff about *him* killing Jubiter Dunlap, I catched that glimpse again, and this time I jumped up and shut down the proceedings, because I *knowed* Jubiter Dunlap was a-setting here before me. I knowed him by a thing which I seen him do—and I remembered it. I'd seen him do it when I was here a year ago."

He stopped then, and studied a minute—laying for an "effect"—I knowed it perfectly well. Then he turned off like he was going to leave the platform, and says, kind of lazy and indifferent—

"Well, I believe that is all."

Why, you never heard such a howl!—and it come from the whole house:

"What *was* it you seen him do? Stay where you are, you little devil! You think you are going to work a body up till his mouth's a-watering and stop there? What *was* it he done?"

That was it, you see—he just done it to get an "effect"; you couldn't 'a' pulled him off of that platform with a yoke of oxen.

"Oh, it wasn't anything much," he says. "I seen him looking a little excited when he found Uncle Silas was actuly fixing to hang himself for a murder that warn't ever done; and he got more and more nervous and worried, I a-watching him sharp but not seeming to look at him—and all of a sudden his hands begun to work and fidget, and pretty soon his left crept up and *his finger drawed a cross on his cheek*, and then I *had* him!"

Well, then they ripped and howled and stomped and clapped their hands till Tom Sawyer was that proud and happy he didn't know what to do with himself. And then the judge he looked down over his pulpit and says—

"My boy, did you *see* all the various details of this strange conspiracy and tragedy that you've been describing?"

"No, your honor, I didn't see any of them."

"Didn't see any of them! Why, you've told the whole history straight through, just the same as if you'd seen it with your eyes. How did you manage that?"

Tom says, kind of easy and comfortable—

"Oh, just noticing the evidence and piecing this and that together, your honor; just an ordinary little bit of detective work; anybody could 'a' done it."

"Nothing of the kind! Not two in a million could 'a' done it. You are a very remarkable boy."

Then they let go and give Tom another smashing round, and he—well, he wouldn't 'a' sold out for a silver mine. Then the judge says—

"But are you certain you've got this curious history straight?"

"Perfectly, your honor. Here is Brace Dunlap—let him deny his share of it if he wants to take the chance; I'll engage to make him wish he hadn't said anything. . . . Well, you see *he's* pretty quiet. And his brother's pretty quiet, and them four witnesses that lied so and got paid for it, they're pretty quiet. And as for Uncle Silas, it ain't any use for him to put in his oar, I wouldn't believe him under oath!"

Well, sir, that fairly made them shout; and even the judge he let go and laughed. Tom he was just feeling like a rainbow. When they was done laughing he looks up at the judge and says—

"Your honor, there's a thief in this house."

"A thief?"

"Yes, sir. And he's got them twelve-thousand-dollar di'monds on him."

By gracious, but it made a stir! Everybody went shouting—

"Which is him? which is him? p'int him out!"

And the judge says—

"Point him out, my lad. Sheriff, you will arrest him. Which one is it?"

Tom says—

"This late dead man here—Jubiter Dunlap."

Then there was another thundering let-go of astonishment and excitement; but Jubiter, which was astonished enough before, was just fairly putrefied with astonishment this time. And he spoke up, about half crying, and says—

"Now *that's* a lie! Your honor, it ain't fair; I'm plenty bad enough without that. I done the other things—Brace he put me up to it, and persuaded me, and promised he'd make me rich, some day, and I done it, and I'm sorry I done it, and I wisht I hadn't; but I hain't stole no di'monds, and I hain't *got* no di'monds; I wisht I may never stir if it ain't so. The sheriff can search me and see."

Tom says—

"Your honor, it wasn't right to call him a thief, and I'll let up on that a little. He did steal the di'monds, but he didn't know it. He stole them from his brother Jake when he was laying dead, after Jake had stole them from the other thieves; but Jubiter didn't know he was stealing them; and he's been swelling around here with them a month; yes, sir, twelve thousand dollars' worth of di'monds on him—all that riches, and going around here every day just like a poor man. Yes, your honor, he's got them on him now."

The judge spoke up and says—

"Search him, sheriff."

Well, sir, the sheriff he ransacked him high and low, and everywhere: searched his hat, socks, seams, boots, everything—and Tom he stood there quiet, laying for another of them effects of hisn. Finally the sheriff he give it up, and everybody looked disappointed, and Jubiter says—

"There, now! what'd I tell you?"

And the judge says—

"It appears you were mistaken this time, my boy."

Then Tom he took an attitude and let on to be studying with all his might, and scratching his head. Then all of a sudden he glanced up chipper, and says—

"Oh, now I've got it! I'd forgot."

Which was a lie, and I knowed it. Then he says—

"Will somebody be good enough to lend me a little small screw-driver? There was one in your brother's hand-bag that you smouched, Jubiter, but I reckon you didn't fetch it with you."

"No, I didn't. I didn't want it, and I give it away."

"That was because you didn't know what it was for."

Jubiter had his boots on again, by now, and when the thing Tom wanted was passed over the people's heads till it got to him, he says to Jubiter—

"Put up your foot on this chair." And he kneeled down and begun to unscrew the heel-plate, everybody watching; and when he got that big di'mond out of that boot-heel and held it up and let it flash and blaze and squirt sunlight everwhichaway, it just took everybody's breath; and Jubiter he looked so sick and sorry you never see the like of it. And when Tom held up the other di'mond he looked sorrier than ever. Land! he was thinking how he would 'a' skipped out and been rich and independent in a foreign land if he'd

"'TOM, GIVE HALF OF IT TO ME'"

only had the luck to guess what the screw-driver was in the carpet-bag for.

Well, it was a most exciting time, take it all around, and Tom got cords of glory. The judge took the di'monds, and stood up in his pulpit, and cleared his throat, and shoved his spectacles back on his head, and says—

"I'll keep them and notify the owners; and when they send for them it will be a real pleasure to me to hand you the two thousand dollars, for you've earned the money—yes, and you've earned the deepest and most sincerest thanks of this community besides, for lifting a wronged and innocent family out of ruin and shame, and saving a good and honorable man from a felon's death, and for exposing to infamy and the punishment of the law a cruel and odious scoundrel and his miserable creatures!"

Well, sir, if there'd been a brass band to bust out some music, then, it would 'a' been just the perfectest thing I ever see, and Tom Sawyer he said the same.

Then the sheriff he nabbed Brace Dunlap and his crowd, and by-and-by next month the judge had them up for trial and jailed the whole lot. And everybody crowded back to Uncle Silas's little old church, and was ever so loving and kind to him and the family and couldn't do enough for them; and Uncle Silas he preached them the blamedest jumbledest idiotic sermons you ever struck, and would tangle you up so you couldn't find your way home in daylight; but the people never let on but what they thought it was the clearest and brightest and elegantest sermons that ever was; and they would set there and cry, for love and pity; but, by George, they give me the jim-jams and the fan-tods and caked up what brains I had, and turned them solid; but by-and-by they loved the old man's intellects back into him again and he was as sound in his skull as ever he was, which ain't no flattery, I reckon. And so the whole family was as happy as birds, and nobody could be gratefuler and lovinger than what they was to Tom Sawyer; and the same to me, though I hadn't done nothing. And when the two thousand dollars come, Tom give half of it to me, and never told anybody so, which didn't surprise me, because I knowed him.

A DOUBLE BARRELLED DETECTIVE STORY

A DOUBLE
BARRELLED
DETECTIVE
STORY
WE OUGHT NEVER TO DO WRONG
WHEN PEOPLE ARE LOOKING
Mark Twain

[See page 12

"IT'S A BIRTH-MARK!"

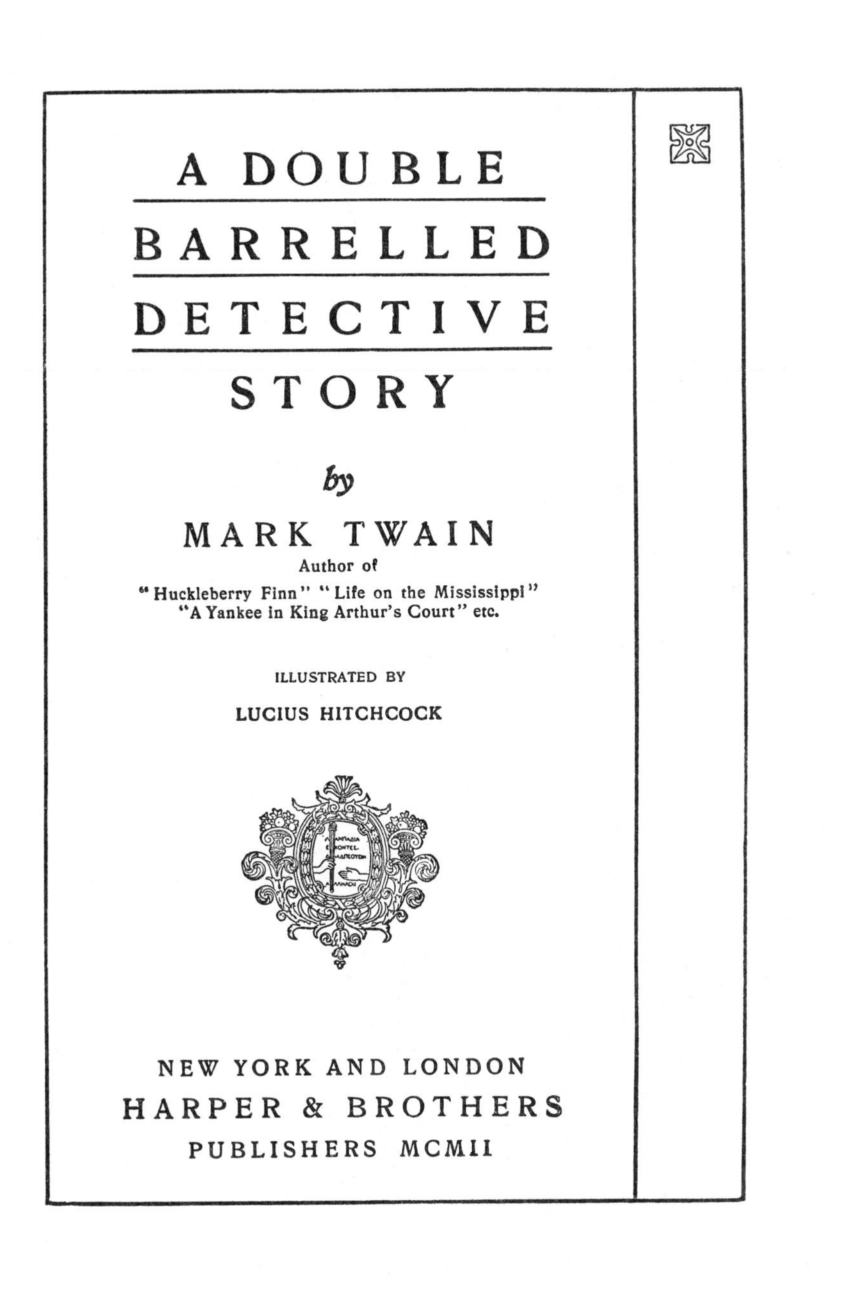

A DOUBLE BARRELLED DETECTIVE STORY

by

MARK TWAIN

Author of

"Huckleberry Finn" "Life on the Mississippi"
"A Yankee in King Arthur's Court" etc.

ILLUSTRATED BY

LUCIUS HITCHCOCK

NEW YORK AND LONDON
HARPER & BROTHERS
PUBLISHERS MCMII

Published April, 1902.

ILLUSTRATIONS

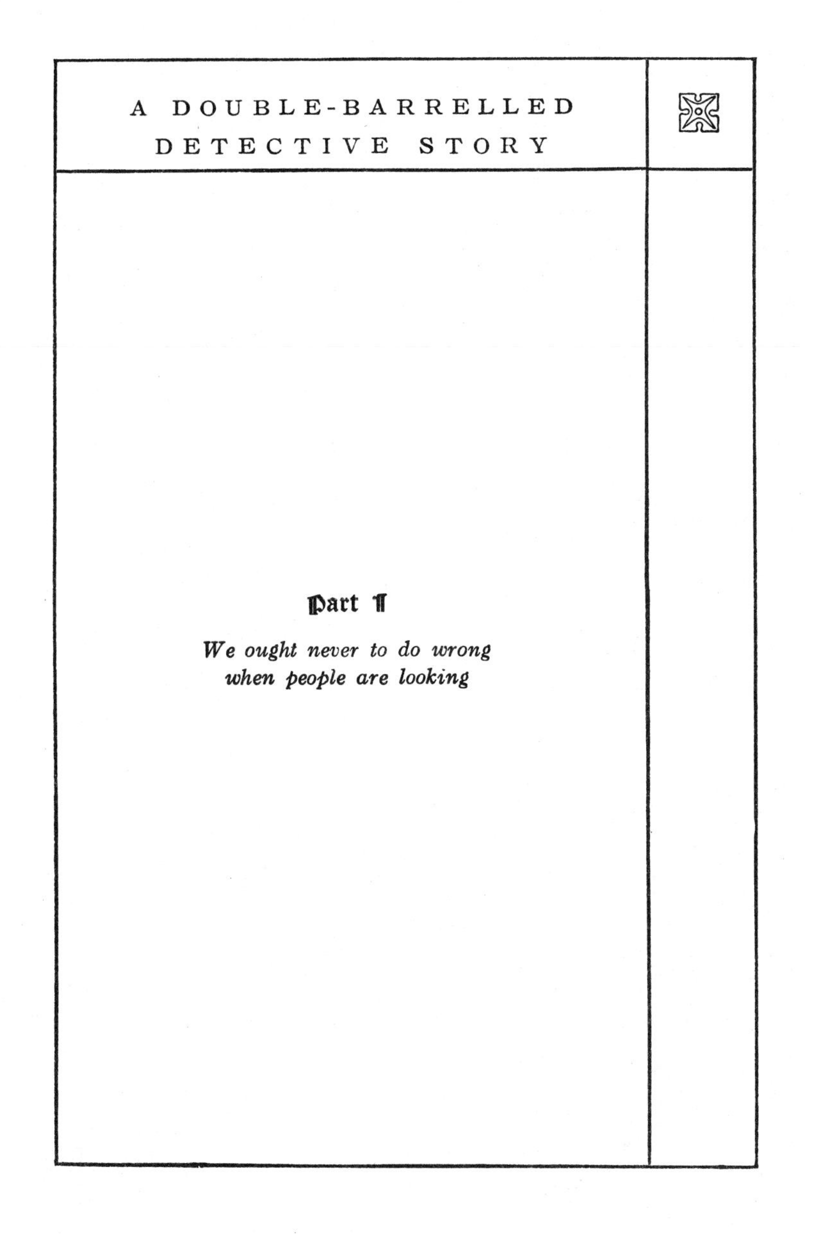

A DOUBLE-BARRELLED DETECTIVE STORY

Part II

We ought never to do wrong when people are looking

I

A wedding

HE first scene is in the country, in Virginia; the time, 1880. There has been a wedding, between a handsome young man of slender means and a rich young girl—a case of love at first sight and a precipitate marriage; a marriage bitterly opposed by the girl's widowed father.

Jacob Fuller, the bridegroom, is twenty-six years old, is of an old but unconsidered family which had

The bride

by compulsion emigrated from Sedgemoor, and for King James's purse's profit, so everybody said—some maliciously, the rest merely because they believed it. The bride is nineteen and beautiful. She is intense, high-strung, romantic, immeasurably proud of her Cavalier blood, and passionate in her love for her young husband. For its sake she braved her father's displeasure, endured his reproaches, listened with loyalty unshaken to his warning predictions, and went from his house without his blessing, proud and happy in the proofs she was thus giving of the quality of the affection which had made its home in her heart.

The morning after the marriage

there was a sad surprise for her. Her husband put aside her proffered caresses, and said:

The wife's surprise

"Sit down. I have something to say to you. I loved you. That was before I asked your father to give you to me. His refusal is not my grievance—I could have endured that. But the things he said of me to you—that is a different matter. There—you needn't speak; I know quite well what they were; I got them from authentic sources. Among other things he said that my character was written in my face; that I was treacherous, a dissembler, a coward, and a brute without sense of pity or compassion: the 'Sedgemoor trade-mark,' he called it—and 'white-sleeve

The Sedgemoor trade mark

Scheme of revenge

badge.' Any other man in my place would have gone to his house and shot him down like a dog. I wanted to do it, and was minded to do it, but a better thought came to me: to put him to shame; to break his heart; to kill him by inches. How to do it? Through my treatment of you, his idol! I would marry you; and then— Have patience. You will see."

From that moment onward, for three months, the young wife suffered all the humiliations, all the insults, all the miseries that the diligent and inventive mind of the husband could contrive, save physical injuries only. Her strong pride stood by her, and she kept the secret of her troubles.

New tortures

Now and then the husband said, "Why don't you go to your father and tell him?" Then he invented new tortures, applied them, and asked again. She always answered, "He shall never know by my mouth," and taunted him with his origin; said she was the lawful slave of a scion of slaves, and must obey, and would — up to that point, but no further; he could kill her if he liked, but he could not break her; it was not in the Sedgemoor breed to do it. At the end of the three months he said, with a dark significance in his manner, "I have tried all things but one" —and waited for her reply. "Try that," she said, and curled her lip in mockery.

A fiend's frenzy

That night he rose at midnight and put on his clothes, then said to her,

"Get up and dress!"

She obeyed—as always, without a word. He led her half a mile from the house, and proceeded to lash her to a tree by the side of the public road; and succeeded, she screaming and struggling. He gagged her then, struck her across the face with his cowhide, and set his blood-hounds on her. They tore the clothes off her, and she was naked. He called the dogs off, and said:

The husband dis-appears

"You will be found—by the passing public. They will be dropping along about three hours from now, and will spread the news—do you hear? Good-by. You have seen the last of me."

HE PROCEEDED TO LASH HER TO A TREE

He went away then. She moaned to herself:

Wife's cry for vengeance

"I shall bear a child—to *him!* God grant it may be a boy!"

The farmers released her by-and-by—and spread the news, which was natural. They raised the country with lynching intentions, but the bird had flown. The young wife shut herself up in her father's house; he shut himself up with her, and thenceforth would see no one. His pride was broken, and his heart; so he wasted away, day by day, and even his daughter rejoiced when death relieved him.

Then she sold the estate and disappeared.

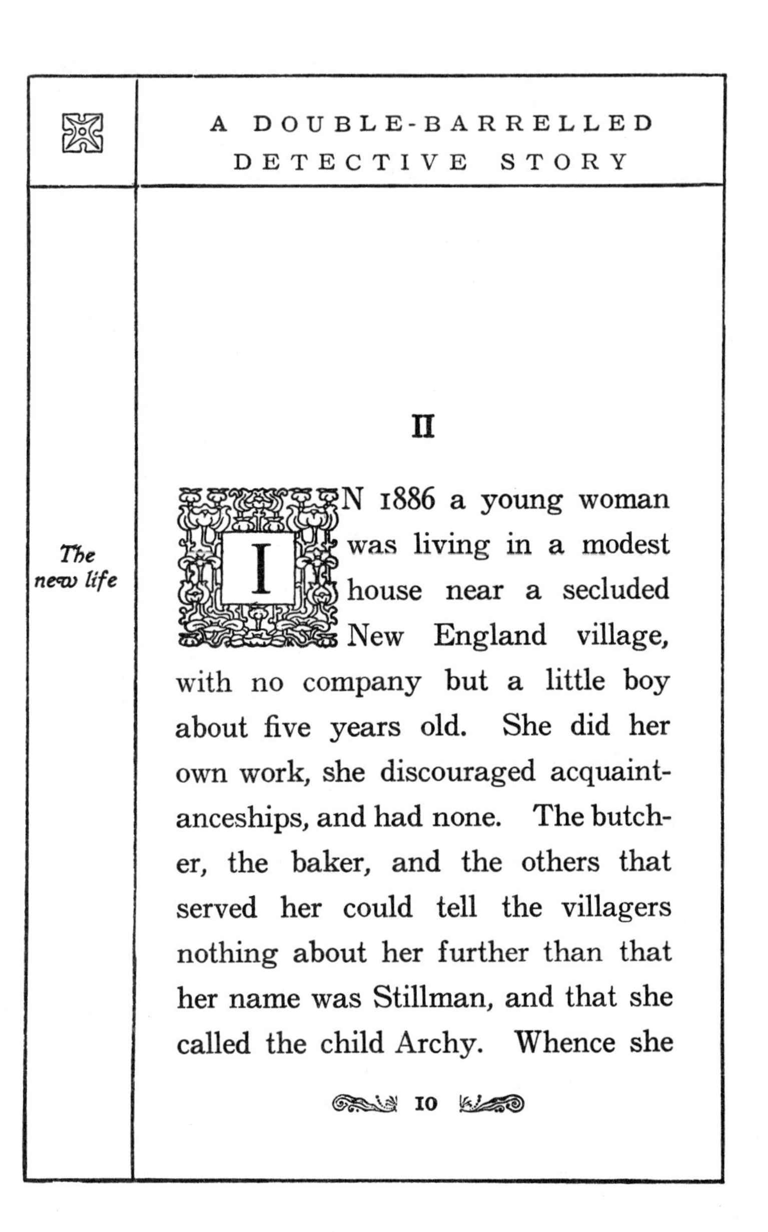

II

The new life

IN 1886 a young woman was living in a modest house near a secluded New England village, with no company but a little boy about five years old. She did her own work, she discouraged acquaintanceships, and had none. The butcher, the baker, and the others that served her could tell the villagers nothing about her further than that her name was Stillman, and that she called the child Archy. Whence she

A lonely child

came they had not been able to find out, but they said she talked like a Southerner. The child had no playmates and no comrade, and no teacher but the mother. She taught him diligently and intelligently, and was satisfied with the results — even a little proud of them. One day Archy said,

"Mamma, am I different from other children?"

"Well, I suppose not. Why?"

"There was a child going along out there and asked me if the postman had been by and I said yes, and she said how long since I saw him and I said I hadn't seen him at all, and she said how did I know he'd been by, then, and I said because I smelt his track on the sidewalk, and

He smelt a man's tracks

The scent of the blood-hound

she said I was a dum fool and made a mouth at me. What did she do that for?"

The young woman turned white, and said to herself, "It's a birth-mark! The gift of the blood-hound is in him." She snatched the boy to her breast and hugged him passionately, saying, "God has appointed the way!" Her eyes were burning with a fierce light and her breath came short and quick with excitement. She said to herself: "The puzzle is solved now; many a time it has been a mystery to me, the impossible things the child has done in the dark, but it is all clear to me now." She set him in his small chair, and said,

"Wait a little till I come, dear; then we will talk about the matter."

She went up to her room and took from her dressing-table several small articles and put them out of sight: a nail-file on the floor under the bed; a pair of nail-scissors under the bureau; a small ivory paper-knife under the wardrobe. Then she returned, and said:

Testing the blood gift

"There! I have left some things which I ought to have brought down." She named them, and said, "Run up and bring them, dear."

The child hurried away on his errand and was soon back again with the things.

"Did you have any difficulty, dear?"

"No, mamma; I only went where you went."

The trial at home

During his absence she had stepped to the bookcase, taken several books from the bottom shelf, opened each, passed her hand over a page, noting its number in her memory, then restored them to their places. Now she said:

"I have been doing something while you have been gone, Archy. Do you think you can find out what it was?"

The boy went to the bookcase and got out the books that had been touched, and opened them at the pages which had been stroked.

The mother took him in her lap, and said:

"I will answer your question now, dear. I have found out that in one

An odd child

way you are quite different from other people. You can see in the dark, you can smell what other people cannot, you have the talents of a bloodhound. They are good and valuable things to have, but you must keep the matter a secret. If people found it out, they would speak of you as an odd child, a strange child, and children would be disagreeable to you, and give you nicknames. In this world one must be like everybody else if he doesn't want to provoke scorn or envy or jealousy. It is a great and fine distinction which has been born to you, and I am glad; but you will keep it a secret, for mamma's sake, won't you?"

The child promised, without understanding.

Uncanny plans

All the rest of the day the mother's brain was busy with excited thinkings; with plans, projects, schemes, each and all of them uncanny, grim, and dark. Yet they lit up her face; lit it with a fell light of their own; lit it with vague fires of hell. She was in a fever of unrest; she could not sit, stand, read, sew; there was no relief for her but in movement. She tested her boy's gift in twenty ways, and kept saying to herself all the time, with her mind in the past: "He broke my father's heart, and night and day all these years I have tried, and all in vain, to think out a way to break his. I have found it now—I have found it now."

To break his heart

When night fell, the demon of unrest still possessed her. She went on with her tests; with a candle she traversed the house from garret to cellar, hiding pins, needles, thimbles, spools, under pillows, under carpets, in cracks in the walls, under the coal in the bin; then sent the little fellow in the dark to find them; which he did, and was happy and proud when she praised him and smothered him with caresses.

Further tests

From this time forward life took on a new complexion for her. She said, "The future is secure—I can wait, and enjoy the waiting." The most of her lost interests revived. She took up music again, and languages, drawing, painting, and the other long-discarded delights of her

A heart too soft

maidenhood. She was happy once more, and felt again the zest of life. As the years drifted by she watched the development of her boy, and was contented with it. Not altogether, but nearly that. The soft side of his heart was larger than the other side of it. It was his only defect, in her eyes. But she considered that his love for her and worship of her made up for it. He was a good hater—that was well; but it was a question if the materials of his hatreds were of as tough and enduring a quality as those of his friendships—and that was not so well.

The years drifted on. Archy was become a handsome, shapely, athletic youth, courteous, dignified, companionable, pleasant in his ways, and looking perhaps a trifle older than he was, which was sixteen. One evening his mother said she had something of grave importance to say to him, adding that he was old enough to hear it now, and old enough and possessed of character enough and stability enough to carry out a stern plan which she had been for years contriving and maturing. Then she told him her bitter story, in all its naked atrociousness. For a while the boy was paralyzed; then he said:

The mother's plan

"I understand. We are Southerners; and by our custom and nature there is but one atonement. I will search him out and kill him."

Inadequate atonement

Worse than death

"Kill him? No! Death is release, emancipation; death is a favor. Do I owe him favors? You must not hurt a hair of his head."

The boy was lost in thought awhile; then he said:

"You are all the world to me, and your desire is my law and my pleasure. Tell me what to do and I will do it."

The mother's eyes beamed with satisfaction, and she said:

Jacob Fuller

"You will go and find him. I have known his hiding-place for eleven years; it cost me five years and more of inquiry, and much money, to locate it. He is a quartz-miner in Colorado, and well-to-do. He lives in Denver. His name is Jacob Fuller. There—it is the first time I have spoken it since

Another wandering Jew

that unforgettable night. Think! That name could have been yours if I had not saved you that shame and furnished you a cleaner one. You will drive him from that place; you will hunt him down and drive him again; and yet again, and again, and again, persistently, relentlessly, poisoning his life, filling it with mysterious terrors, loading it with weariness and misery, making him wish for death, and that he had a suicide's courage; you will make of him another wandering Jew; he shall know no rest any more, no peace of mind, no placid sleep; you shall shadow him, cling to him, persecute him, till you break his heart, as he broke my father's and mine."

Money and disguises

"I will obey, mother."

"I believe it, my child. The preparations are all made; everything is ready. Here is a letter of credit; spend freely, there is no lack of money. At times you may need disguises. I have provided them; also some other conveniences." She took from the drawer of the type-writer table several squares of paper. They all bore these type-written words:

$10,000 REWARD.

It is believed that a certain man who is wanted in an Eastern State is sojourning here. In 1880, in the night, he tied his young wife to a tree by the public road, cut her across

the face with a cowhide, and
made his dogs tear her clothes
from her, leaving her naked.
He left her there, and fled
the country. A blood-relative
of hers has searched for him
for seventeen years. Address
...., ..., Post-office. The
above reward will be paid in
cash to the person who will
furnish the seeker, in a personal interview, the criminal's address.

The first placard

"When you have found him and acquainted yourself with his scent, you will go in the night and placard one of these upon the building he occupies, and another one upon the post-office or in some other prominent

Careful cruelty

place. It will be the talk of the region. At first you must give him several days in which to force a sale of his belongings at something approaching their value. We will ruin him by-and-by, but gradually; we must not impoverish him at once, for that could bring him to despair and injure his health, possibly kill him."

She took three or four more type-written forms from the drawer—duplicates—and read one:

........, ..., 18....

To Jacob Fuller:

You have ...... days in which to settle your affairs. You will not be disturbed during that limit, which will expire at ...... M., on the ...... of

The second placard

...... You must then MOVE ON. If you are still in the place after the named hour, I will placard you on all the dead walls, detailing your crime once more, and adding the date, also the scene of it, with all names concerned, including your own. Have no fear of bodily injury—it will in no circumstances ever be inflicted upon you. You brought misery upon an old man, and ruined his life and broke his heart. What he suffered, you are to suffer.

"You will add no signature. He must receive this before he learns of the reward-placard—before he rises

Moving on

in the morning—lest he lose his head and fly the place penniless."

"I shall not forget."

"You will need to use these forms only in the beginning—once may be enough. Afterward, when you are ready for him to vanish out of a place, see that he gets a copy of *this* form, which merely says:

MOVE ON. You have ...... days.

"He will obey. That is sure."

III

Extracts from Letters to the Mother.

DENVER, *April* 3, 1897.

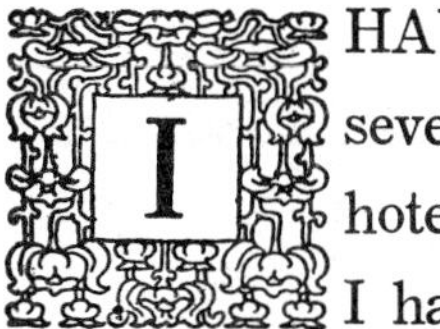

I HAVE now been living several days in the same hotel with Jacob Fuller. I have his scent; I could track him through ten divisions of infantry and find him. I have often been near him and heard him talk. He owns a good mine, and has a fair income from it; but he is not rich. He learned mining in a good way—by working at it for wages. He is a cheerful

Close to his prey

A pleasing personality

creature, and his forty-three years sit lightly upon him; he could pass for a younger man — say thirty-six or thirty-seven. He has never married again—passes himself off for a widower. He stands well, is liked, is popular, and has many friends. Even I feel a drawing toward him—the paternal blood in me making its claim. How blind and unreasoning and arbitrary are some of the laws of nature—the most of them, in fact! My task is become hard now—you realize it? you comprehend, and make allowances?—and the fire of it has cooled, more than I like to confess to myself. But I will carry it out. Even with the pleasure paled, the duty remains, and I will not spare him.

And for my help, a sharp resentment rises in me when I reflect that he who committed that odious crime is the only one who has not suffered by it. The lesson of it has manifestly reformed his character, and in the change he is happy. He, the guilty party, is absolved from all suffering; you, the innocent, are borne down with it. But be comforted—he shall harvest his share.

Happy criminal

SILVER GULCH, *May* 19.

I placarded Form No. 1 at midnight of April 3; an hour later I slipped Form No. 2 under his chamber door, notifying him to leave Denver at or before 11.50 the night of the 14th.

The Warning

Some late bird of a reporter stole

The "scoop"

one of my placards, then hunted the town over and found the other one, and stole that. In this manner he accomplished what the profession call a "scoop"—that is, he got a valuable item, and saw to it that no other paper got it. And so his paper—the principal one in the town—had it in glaring type on the editorial page in the morning, followed by a Vesuvian opinion of our wretch a column long, which wound up by adding a thousand dollars to our reward on the *paper's* account! The journals out here know how to do the noble thing—when there's business in it.

At breakfast I occupied my usual seat—selected because it afforded a view of papa Fuller's face, and was near enough for me to hear the talk that went on at his table. Seventy-five or a hundred people were in the room, and all discussing that item, and saying they hoped the seeker would find that rascal and remove the pollution of his presence from the town—with a rail, or a bullet, or something.

Removal by rail or bullet

When Fuller came in he had the Notice to Leave—folded up—in one hand, and the newspaper in the other; and it gave me more than half a pang to see him. His cheerfulness was all gone, and he looked old and pinched and ashy. And then—only think of the things he had to listen to! Mamma, he heard his own unsuspecting friends describe him with epithets

Calls himself names

and characterizations drawn from the very dictionaries and phrase-books of Satan's own authorized editions down below. And more than that, he had to *agree* with the verdicts and applaud them. His applause tasted bitter in his mouth, though; he could not disguise that from me; and it was observable that his appetite was gone; he only nibbled; he couldn't eat. Finally a man said:

"It is quite likely that that relative is in the room and hearing what this town thinks of that unspeakable scoundrel. I hope so."

Ah, dear, it was pitiful the way Fuller winced, and glanced around scared! He couldn't endure any more, and got up and left.

He sells out

During several days he gave out that he had bought a mine in Mexico, and wanted to sell out and go down there as soon as he could, and give the property his personal attention. He played his cards well; said he would take $40,000—a quarter in cash, the rest in safe notes; but that as he greatly needed money on account of his new purchase, he would diminish his terms for cash in full. He sold out for $30,000. And then, what do you think he did? He asked for *greenbacks*, and took them, saying the man in Mexico was a New-Englander, with a head full of crotchets, and preferred greenbacks to gold or drafts. People thought it queer, since a draft on New York could pro-

Sticking to his trail

duce greenbacks quite conveniently. There was talk of this odd thing, but only for a day; that is as long as any topic lasts in Denver.

I was watching, all the time. As soon as the sale was completed and the money paid—which was on the 11th—I began to stick to Fuller's track without dropping it for a moment. That night—no, 12th, for it was a little past midnight—I tracked him to his room, which was four doors from mine in the same hall, then I went back and put on my muddy day-laborer disguise, darkened my complexion, and sat down in my room in the gloom, with a gripsack handy, with a change in it, and my door ajar. For I suspected that the bird would

I CAUGHT THE FAMILIAR WHIFF

Disguised as a woman

take wing now. In half an hour an old woman passed by, carrying a grip; I caught the familiar whiff and followed, with my grip, for it was Fuller. He left the hotel by a side entrance, and at the corner he turned up an unfrequented street and walked three blocks in a light rain and a heavy darkness, and got into a two-horse hack, which, of course, was waiting for him by appointment. I took a seat (uninvited) on the trunk platform behind, and we drove briskly off. We drove ten miles, and the hack stopped at a way station and was discharged. Fuller got out and took a seat on a barrow under the awning, as far as he could get from the light; I went inside, and watched the ticket-

On the train

office. Fuller bought no ticket; I bought none. Presently the train came along, and he boarded a car; I entered the same car at the other end, and came down the aisle and took the seat behind him. When he paid the conductor and named his objective point, I dropped back several seats, while the conductor was changing a bill, and when he came to me I paid to the same place —about a hundred miles westward.

False whiskers

From that time for a week on end he led me a dance. He travelled here and there and yonder—always on a general westward trend—but he was not a woman after the first day. He was a laborer, like myself, and wore bushy false whiskers. His outfit was perfect, and he could do the character without thinking about it, for he had served the trade for wages. His nearest friend could not have recognized him. At last he located himself here, the obscurest little mountain camp in Montana; he has a shanty, and goes out prospecting daily; is gone all day, and avoids society. I am living at a miner's boarding-house, and it is an awful place: the bunks, the food, the dirt—everything.

Living in a shanty

We have been here four weeks, and in that time I have seen him but once; but every night I go over his track and post myself. As soon as he engaged a shanty here I went to a town fifty miles away and telegraphed that Denver hotel to keep my baggage

till I should send for it. I need nothing here but a change of army shirts, and I brought that with me.

SILVER GULCH, *June* 12.

Fuller feels safe

The Denver episode has never found its way here, I think. I know the most of the men in camp, and they have never referred to it, at least in my hearing. Fuller doubtless feels quite safe in these conditions. He has located a claim, two miles away, in an out-of-the-way place in the mountains; it promises very well, and he is working it diligently. Ah, but the change in him! He never smiles, and he keeps quite to himself, consorting with no one—he who was so fond of company and so cheery only two months ago. I have seen him passing along several times recently—drooping, forlorn, the spring gone from his step, a pathetic figure. He calls himself David Wilson.

Drooping and forlorn

I can trust him to remain here until we disturb him. Since you insist, I will banish him again, but I do not see how he can be unhappier than he already is. I will go back to Denver and treat myself to a little season of comfort, and edible food, and endurable beds, and bodily decency; then I will fetch my things, and notify poor papa Wilson to move on.

DENVER, *June* 19.

They miss him here. They all hope he is prospering in Mexico, and they

They are all sorry

do not say it just with their mouths, but out of their hearts. You know you can always tell. I am loitering here overlong, I confess it. But if you were in my place you would have charity for me. Yes, I know what you will say, and you are right: if I were in *your* place, and carried your scalding memories in my heart—

I will take the night train back to-morrow.

DENVER, *June* 20.

Hunting the wrong man

God forgive us, mother, we are hunting the *wrong man!* I have not slept any all night. I am now waiting, at dawn, for the *morning* train—and how the minutes drag, how they drag!

This Jacob Fuller is a *cousin* of

Not the criminal

the guilty one. How stupid we have been not to reflect that the guilty one would never again wear his own name after that fiendish deed! The Denver Fuller is four years younger than the other one; he came here a young widower in '79, aged twenty-one—a year before you were married; and the documents to prove it are innumerable. Last night I talked with familiar friends of his who have known him from the day of his arrival. I said nothing, but a few days from now I will land him in this town again, with the loss upon his mine made good; and there will be a banquet, and a torch-light procession, and there will not be any expense on anybody but me. Do you call this "gush"?

I am only a boy, as you well know; it is my privilege. By-and-by I shall not be a boy any more.

SILVER GULCH, *July* 3.

Gone and left no clew

Mother, he is gone! Gone, and left no trace. The scent was cold when I came. To-day I am out of bed for the first time since. I wish I were not a boy; then I could stand shocks better. They all think he went west. I start to-night, in a wagon—two or three hours of that, then I get a train. I don't know where I'm going, but I must go; to try to keep still would be torture.

Of course he has effaced himself with a new name and a disguise. This means that *I may have to search*

The hunter hunted

the whole globe to find him. Indeed it is what I expect. Do you see, mother? It is *I* that am the Wandering Jew. The irony of it! We arranged that for another.

Think of the difficulties! And there would be none if I only could advertise for him. But if there is any way to do it that would not frighten him, I have not been able to think it out, and I have tried till my brains are addled. "If the gentleman who lately bought a mine in Mexico and sold one in Denver will send his address to" (to whom, mother?), "it will be explained to him that it was all a mistake; his forgiveness will be asked, and full reparation made for a loss which he sustained in a

Not the man wanted

certain matter." Do you see? He would think it a trap. Well, any one would. If I should say, "It is now known that he was not the man wanted, but another man—a man who once bore the same name, but discarded it for good reasons"—would that answer? But the Denver people would wake up then and say "Oho!" and they would remember about the suspicious greenbacks, and say, "Why did he run away if he wasn't the right man?—it is too thin." If I failed to find him he would be ruined there—there where there is no taint upon him now. You have a better head than mine. Help me.

I have one clew, and only one. I know his handwriting. If he puts his new false name upon a hotel register and does not disguise it too much, it will be valuable to me if I ever run across it.

SAN FRANCISCO, *June* 28, 1898.

You already know how well I have searched the States from Colorado to the Pacific, and how nearly I came to getting him once. Well, I have had another close miss. It was here, yesterday. I struck his trail, *hot*, on the street, and followed it on a run to a cheap hotel. That was a costly mistake; a dog would have gone the other way. But I am only part dog, and can get very humanly stupid when excited. He had been stopping in that house ten days; I

A close miss

Has to keep moving

almost know, now, that he stops long nowhere, the past six or eight months, but is restless and has to keep moving. I understand that feeling! and I know what it is to feel it. He still uses the name he had registered when I came so near catching him nine months ago—"James Walker"; doubtless the same he adopted when he fled from Silver Gulch. An unpretending man, and has small taste for fancy names. I recognized the hand easily, through its slight disguise. A square man, and not good at shams and pretences.

They said he was just gone, on a journey; left no address; didn't say where he was going; looked frightened when asked to leave his address; had no baggage but a cheap valise; carried it off on foot—a "stingy old person, and not much loss to the house." "*Old!*" I suppose he is, now. I hardly heard; I was there but a moment. I rushed along his trail, and it led me to a wharf. Mother, the smoke of the steamer he had taken was just fading out on the horizon! I should have saved half an hour if I had gone in the right direction at first. I could have taken a fast tug, and should have stood a chance of catching that vessel. She is bound for Melbourne.

The aging criminal

HOPE CANYON, CALIFORNIA,
October 3, 1900.

You have a right to complain. "A letter a year" *is* a paucity; I freely acknowledge it; but how can one write when there is nothing to write about but failures? No one can keep it up; it breaks the heart.

Chased over the world

I told you—it seems ages ago, now—how I missed him at Melbourne, and then chased him all over Australasia for months on end.

Well, then, after that I followed him to India; almost *saw* him in Bombay; traced him all around — to Baroda, Rawal-Pindi, Lucknow, Lahore, Cawnpore, Allahabad, Calcutta, Madras—oh, everywhere; week after week, month after month, through the dust and swelter—always approximately on his track, sometimes close upon him, yet never catching him. And down to Ceylon, and then to— Never mind; by-and-by I will write it all out.

Back in California

I chased him home to California, and down to Mexico, and back again to California. Since then I have been hunting him about the State from the first of last January down to a month ago. I feel almost sure he is not far from Hope Canyon; I traced him to a point thirty miles from here, but there I lost the trail; some one gave him a lift in a wagon, I suppose.

I am taking a rest, now—modified by searchings for the lost trail. I was tired to death, mother, and low-spirited, and sometimes coming uncomfortably near to losing hope; but the miners in this little camp are good fellows, and I am used to their sort this long time back; and their breezy ways freshen a person up and

Sammy Hillyer

make him forget his troubles. I have been here a month. I am cabining with a young fellow named "Sammy" Hillyer, about twenty-five, the only son of his mother—like me—and loves her dearly, and writes to her every week—part of which is like me. He is a timid body, and in the matter of intellect—well, he cannot be depended upon to set a river on fire; but no matter, he is well liked; he is good and fine, and it is meat and bread and rest and luxury to sit and talk with him and have a comradeship again. I wish "James Walker" could have it. He had friends; he liked company. That brings up that picture of him, the time that I saw him last. The pathos of it! It comes before me often and often. At that very time, poor thing, I was girding up my conscience to make him move on again!

The black sheep

Hillyer's heart is better than mine, better than anybody's in the community, I suppose, for he is the one friend of the black sheep of the camp —Flint Buckner—and the only man Flint ever talks with or allows to talk with him. He says he knows Flint's history, and that it is trouble that has made him what he is, and so one ought to be as charitable toward him as one can. Now none but a pretty large heart could find space to accommodate a lodger like Flint Buckner, from all I hear about him outside. I think that this one de-

A man of misery

tail will give you a better idea of Sammy's character than any labored-out description I could furnish you of him. In one of our talks he said something about like this: "Flint's a kinsman of mine, and he pours out all his troubles to me—empties his breast from time to time, or I reckon it would burst. There couldn't be any unhappier man, Archy Stillman; his life has been made up of misery of mind—he isn't near as old as he looks. He has lost the feel of reposefulness and peace—oh, years and years ago! He doesn't know what good luck is—never has had any; often says he wishes he was in the other hell, he is so tired of this one."

IV

No real gentleman will tell the naked truth in the presence of ladies

IT was a crisp and spicy morning in early October. The lilacs and laburnums, lit with the glory-fires of autumn, hung burning and flashing in the upper air, a fairy bridge provided by kind Nature for the wingless wild things that have their homes in the tree-tops and would visit together; the larch and the pomegranate flung their purple and yel-

Fine days and words

More fine days and words

low flames in brilliant broad splashes along the slanting sweep of the woodland; the sensuous fragrance of innumerable deciduous flowers rose upon the swooning atmosphere; far in the empty sky a solitary œsophagus slept upon motionless wing; everywhere brooded stillness, serenity, and the peace of God.

October is the time—1900; Hope Canyon is the place, a silver-mining camp away down in the Esmeralda region. It is a secluded spot, high and remote; recent as to discovery; thought by its occupants to be rich in metal—a year or two's prospecting will decide that matter one way or the other. For inhabitants, the camp has about two hundred miners,

A new camp

one white woman and child, several Chinese washermen, five squaws, and a dozen vagrant buck Indians in rabbit-skin robes, battered plug hats, and tin-can necklaces. There are no mills as yet; there is no church, no newspaper. The camp has existed but two years; it has made no big strike; the world is ignorant of its name and place.

On both sides of the canyon the mountains rise wall-like, three thousand feet, and the long spiral of straggling huts down in its narrow bottom gets a kiss from the sun only once a day, when he sails over at noon. The village is a couple of miles long; the cabins stand well apart from each other. The tavern is the only "frame"

The tavern

house — the only house, one might say. It occupies a central position, and is the evening resort of the population. They drink there, and play seven-up and dominoes; also billiards, for there is a table, crossed all over with torn places repaired with court-plaster; there are some cues, but no leathers; some chipped balls which clatter when they run, and do not slow up gradually, but stop suddenly and sit down; there is part of a cube of chalk, with a projecting jag of flint in it; and the man who can score six on a single break can set up the drinks at the bar's expense.

Flint Buckner's cabin was the last one of the village, going south; his silver claim was at the other end of the village, northward, and a little beyond the last hut in that direction. He was a sour creature, unsociable, and had no companionships. People who had tried to get acquainted with him had regretted it and dropped him. His history was not known. Some believed that Sammy Hillyer knew it; others said no. If asked, Hillyer said no, he was not acquainted with it. Flint had a meek English youth of sixteen or seventeen with him, whom he treated roughly, both in public and in private, and of course this lad was applied to for information, but with no success. Fetlock Jones—name of the youth—said that Flint picked him up on a prospecting

A sour creature

Salary, bacon and beans

tramp, and as he had neither home nor friends in America, he had found it wise to stay and take Buckner's hard usage for the sake of the salary, which was bacon and beans. Further than this he could offer no testimony.

Fetlock had been in this slavery for a month now, and under his meek exterior he was slowly consuming to a cinder with the insults and humiliations which his master had put upon him. For the meek suffer bitterly from these hurts; more bitterly, perhaps, than do the manlier sort, who can burst out and get relief with words or blows when the limit of endurance has been reached. Good-hearted people wanted to help Fetlock out of his trouble, and tried to get him to leave

Buckner; but the boy showed fright at the thought, and said he "dasn't." Pat Riley urged him, and said:

Afraid to quit

"You leave the damned hunks and come with me; don't you be afraid. I'll take care of *him.*"

The boy thanked him with tears in his eyes, but shuddered and said he "dasn't risk it"; he said Flint would catch him alone, some time, in the night, and then— "Oh, it makes me sick, Mr. Riley, to think of it."

Others said, "Run away from him; we'll stake you; skip out for the coast some night." But all these suggestions failed; he said Flint would hunt him down and fetch him back, just for meanness.

The people could not understand

The problem of murder

this. The boy's miseries went steadily on, week after week. It is quite likely that the people would have understood if they had known how he was employing his spare time. He slept in an out-cabin near Flint's; and there, nights, he nursed his bruises and his humiliations, and studied and studied over a single problem—how he could murder Flint Buckner and not be found out. It was the only joy he had in life; these hours were the only ones in the twenty-four which he looked forward to with eagerness and spent in happiness.

He thought of poison. No—that would not serve; the inquest would reveal where it was procured and who had procured it. He thought of a shot in the back in a lonely place when Flint would be homeward-bound at midnight — his unvarying hour for the trip. No — somebody might be near, and catch him. He thought of stabbing him in his sleep. No— he might strike an inefficient blow, and Flint would seize him. He examined a hundred different ways— none of them would answer; for in even the very obscurest and secretest of them there was always the fatal defect of a *risk*, a chance, a possibility that he might be found out. He would have none of that.

The risk in killing

But he was patient, endlessly patient. There was no hurry, he said to himself. He would never leave Flint till he left him a corpse;

Revenge is slow

there was no hurry — he would find the way. It was somewhere, and he would endure shame and pain and misery until he found it. Yes, somewhere there was a way which would leave not a trace, not even the faintest clew to the murderer—there was no hurry—he would find that way, and then—oh, then, it would just be good to be alive! Meantime he would diligently keep up his reputation for meekness; and also, as always theretofore, he would allow no one to hear him say a resentful or offensive thing about his oppressor.

Two days before the before-mentioned October morning Flint had bought some things, and he and Fetlock had brought them home to Flint's

Mining material

cabin: a fresh box of candles, which they put in the corner; a tin can of blasting-powder, which they placed upon the candle-box; a keg of blasting-powder, which they placed under Flint's bunk; a huge coil of fuse, which they hung on a peg. Fetlock reasoned that Flint's mining operations had outgrown the pick, and that blasting was about to begin now. He had seen blasting done, and he had a notion of the process, but he had never helped in it. His conjecture was right — blasting-time had come. In the morning the pair carried fuse, drills, and the powder-can to the shaft; it was now eight feet deep, and to get into it and out of it a short ladder was used. They

First lessons

descended, and by command Fetlock held the drill—without any instructions as to the right way to hold it—and Flint proceeded to strike. The sledge came down; the drill sprang out of Fetlock's hand, almost as a matter of course.

"You mangy son of a nigger, is that any way to hold a drill? Pick it up! Stand it up! There—hold fast. D—— you! *I'll* teach you!"

At the end of an hour the drilling was finished.

"Now, then, charge it."

The boy started to pour in the powder.

"Idiot!"

A heavy bat on the jaw laid the lad out.

"Get up! You can't lie snivelling there. Now, then, stick in the fuse *first*. *Now* put in the powder. Hold on, hold on! Are you going to fill the hole *all* up? Of all the sap-headed milksops I— Put in some dirt! Put in some gravel! Tamp it down! Hold on, hold on! Oh, great Scott! get out of the way!" He snatched the iron and tamped the charge himself, meantime cursing and blaspheming like a fiend. Then he fired the fuse, climbed out of the shaft, and ran fifty yards away, Fetlock following. They stood waiting a few minutes, then a great volume of smoke and rocks burst high into the air with a thunderous explosion; after a little there was a shower of de-

The first blast

scending stones; then all was serene again.

"I wish to God you'd been in it!" remarked the master.

They went down the shaft, cleaned it out, drilled another hole, and put in another charge.

Timing the fuse

"Look here! How much fuse are you proposing to waste? Don't you know how to time a fuse?"

"No, sir."

"You *don't!* Well, if you don't beat anything *I* ever saw!"

He climbed out of the shaft and spoke down:

"Well, idiot, are you going to be all day? Cut the fuse and light it!"

The trembling creature began,

"If you please, sir, I—"

"You talk back to *me?* Cut it and light it!"

The boy cut and lit.

"Ger-reat Scott! a one-minute fuse! I wish you were in—"

In his rage he snatched the ladder out of the shaft and ran. The boy was aghast.

"Oh, my God! Help! Help! Oh, save me!" he implored. "Oh, what can I do! What *can* I do!"

He backed against the wall as tightly as he could; the sputtering fuse frightened the voice out of him; his breath stood still; he stood gazing and impotent; in two seconds, three seconds, four, he would be flying toward the sky torn to fragments. Then he had an inspiration. He

Close to death

sprang at the fuse and severed the inch of it that was left above ground, and was saved.

He sank down limp and half lifeless with fright, his strength all gone; but he muttered with a deep joy:

Fetlock finds a way

"He has learnt me! I knew there was a way, if I would wait."

After a matter of five minutes Buckner stole to the shaft, looking worried and uneasy, and peered down into it. He took in the situation; he saw what had happened. He lowered the ladder, and the boy dragged himself weakly up it. He was very white. His appearance added something to Buckner's uncomfortable state, and he said, with a show of

HE BACKED AGAINST THE WALL AS TIGHTLY AS HE COULD

regret and sympathy which sat upon him awkwardly from lack of practice:

"It was an accident, you know. Don't say anything about it to anybody; I was excited, and didn't notice what I was doing. You're not looking well; you've worked enough for to-day; go down to my cabin and eat what you want, and rest. It's just an accident, you know, on account of my being excited."

Something learned

"It scared me," said the lad, as he started away; "but I learnt something, so I don't mind it."

"Damned easy to please!" muttered Buckner, following him with his eye. "I wonder if he'll tell? Mightn't he? . . . I wish it *had* killed him."

Feverish work

The boy took no advantage of his holiday in the matter of resting; he employed it in work, eager and feverish and happy work. A thick growth of chaparral extended down the mountain-side clear to Flint's cabin; the most of Fetlock's labor was done in the dark intricacies of that stubborn growth; the rest of it was done in his own shanty. At last all was complete, and he said:

"If he's got any suspicions that I'm going to tell on him, he won't keep them long, to-morrow. He will see that I am the same milksop as I always was—all day and the next. And the day after to-morrow night there'll be an end of him, and nobody will ever guess who finished him up nor how it was done. He dropped me the idea his own self, and that's odd."

The certain way

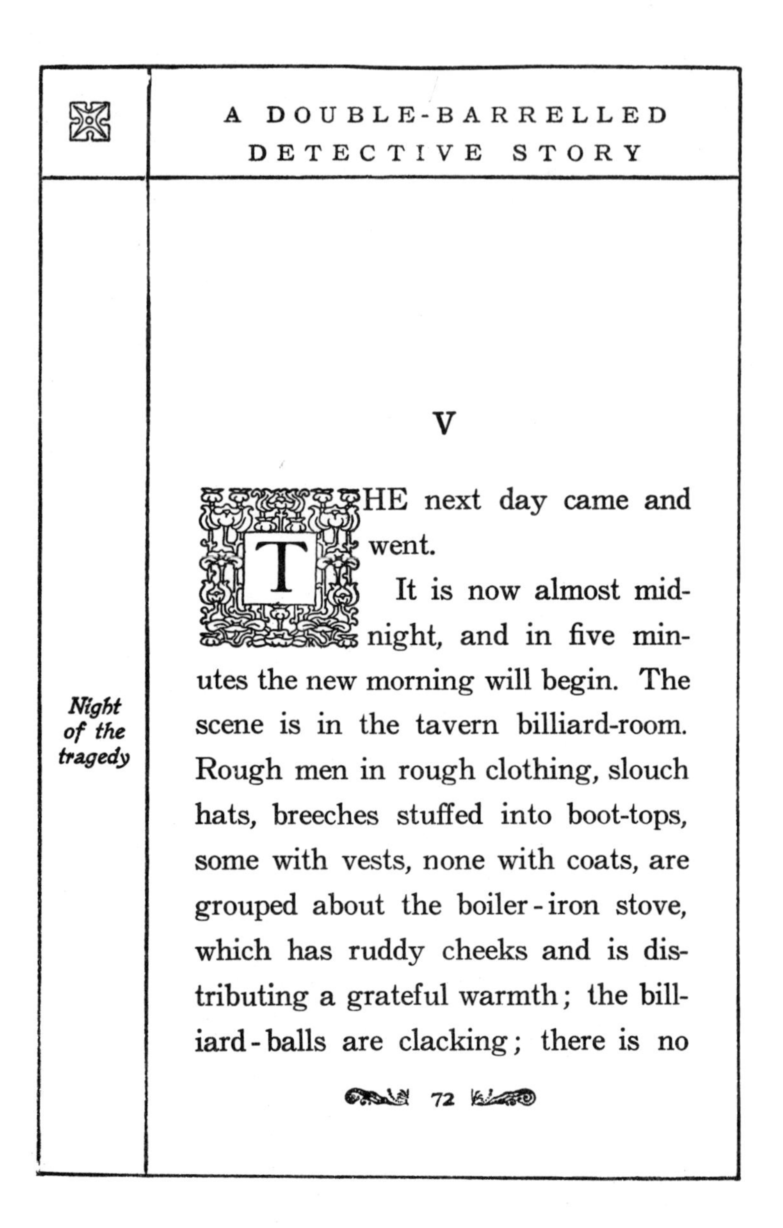

V

THE next day came and went.

Night of the tragedy

It is now almost midnight, and in five minutes the new morning will begin. The scene is in the tavern billiard-room. Rough men in rough clothing, slouch hats, breeches stuffed into boot-tops, some with vests, none with coats, are grouped about the boiler-iron stove, which has ruddy cheeks and is distributing a grateful warmth; the billiard-balls are clacking; there is no

other sound—that is, within; the wind is fitfully moaning without. The men look bored; also expectant. A hulking, broad-shouldered miner, of middle age, with grizzled whiskers, and an unfriendly eye set in an unsociable face, rises, slips a coil of fuse upon his arm, gathers up some other personal properties, and departs without word or greeting to anybody. It is Flint Buckner. As the door closes behind him a buzz of talk breaks out.

Regular method

"The regularest man that ever was," said Jake Parker, the blacksmith; "you can tell when it's twelve just by him leaving, without looking at your Waterbury."

"And it's the only virtue he's got,

A blight on society

as fur as I know," said Peter Hawes, miner.

"He's just a blight on this society," said Wells-Fargo's man, Ferguson. "If I was running this shop I'd make him say something, *some* time or other, or vamos the ranch." This with a suggestive glance at the barkeeper, who did not choose to see it, since the man under discussion was a good customer, and went home pretty well set up, every night, with refreshments furnished from the bar.

"Say," said Ham Sandwich, miner, "does any of you boys ever recollect of him asking you to take a drink?"

"*Him?* Flint *Buckner?* Oh, Laura!"

This sarcastic rejoinder came in a

spontaneous general outburst in one form of words or another from the crowd. After a brief silence, Pat Riley, miner, said:

"He's the 15-puzzle, that cuss. And his boy's another one. *I* can't make them out."

The 15-puzzle

"Nor anybody else," said Ham Sandwich; "and if they are 15-puzzles, how are you going to rank up that other one? When it comes to A 1 right-down solid mysteriousness, he lays over both of them. *Easy*—don't he?"

"You bet!"

Everybody said it. Every man but one. He was the new-comer—Peterson. He ordered the drinks all round, and asked who No. 3 might

A boy mystery

be. All answered at once, "Archy Stillman!"

"Is he a mystery?" asked Peterson.

"Is *he* a mystery? Is Archy *Stillman* a mystery?" said Wells-Fargo's man, Ferguson. "Why, the fourth dimension's foolishness to *him.*"

For Ferguson was learned.

Peterson wanted to hear all about him; everybody wanted to tell him; everybody began. But Billy Stevens, the barkeeper, called the house to order, and said one at a time was best. He distributed the drinks, and appointed Ferguson to lead. Ferguson said:

"Well, he's a boy. And that is just about all we know about him.

You can pump him till you are tired; it ain't any use; you won't get anything. At least about his intentions, or line of business, or where he's from, and such things as that. And as for getting at the nature and get-up of his main big chief mystery, why, he'll just change the subject, that's all. You can *guess* till you're black in the face—it's your privilege—but suppose you do, where do you arrive at? Nowhere, as near as I can make out."

Not to be pumped

"What *is* his big chief one?"

"Sight, maybe. Hearing, maybe. Instinct, maybe. Magic, maybe. Take your choice—grown-ups, twenty-five; children and servants, half price. Now I'll tell you what he can do. You

Can't hide from him

can start here, and just disappear; you can go and hide wherever you want to, I don't care where it is, nor how far — and he'll go straight and put his finger on you."

"You don't mean it!"

"I just do, though. Weather's nothing to him — elemental conditions is nothing to him — he don't even take notice of them."

"Oh, come! Dark? Rain? Snow? Hey?"

"It's all the same to *him*. *He* don't give a damn."

"Oh, *say*—including *fog*, per'aps?"

"*Fog!* he's got an eye 't can plunk through it like a bullet."

"Now, boys, honor bright, what's he giving me?"

"It's a fact!" they all shouted. "Go on, Wells-Fargo."

"Well, sir, you can leave him here, chatting with the boys, and you can slip out and go to any cabin in this camp and open a book — yes, sir, a dozen of them—and take the page in your memory, and he'll start out and go straight to that cabin and open every one of them books at the right page, and call it off, and never make a mistake."

"He must be the devil!"

Is he the devil?

"More than one has thought it. Now I'll tell you a perfectly wonderful thing that he done. The other night he—"

There was a sudden great murmur of sounds outside, the door flew open,

Child lost at night

and an excited crowd burst in, with the camp's one white woman in the lead and crying:

"My child! my child! she's lost and gone! For the love of God help me to find Archy Stillman; we've hunted everywhere!"

Said the barkeeper:

"Sit down, sit down, Mrs. Hogan, and don't worry. He asked for a bed three hours ago, tuckered out tramping the trails the way he's always doing, and went up stairs. Ham Sandwich, run up and roust him out; he's in No. 14."

The youth was soon downstairs and ready. He asked Mrs. Hogan for particulars.

"Bless you, dear, there ain't any;

I wish there was. I put her to sleep at seven in the evening, and when I went in there an hour ago to go to bed myself, she was gone. I rushed for your cabin, dear, and you wasn't there, and I've hunted for you ever since, at every cabin down the gulch, and now I've come up again, and I'm that distracted and scared and heart-broke; but, thanks to God, I've found you at last, dear heart, and you'll find my child. Come on! come quick!"

Strange disappearance

"Move right along; I'm with you, madam. Go to your cabin first."

The whole company streamed out to join the hunt. All the southern half of the village was up, a hundred men strong, and waiting outside, a

The search begins

vague dark mass sprinkled with twinkling lanterns. The mass fell into columns by threes and fours to accommodate itself to the narrow road, and strode briskly along southward in the wake of the leaders. In a few minutes the Hogan cabin was reached.

"There's the bunk," said Mrs. Hogan; "there's where she was; it's where I laid her at seven o'clock; but where she is now, God only knows."

"Hand me a lantern," said Archy. He set it on the hard earth floor and knelt by it, pretending to examine the ground closely. "Here's her track," he said, touching the ground here and there and yonder with his finger. "Do you see?"

On the trail

Several of the company dropped upon their knees and did their best to see. One or two thought they discerned something like a track; the others shook their heads and confessed that the smooth hard surface had no marks upon it which their eyes were sharp enough to discover. One said, "Maybe a child's foot could make a mark on it, but *I* don't see how."

Young Stillman stepped outside, held the light to the ground, turned leftward, and moved along three steps, closely examining; then said, "I've got the direction—come along; take the lantern, somebody."

He strode off swiftly southward, the files following, swaying and bend-

Following an invisible clew

ing in and out with the deep curves of the gorge. Thus a mile, and the mouth of the gorge was reached; before them stretched the sage-brush plain, dim, vast, and vague. Stillman called a halt, saying, "We mustn't start wrong, now; we must take the direction again." He took a lantern and examined the ground for a matter of twenty yards; then said, "Come on; it's all right," and gave up the lantern. In and out among the sage-bushes he marched, a quarter of a mile, bearing gradually to the right; then took a new direction and made another great semicircle; then changed again and moved due west nearly half a mile—and stopped.

"She gave it up, here, poor little

chap. Hold the lantern. You can see where she sat."

But this was in a slick alkali flat which was surfaced like steel, and no person in the party was quite hardy enough to claim an eyesight that could detect the track of a cushion on a veneer like that. The bereaved mother fell upon her knees and kissed the spot, lamenting.

"But where is she, then?" some one said. "She didn't stay here. We can see *that* much, anyway."

Stillman moved about in a circle around the place, with the lantern, pretending to hunt for tracks. "Well!" he said presently, in an annoyed tone, "I don't understand it." He examined again. "No use. She was

Vanishing trail

here—that's certain; she never *walked* away from here—and that's certain. It's a puzzle; I can't make it out."

The mother lost heart then.

"Oh, my God! oh, blessed Virgin! some flying beast has got her. I'll never see her again!"

"Ah, *don't* give up," said Archy. "We'll find her—don't give up."

"God bless you for the words, Archy Stillman!" and she seized his hand and kissed it fervently.

Peterson, the new-comer, whispered satirically in Ferguson's ear:

"Wonderful performance to find this place, wasn't it? Hardly worth while to come so far, though; any other supposititious place would have answered just as well—hey?"

Faith in Still-man

Ferguson was not pleased with the innuendo. He said, with some warmth:

"Do you mean to insinuate that the child hasn't been here? I tell you the child *has* been here! Now if you want to get yourself into as tidy a little fuss as—"

"All right!" sang out Stillman. "Come, everybody, and look at this! It was right under our noses all the time, and we didn't see it."

There was a general plunge for the ground at the place where the child was alleged to have rested, and many eyes tried hard and hopefully to see the thing that Archy's finger was resting upon. There was a pause, then a several-barrelled sigh of dis-

A new scent

appointment. Pat Riley and Ham Sandwich said, in the one breath:

"What is it, Archy? There's nothing here."

"Nothing? Do you call *that* nothing?" and he swiftly traced upon the ground a form with his finger. "There—don't you recognize it now? It's Injun Billy's track. He's got the child."

"God be praised!" from the mother.

"Take away the lantern. I've got the direction. Follow!"

He started on a run, racing in and out among the sage-bushes a matter of three hundred yards, and disappeared over a sand-wave; the others struggled after him, caught him up, and found him waiting. Ten steps

away was a little wickieup, a dim and formless shelter of rags and old horse-blankets, a dull light showing through its chinks.

"You lead, Mrs. Hogan," said the lad. "It's your privilege to be first."

The lost one found

All followed the sprint she made for the wickieup, and saw, with her, the picture its interior afforded. Injun Billy was sitting on the ground; the child was asleep beside him. The mother hugged it with a wild embrace, which included Archy Stillman, the grateful tears running down her face, and in a choked and broken voice she poured out a golden stream of that wealth of worshipping endearments which has its home in full

At Injun Billy's

richness nowhere but in the Irish heart.

"I find her bymeby it is ten o'clock," Billy explained. "She 'sleep out yonder, ve'y tired — face wet, been cryin', 'spose; fetch her home, feed her, she heap much hungry — go 'sleep 'gin."

In her limitless gratitude the happy mother waived rank and hugged him too, calling him "the angel of God in disguise."

And he probably was in disguise if he was that kind of an official. He was dressed for the character.

At half past one in the morning the procession burst into the village, singing "When Johnny Comes Marching Home," waving its lanterns, and swallowing the drinks that were brought out all along its course. It concentrated at the tavern, and made a night of what was left of the morning.

Made a night of it

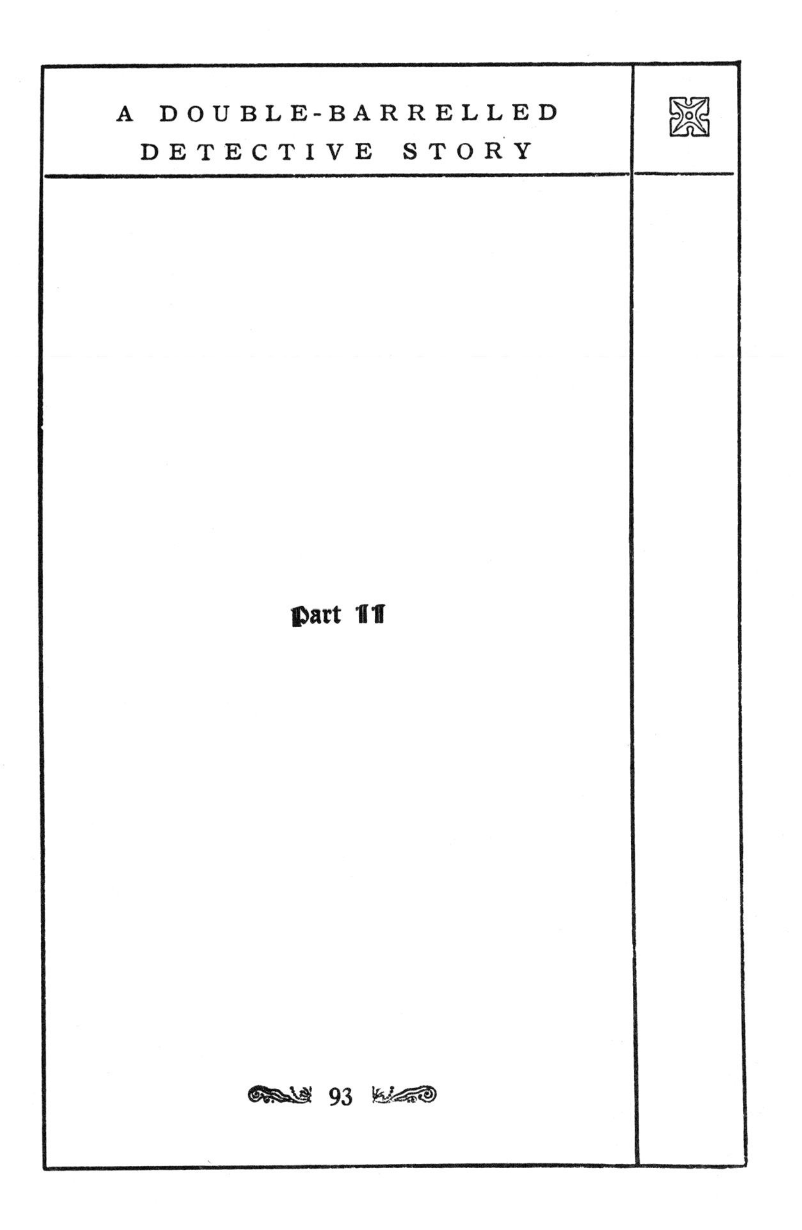

Part II

I

THE next afternoon the village was electrified with an immense sensation. A grave and dignified foreigner of distinguished bearing and appearance had arrived at the tavern, and entered this formidable name upon the register:

Sherlock Holmes!

Sherlock Holmes.

The news buzzed from cabin to cabin, from claim to claim; tools were dropped, and the town swarmed

Fet-lock's uncle

toward the centre of interest. A man passing out at the northern end of the village shouted it to Pat Riley, whose claim was the next one to Flint Buckner's. At that time Fetlock Jones seemed to turn sick. He muttered to himself:

"Uncle *Sherlock!* The mean luck of it!—that *he* should come just when . . ." He dropped into a reverie, and presently said to himself: "But what's the use of being afraid of *him?* Anybody that knows him the way I do knows he can't detect a crime, except when he plans it all out beforehand and arranges the clews and hires some fellow to commit it according to instructions. . . . Now there ain't going to *be* any clews this time—so, what show has he got? None at all. No, sir; everything's ready. If I was to risk putting it off. . . . No, I won't run any risk like that. Flint Buckner goes out of this world to-night, for sure." Then another trouble presented itself. "Uncle Sherlock 'll be wanting to talk home matters with me this evening, and how am I going to get rid of him? for I've *got* to be at my cabin a minute or two about eight o'clock." This was an awkward matter, and cost him much thought. But he found a way to beat the difficulty. "We'll go for a walk, and I'll leave him in the road a minute, so that he won't see what it is I do: the best way to throw a detective off the track. any-

Planning for a detective

way, is to have him along when you are preparing the thing. Yes, that's the safest—I'll take him with me."

The shadow of great-ness

Meantime the road in front of the tavern was blocked with villagers waiting and hoping for a glimpse of the great man. But he kept his room, and did not appear. None but Ferguson, Jake Parker the black-smith, and Ham Sandwich had any luck. These enthusiastic admirers of the great scientific detective hired the tavern's detained-baggage lockup, which looked into the detective's room across a little alleyway ten or twelve feet wide, ambushed themselves in it, and cut some peep-holes in the win-dow-blind. Mr. Holmes's blinds were down; but by-and-by he raised them. It gave the spies a hair-lifting but pleasurable thrill to find themselves face to face with the Extraordinary Man who had filled the world with the fame of his more than human ingenuities. There he sat — not a myth, not a shadow, but real, alive, compact of substance, and almost within touching distance with the hand.

Sherlock inspected

"Look at that head!" said Ferguson, in an awed voice. "By gracious! *that's* a head!"

"You bet!" said the blacksmith, with deep reverence. "Look at his nose! look at his eyes! Intellect? Just a battery of it!"

"And that paleness," said Ham

Sherlock thinking

Sandwich. "Comes from thought—that's what it comes from. Hell! duffers like us don't know what real thought *is*."

"No more we don't," said Ferguson. "What we take for thinking is just blubber-and-slush."

"Right you are, Wells-Fargo. And look at that frown—that's *deep* thinking—away down, down, forty fathom into the bowels of things. He's on the track of something."

"Well, he is, and don't you forget it. Say—look at that awful gravity—look at that pallid solemness—there ain't any corpse can lay over it."

"No, sir, not for dollars! And it's his'n by hereditary rights, too; he's been dead four times a'ready, and there's history for it. Three times natural, once by accident. I've heard say he smells damp and cold, like a grave. And he—"

" 'Sh! Watch him! There — he's got his thumb on the bump on the near corner of his forehead, and his forefinger on the off one. His think-works is just a-*grinding* now, you bet your other shirt."

Grinding his mind

"That's so. And now he's gazing up toward heaven and stroking his mustache slow, and—"

"Now he has rose up standing, and is putting his clews together on his left fingers with his right finger. See? he touches the forefinger—now middle finger—now ring-finger—"

"Stuck!"

"Look at him scowl! He can't seem to make out *that* clew. So he—"

"See him smile!—like a tiger—and tally off the other fingers like nothing! He's got it, boys; he's got it sure!"

Real thinking

"Well, I should *say!* I'd hate to be in that man's place that he's after."

Mr. Holmes drew a table to the window, sat down with his back to the spies, and proceeded to write. The spies withdrew their eyes from the peep-holes, lit their pipes, and settled themselves for a comfortable smoke and talk. Ferguson said, with conviction:

"Boys, it's no use talking, he's a wonder! He's got the signs of it all over him."

"You hain't ever said a truer word than that, Wells-Fargo," said Jake Parker. "Say, wouldn't it 'a' been nuts if he'd a-been here last night?"

"Oh, by George, but wouldn't it!" said Ferguson. "Then we'd have seen *scientific* work. Intellect—just pure intellect—away up on the upper levels, dontchuknow. Archy is all right, and it don't become anybody to belittle *him,* I can tell you. But his gift is only just eyesight, sharp as an owl's, as near as I can make it out just a grand natural animal talent, no more, no less, and prime as far as it goes, but no intellect in it, and for awfulness and marvellousness no more to be compared to what

Scientific detecting

Sher-lock's method

this man does than—than— Why, let me tell you what *he'd* have done. He'd have stepped over to Hogan's and glanced—just *glanced*, that's all—at the premises, and that's enough. See everything? Yes, sir, to the last little *de*tail; and he'd know more about that place than the Hogans would know in seven years. Next, he would sit down on the bunk, just as ca'm, and say to Mrs. Hogan—*Say*, Ham, consider that you are Mrs. Hogan. I'll ask the questions; you answer them."

"All right; go on."

"Madam, if you please—attention—do not let your mind wander. Now, then—sex of the child?"

"Female, your Honor."

Just questions

"Um — female. Very good, very good. Age?"

"Turned six, your Honor."

"Um — young, weak — two miles. Weariness will overtake it then. It will sink down and sleep. We shall find it two miles away, or less. Teeth?"

"Five, your Honor, and one a-coming."

"Very good, very good, *very* good indeed. You see, boys, *he* knows a clew when he sees it, when it wouldn't mean a dern thing to anybody else. Stockings, madam? Shoes?"

"Yes, your Honor—both."

"Yarn, perhaps? Morocco?"

"Yarn, your Honor. And kip."

"Um—kip. This complicates the

And then!

matter. However, let it go—we shall manage. Religion?"

"Catholic, your Honor."

"Very good. Snip me a bit from the bed blanket, please. Ah, thanks. Part wool—foreign make. Very well. A snip from some garment of the child's, please. Thanks. Cotton. Shows wear. An excellent clew, excellent. Pass me a pellet of the floor dirt, if you'll be so kind. Thanks, many thanks. Ah, admirable, admirable! *Now* we know where we are, I think. You see, boys, he's got all the clews he wants now; he don't need anything more. Now, then, what does this Extraordinary Man do? He lays those snips and that dirt out on the table and leans over them on his elbows, and puts them together side by side and studies them—mumbles to himself, 'Female'; changes them around—mumbles, 'Six years old'; changes them this way and that—again mumbles: 'Five teeth — one a-coming — Catholic — yarn — cotton — kip — damn that kip.' Then he straightens up and gazes toward heaven, and ploughs his hands through his hair—ploughs and ploughs, muttering, 'Damn that kip!' Then he stands up and frowns, and begins to tally off his clews on his fingers—and gets stuck at the ring-finger. But only just a minute—then his face glares all up in a smile like a house afire, and he straightens up stately and

More thinking

majestic, and says to the crowd, 'Take a lantern, a couple of you, and go down to Injun Billy's and fetch the child—the rest of you go 'long home to bed; good-night, madam; good-night, gents.' And he bows like the Matterhorn, and pulls out for the tavern. That's *his* style, and the *Only*—scientific, intellectual—all over in fifteen minutes—no poking around all over the sage-brush range an hour and a half in a mass-meeting crowd for *him*, boys—you hear *me!*"

Easy style

"By Jackson, it's grand!" said Ham Sandwich. "Wells-Fargo, you've got him down to a dot. He ain't painted up any exacter to the life in the books. By George, I can just *see* him—can't you, boys?"

"You bet you! It's just a photograft, that's what it is."

Ferguson was profoundly pleased with his success, and grateful. He sat silently enjoying his happiness a little while, then he murmured, with a deep awe in his voice,

"I wonder if God made him?"

There was no response for a moment; then Ham Sandwich said, reverently,

"Not all at one time, I reckon."

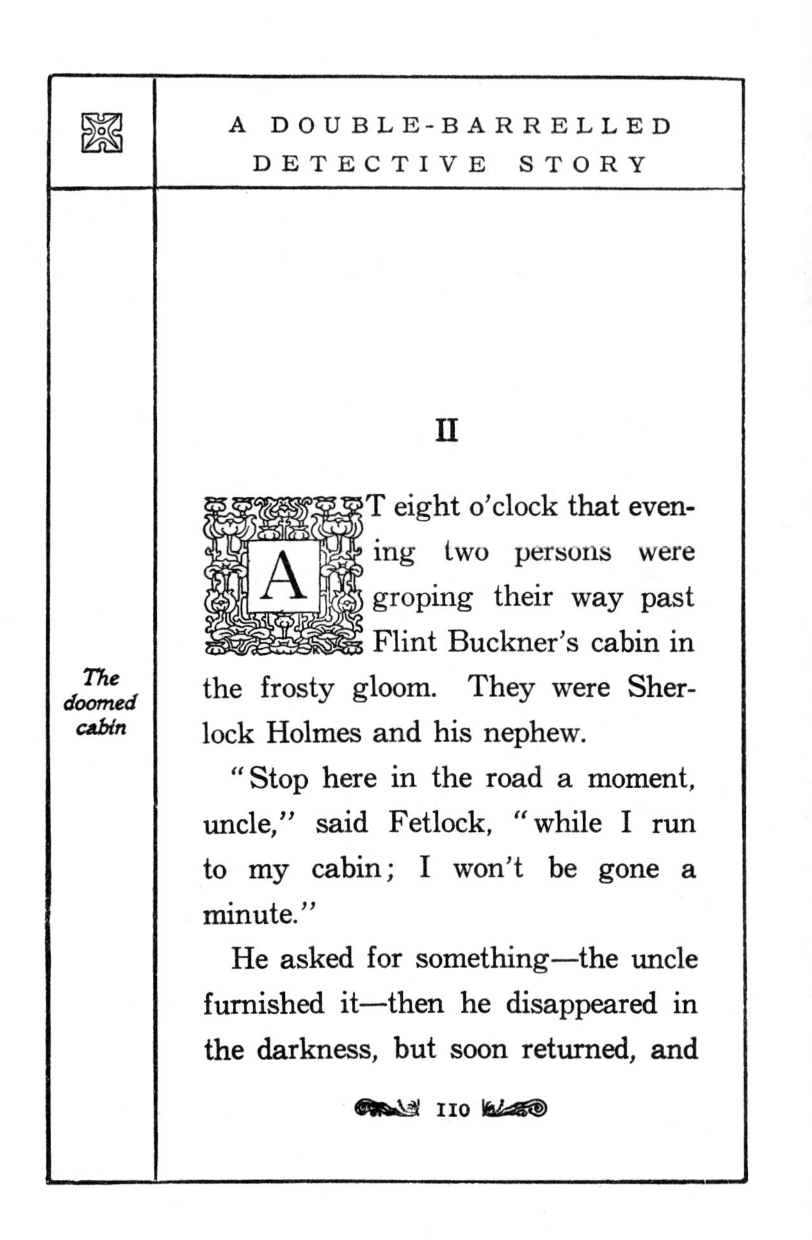

II

The doomed cabin

AT eight o'clock that evening two persons were groping their way past Flint Buckner's cabin in the frosty gloom. They were Sherlock Holmes and his nephew.

"Stop here in the road a moment, uncle," said Fetlock, "while I run to my cabin; I won't be gone a minute."

He asked for something—the uncle furnished it—then he disappeared in the darkness, but soon returned, and

Back to the tavern

the talking-walk was resumed. By nine o'clock they had wandered back to the tavern. They worked their way through the billiard-room, where a crowd had gathered in the hope of getting a glimpse of the Extraordinary Man. A royal cheer was raised. Mr. Holmes acknowledged the compliment with a series of courtly bows, and as he was passing out his nephew said to the assemblage,

"Uncle Sherlock's got some work to do, gentlemen, that 'll keep him till twelve or one, but he'll be down again then, or earlier if he can, and hopes some of you'll be left to take a drink with him."

"By George, he's just a duke, boys! Three cheers for Sherlock

Holmes, the greatest man that ever lived!" shouted Ferguson. "Hip, hip, hip—"

"Hurrah! hurrah! hurrah! Tiger!"

To treat his admirers

The uproar shook the building, so hearty was the feeling the boys put into their welcome. Upstairs the uncle reproached the nephew gently, saying,

"What did you get me into that engagement for?"

"I reckon you don't want to be unpopular, do you, uncle? Well, then, don't you put on any exclusiveness in a mining-camp, that's all. The boys admire you; but if you was to leave without taking a drink with them, they'd set you down for a snob. And, besides, you said you had home talk enough in stock to keep us up and at it half the night."

The boy was right, and wise—the uncle acknowledged it. The boy was wise in another detail which he did not mention,—except to himself: "Uncle and the others will come handy—in the way of nailing an *alibi* where it can't be budged."

An alibi

He and his uncle talked diligently about three hours. Then, about midnight, Fetlock stepped down stairs and took a position in the dark a dozen steps from the tavern, and waited. Five minutes later Flint Buckner came rocking out of the billiard-room and almost brushed him as he passed.

"I've *got* him!" muttered the boy.

Buckner's last walk

He continued to himself, looking after the shadowy form: "Good-by—good-by for good, Flint Buckner; you called my mother a—well, never mind what; it's all right, now; you're taking your last walk, friend."

He went musing back into the tavern. "From now till one is an hour. We'll spend it with the boys; it's good for the *alibi*."

He brought Sherlock Holmes to the billiard-room, which was jammed with eager and admiring miners; the guest called the drinks, and the fun began. Everybody was happy; everybody was complimentary; the ice was soon broken; songs, anecdotes, and more drinks followed, and the pregnant minutes flew. At six minutes to one, when the jollity was at its highest—

Boom!

There was silence instantly. The deep sound came rolling and rumbling from peak to peak up the gorge, then died down, and ceased. The spell broke, then, and the men made a rush for the door, saying,

The explosion

"Something's blown up!"

Outside, a voice in the darkness said,

"It's away down the gorge; I saw the flash."

The crowd poured down the canyon—Holmes, Fetlock, Archy Stillman, everybody. They made the mile in a few minutes. By the light of a lantern they found the smooth and solid dirt floor of Flint Buckner's cabin; of the cabin itself not a vestige remained, not a rag nor a splinter. Nor any sign of Flint. Search-parties sought here and there and yonder, and presently a cry went up.

"Here he is!"

Some remains

It was true. Fifty yards down the gulch they had found him—that is, they had found a crushed and lifeless mass which represented him. Fetlock Jones hurried thither with the others and looked.

The inquest was a fifteen-minute affair. Ham Sandwich, foreman of the jury, handed up the verdict, which was phrased with a certain unstudied literary grace, and closed with this finding, to wit: that "deceased came to his death by his own act or some other person or persons unknown to this jury not leaving any family or similar effects behind but his cabin which was blown away and God have mercy on his soul amen."

Impatient jury

Then the impatient jury rejoined the main crowd, for the storm-centre of interest was there — Sherlock Holmes. The miners stood silent and reverent in a half-circle, enclosing a large vacant space which included the front exposure of the site of the late premises. In this considerable space the Extraordinary Man was moving about, attended by his nephew with a lantern. With a tape he took measurements of the cabin site; of the distance from the wall of chapar-

Sherlock's real job

ral to the road; of the height of the chaparral bushes; also various other measurements. He gathered a rag here, a splinter there, and a pinch of earth yonder, inspected them profoundly, and preserved them. He took the "lay" of the place with a pocket-compass, allowing two seconds for magnetic variation. He took the time (Pacific) by his watch, correcting it for local time. He paced off the distance from the cabin site to the corpse, and corrected that for tidal differentiation. He took the altitude with a pocket-aneroid, and the temperature with a pocket-thermometer. Finally he said, with a stately bow:

"It is finished. Shall we return, gentlemen?"

He took up the line of march for the tavern, and the crowd fell into his wake, earnestly discussing and admiring the Extraordinary Man, and interlarding guesses as to the origin of the tragedy and who the author of it might be.

Luck for the camp

"My, but it's grand luck having him here—hey, boys?" said Ferguson.

"It's the biggest thing of the century," said Ham Sandwich. "It 'll go all over the world; you mark my words."

"*You* bet!" said Jake Parker the blacksmith. "It 'll boom this camp. Ain't it so, Wells-Fargo?"

"Well, as you want my opinion—if it's any sign of how *I* think about

Bunches of clews

it, I can tell you this: yesterday I was holding the Straight Flush claim at two dollars a foot; I'd like to see the man that can get it at sixteen to-day."

"Right you are, Wells-Fargo! It's the grandest luck a new camp ever struck. Say, did you see him collar them little rags and dirt and things? What an eye! He just can't over-look a clew—'tain't *in* him."

"That's so. And they wouldn't mean a thing to anybody else; but to him, why, they're just a book—large print at that."

"Sure's you're born! Them odds and ends have got their little old secret, and they think there ain't anybody can pull it; but, land! when he sets his grip there they've got to squeal, and don't you forget it."

"Boys, I ain't sorry, now, that he wasn't here to roust out the child; this is a bigger thing, by a long sight. Yes, sir, and more tangled up and scientific and intellectual."

A big thing

"I reckon we're all of us glad it's turned out this way. Glad? 'George! it ain't any name for it. Dontchu-know, Archy could 've *learnt* some-thing if he'd had the nous to stand by and take notice of how that man works the system. But no; he went poking up into the chaparral and just missed the whole thing."

"It's true as gospel; I seen it myself. Well, Archy's young. He'll know better one of these days."

Who did it?

"Say, boys, who do you reckon done it?"

That was a difficult question, and brought out a world of unsatisfying conjecture. Various men were mentioned as possibilities, but one by one they were discarded as not being eligible. No one but young Hillyer had been intimate with Flint Buckner; no one had really had a quarrel with him; he had affronted every man who had tried to make up to him, although not quite offensively enough to require bloodshed. There was one name that was upon every tongue from the start, but it was the last to get utterance —Fetlock Jones's. It was Pat Riley that mentioned it.

"Oh, well," the boys said, "of

course we've all thought of him, because he had a million rights to kill Flint Buckner, and it was just his plain duty to do it. But all the same there's two things we can't get around: for one thing, he hasn't got the sand; and for another, he wasn't anywhere near the place when it happened."

Fetlock's alibi

"I know it," said Pat. "He was there in the billiard-room with us when it happened."

"Yes, and was there all the time for an hour *before* it happened."

"It's so. And lucky for him, too. He'd have been suspected in a minute if it hadn't been for that."

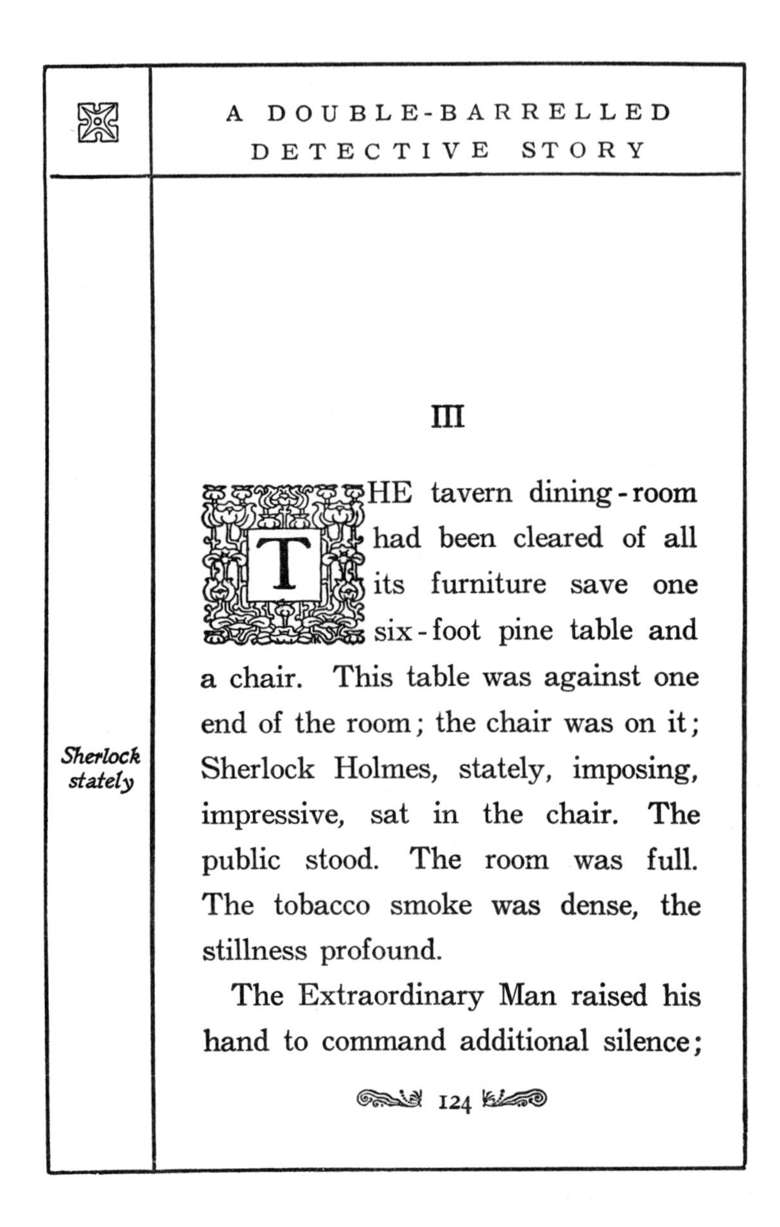

III

THE tavern dining-room had been cleared of all its furniture save one six-foot pine table and a chair. This table was against one end of the room; the chair was on it; Sherlock Holmes, stately, imposing, impressive, sat in the chair. The public stood. The room was full. The tobacco smoke was dense, the stillness profound.

Sherlock stately

The Extraordinary Man raised his hand to command additional silence;

held it in the air a few moments; then, in brief, crisp terms he put forward question after question, and noted the answers with "Um-ums," nods of the head, and so on. By this process he learned all about Flint Buckner, his character, conduct, and habits, that the people were able to tell him. It thus transpired that the Extraordinary Man's nephew was the only person in the camp who had a killing-grudge against Flint Buckner. Mr. Holmes smiled compassionately upon the witness, and asked, languidly—

Thinks with his voice

"Do any of you gentlemen chance to know where the lad Fetlock Jones was at the time of the explosion?"

A thunderous response followed—

"In the billiard-room of this house!"

Just questions

"Ah. And had he just come in?"

"Been there all of an hour!"

"Ah. It is about—about—well, about how far might it be to the scene of the explosion?"

"All of a mile!"

"Ah. It isn't *much* of an alibi, 'tis true, but—"

A storm-burst of laughter, mingled with shouts of, "By jiminy, but he's chain-lightning!" and, "Ain't you sorry you spoke, Sandy?" shut off the rest of the sentence, and the crushed witness drooped his blushing face in pathetic shame. The inquisitor resumed:

"The lad Jones's somewhat *distant* connection with the case" (*laughter*) "having been disposed of, let us now call the *eye*-witnesses of the tragedy, and listen to what they have to say."

He got out his fragmentary clews and arranged them on a sheet of cardboard on his knee. The house held its breath and watched.

A few words

"We have the longitude and the latitude, corrected for magnetic variation, and this gives us the exact location of the tragedy. We have the altitude, the temperature, and the degree of humidity prevailing — inestimably valuable, since they enable us to estimate with precision the degree of influence which they would exercise upon the mood and disposition of the assassin at that time of the night." (*Buzz of admiration; muttered remark, "By George, but he's deep!"*) He

More words

fingered his clews. "And now let us ask these mute witnesses to speak to us.

"Here we have an empty linen shot-bag. What is its message? This: that robbery was the motive, not revenge. What is its further message? This: that the assassin was of inferior intelligence—shall we say light-witted, or perhaps approaching that? How do we know this? Because a person of sound intelligence would not have proposed to rob the man Buckner, who never had much money with him. But the assassin might have been a stranger? Let the bag speak again. I take from it this article. It is a bit of silver-bearing quartz. It is peculiar. Examine it, please—you—and

you—and you. Now pass it back, please. There is but one lode on this coast which produces just that character and color of quartz; and that is a lode which crops out for nearly two miles on a stretch, and in my opinion is destined, at no distant day, to confer upon its locality a globe-girdling celebrity, and upon its two hundred owners riches beyond the dreams of avarice. Name that lode, please."

Like a clew

"The Consolidated Christian Science and Mary Ann!" was the prompt response.

A wild crash of hurrahs followed, and every man reached for his neighbor's hand and wrung it, with tears in his eyes; and Wells-Fargo Ferguson shouted, "The Straight Flush is on the lode, and up she goes to a hundred and fifty a foot—you hear *me!*"

When quiet fell, Mr. Holmes resumed:

Commonplace

"We perceive, then, that three facts are established, to wit: the assassin was approximately light-witted; he was not a stranger; his motive was robbery, not revenge. Let us proceed. I hold in my hand a small fragment of fuse, with the recent smell of fire upon it. What is its testimony? Taken with the corroborative evidence of the quartz, it reveals to us that the assassin was a miner. What does it tell us further? This, gentlemen: that the assassination was consummated by means of an explosive. What else does it say? This: that the explosive was located against the side of the cabin nearest the road—the front side—for within six feet of that spot I found it.

"I hold in my fingers a burnt Swedish match—the kind one rubs on a safety-box. I found it in the road, 622 feet from the abolished cabin. What does it say? This: that the train was fired from that point. What further does it tell us? This: that the assassin was left-handed. How do I know this? I should not be able to explain to you, gentlemen, how I know it, the signs being so subtle that only long experience and deep study can enable one to detect them. But the signs are here, and they are re-enforced by a fact which you must

Subtle signs

Left-handed villany

have often noticed in the great detective narratives—that *all* assassins are left-handed."

"By Jackson, *that's* so!" said Ham Sandwich, bringing his great hand down with a resounding slap upon his thigh; "blamed if I ever thought of it before."

"Nor I!" "Nor I!" cried several. "Oh, there can't anything escape ***him*** —look at his eye!"

"Gentlemen, distant as the murderer was from his doomed victim, he did not wholly escape injury. This fragment of wood which I now exhibit to you struck him. It drew blood. Wherever he is, he bears the telltale mark. I picked it up where he stood when he fired the fatal train."

He looked out over the house from his high perch, and his countenance began to darken; he slowly raised his hand, and pointed—

"There stands the assassin!"

Guess at the criminal

For a moment the house was paralyzed with amazement; then twenty voices burst out with:

"Sammy Hillyer? Oh, *hell,* no! *Him?* It's pure foolishness!"

"Take care, gentlemen — be not hasty. Observe—he has the blood-mark on his brow."

Hillyer turned white with fright. He was near to crying. He turned this way and that, appealing to every face for help and sympathy; and held out his supplicating hands toward Holmes and began to plead:

"*Don't,* oh, don't! I never did it; I give my word I never did it. The way I got this hurt on my forehead was—"

"Arrest him, constable!" cried Holmes. "I will swear out the warrant."

The constable moved reluctantly forward—hesitated—stopped.

Hillyer broke out with another appeal. "Oh, Archy, don't let them do it; it would kill mother! *You* know how I got the hurt. Tell them, and save me, Archy; save me!"

Archy to the rescue

Stillman worked his way to the front, and said:

"Yes, I'll save you. Don't be afraid." Then he said to the house, "Never mind how he got the hurt; it

"YES, I'LL SAVE YOU"

hasn't anything to do with this case, and isn't of any consequence."

"God bless you, Archy, for a true friend!"

"Hurrah for Archy! Go in, boy, and play 'em a knock-down flush to their two pair 'n' a jack!" shouted the house, pride in their home talent and a patriotic sentiment of loyalty to it rising suddenly in the public heart and changing the whole attitude of the situation.

Real detective work

Young Stillman waited for the noise to cease; then he said,

"I will ask Tom Jeffries to stand by that door yonder, and Constable Harris to stand by the other one here, and not let anybody leave the room."

"Said and done. Go on, old man!"

The criminal present

"The criminal is present, I believe. I will show him to you before long, in case I am right in my guess. Now I will tell you all about the tragedy, from start to finish. The motive *wasn't* robbery; it was revenge. The murderer *wasn't* light-witted. He *didn't* stand 622 feet away. He *didn't* get hit with a piece of wood. He *didn't* place the explosive against the cabin. He *didn't* bring a shot-bag with him, and he *wasn't* left-handed. With the exception of these errors, the distinguished guest's statement of the case is substantially correct."

A comfortable laugh rippled over the house; friend nodded to friend, as much as to say, "That's the word, with the bark *on* it. Good lad, good boy. *He* ain't lowering his flag any!"

The guest's serenity was not disturbed. Stillman resumed:

"I also have some witnesses; and I will presently tell you where you can find some more." He held up a piece of coarse wire; the crowd craned their necks to see. "It has a smooth coating of melted tallow on it. And here is a candle which is burned half-way down. The remaining half of it has marks cut upon it an inch apart. Soon I will tell you where I found these things. I will now put aside reasonings, guesses, the impressive hitching of odds and ends of clews together, and the other showy theatricals of the detective trade, and tell you

Telltale evidence

Just how it happened

in a plain, straightforward way just how this dismal thing happened."

He paused a moment, for effect—to allow silence and suspense to intensify and concentrate the house's interest; then he went on:

"The assassin studied out his plan with a good deal of pains. It was a good plan, very ingenious, and showed an intelligent mind, not a feeble one. It was a plan which was well calculated to ward off all suspicion from its inventor. In the first place, he marked a candle into spaces an inch apart, and lit it and timed it. He found it took three hours to burn four inches of it. I tried it myself for half an hour, awhile ago, upstairs here, while the inquiry into Flint Buckner's character

The candle test

and ways was being conducted in this room, and I arrived in that way at the rate of a candle's consumption when sheltered from the wind. Having proved his trial-candle's rate, he blew it out—I have already shown it to you—and put his inch-marks on a fresh one.

"He put the fresh one into a tin candlestick. Then at the five-hour mark he bored a hole through the candle with a red-hot wire. I have already shown you the wire, with a smooth coat of tallow on it — tallow that had been melted and had cooled.

"With labor — very hard labor, I should say—he struggled up through the stiff chaparral that clothes the

Laying the mine

steep hill-side back of Flint Buckner's place, tugging an empty flour-barrel with him. He placed it in that absolutely secure hiding-place, and in the bottom of it he set the candlestick. Then he measured off about thirty-five feet of fuse—the barrel's distance from the back of the cabin. He bored a hole in the side of the barrel—here is the large gimlet he did it with. He went on and finished his work; and when it was done, one end of the fuse was in Buckner's cabin, and the other end, with a notch chipped in it to expose the powder, was in the hole in the candle—timed to blow the place up at one o'clock this morning, provided the candle was lit about eight o'clock yesterday evening—which I am betting it was—and provided there was an explosive in the cabin and connected with that end of the fuse—which I am also betting there was, though I can't prove it. Boys, the barrel is there in the chaparral, the candle's remains are in it in the tin stick; the burnt-out fuse is in the gimlet-hole, the other end is down the hill where the late cabin stood. I saw them all an hour or two ago, when the Professor here was measuring off unimplicated vacancies and collecting relics that hadn't anything to do with the case."

Full details

He paused. The house drew a long, deep breath, shook its strained cords and muscles free and burst into cheers.

"Dang him!" said Ham Sandwich, "that's why he was snooping around in the chaparral, instead of picking up points out of the P'fessor's game. Looky here—*he* ain't no fool, boys."

"No, sir! Why, great Scott—"

But Stillman was resuming:

A second hiding-place

"While we were out yonder an hour or two ago, the owner of the gimlet and the trial-candle took them from a place where he had concealed them—it was not a good place—and carried them to what he probably thought was a better one, two hundred yards up in the pine woods, and hid them there, covering them over with pine needles. It was there that I found them. The gimlet exactly fits the hole in the barrel. And now—"

A pretty fairy-tale

The Extraordinary Man interrupted him. He said, sarcastically:

"We have had a very pretty fairy-tale, gentlemen—very pretty indeed. Now I would like to ask this young man a question or two."

Some of the boys winced, and Ferguson said,

"I'm afraid Archy's going to catch it now."

The others lost their smiles and sobered down. Mr. Holmes said:

"Let us proceed to examine into this fairy-tale in a consecutive and orderly way—by geometrical progression, so to speak—linking detail to detail in a steadily advancing and remorselessly consistent and unassailable march upon this tinsel toy-fortress of error, the dream-fabric of a callow imagination. To begin with, young sir, I desire to ask you but three questions at present—*at present.* Did I understand you to say it was your opinion that the supposititious candle was lighted at about eight o'clock yesterday evening?"

The time test

"Yes, sir—about eight."

"Could you say exactly eight?"

"Well, no, I couldn't be that exact."

"Um. If a person had been passing along there just about that time, he would have been almost sure to encounter that assassin, do you think?"

"Yes, I should think so."

"Thank you, that is all. For the present. I say, all *for the present.*"

"Dern him! he's laying for Archy," said Ferguson.

"It's so," said Ham Sandwich. "I don't like the look of it."

Stillman said, glancing at the guest,

"I was along there myself at half past eight—no, about nine."

On the spot

"In-deed? This is interesting—this is very interesting. Perhaps you encountered the assassin yourself?"

"No, I encountered no one."

"Ah. Then—if you will excuse the remark—I do not quite see the relevancy of the information."

"It has none. At present. I say it has none—at present." He paused. Presently he resumed: "I did not encounter the assassin, but I am on his track, I am sure, for I believe he is in this room. I will ask you all to pass one by one in front of me—here, where there is a good light—so that I can see your feet."

Watching their feet

A buzz of excitement swept the place, and the march began, the guest looking on with an iron attempt at gravity which was not an unqualified success. Stillman stooped, shaded his eyes with his hand, and gazed down intently at each pair of feet as it passed. Fifty men tramped monotonously by—with no result. Sixty. Seventy. The thing was beginning to look absurd. The guest remarked, with suave irony,

"Assassins appear to be scarce this evening."

The house saw the humor of it, and

refreshed itself with a cordial laugh. Ten or twelve more candidates tramped by — no, *danced* by, with airy and ridiculous capers which convulsed the spectators — then suddenly Stillman put out his hand and said,

The real assassin

"This is the assassin!"

"Fetlock Jones, by the great Sanhedrim!" roared the crowd; and at once let fly a pyrotechnic explosion and dazzle and confusion of stirring remarks inspired by the situation.

At the height of the turmoil the guest stretched out his hand, commanding peace. The authority of a great name and a great personality laid its mysterious compulsion upon the house, and it obeyed. Out of the panting calm which succeeded, the guest spoke, saying, with dignity and feeling:

"*This* is serious. It strikes at an innocent life. Innocent beyond suspicion! Innocent beyond peradventure! Hear me *prove* it; observe how simple a fact can brush out of existence this witless lie. Listen. My friends, that lad was never out of my sight yesterday evening at *any* time!"

With the detective

It made a deep impression. Men turned their eyes upon Stillman with grave inquiry in them. His face brightened, and he said,

"I *knew* there was another one!" He stepped briskly to the table and glanced at the guest's feet, then up at his face, and said: "You were *with* him! You were not fifty steps from him when he lit the candle that by-and-by fired the powder!" (*Sensation.*) "And what is more, you furnished the matches yourself!"

He furnished matches

Plainly the guest seemed hit; it looked so to the public. He opened his mouth to speak; the words did not come freely.

"This — er — this is insanity — this—"

Stillman pressed his evident advantage home. He held up a charred match.

"Here is one of them. I found it in the barrel—and there's *another* one there."

The guest found his voice at once.

"*Yes*—and put them there yourself!"

It was recognized as a good shot. Stillman retorted:

Wax matches

"It is *wax*—a breed unknown to this camp. I am ready to be searched for the box. Are you?"

The guest was staggered this time—the dullest eye could see it. He fumbled with his hands; once or twice his lips moved, but the words did not come. The house waited and watched, in tense suspense, the stillness adding effect to the situation. Presently Stillman said, gently,

"We are waiting for your decision."

There was silence again during several moments; then the guest answered, in a low voice,

"I refuse to be searched."

There was no noisy demonstration,

but all about the house one voice after another muttered:

"That settles it! He's Archy's meat."

Archy's meat

What to do now? Nobody seemed to know. It was an embarrassing situation for the moment—merely, of course, because matters had taken such a sudden and unexpected turn that these unpractised minds were not prepared for it, and had come to a standstill, like a stopped clock, under the shock. But after a little the machinery began to work again, tentatively, and by twos and threes the men put their heads together and privately buzzed over this and that and the other proposition. One of

Thanking the assassin

these propositions met with much favor; it was, to confer upon the assassin a vote of thanks for removing Flint Buckner, and let him go. But the cooler heads opposed it, pointing out that addled brains in the Eastern States would pronounce it a scandal, and make no end of foolish noise about it. In the end the cool heads got the upper hand, and obtained general consent to a proposition of their own, and their leader then called the house to order and stated it—to this effect: that Fetlock Jones be jailed and put upon his trial.

The motion was carried. Apparently there was nothing further to do now, and the people were glad, for, privately, they were impatient to get

out and rush to the scene of the tragedy, and see whether that barrel and the other things were really there or not.

But no—the break-up got a check. The surprises were not over yet. For a while Fetlock Jones had been silently sobbing, unnoticed in the absorbing excitements which had been following one another so persistently for some time; but when his arrest and trial were decreed, he broke out despairingly, and said:

Wanted no trial

"No! it's no use. I don't want any jail, I don't want any trial; I've had all the hard luck I want, and all the miseries. Hang me now, and let me out! It would all come out, anyway—there couldn't anything save

The confession

me. He has told it all, just as if he'd been with me and seen it — *I* don't know how he found out; and you'll find the barrel and things, and then I wouldn't have any chance any more. I killed him; and *you'd* have done it too, if he'd treated you like a dog, and you only a boy, and weak and poor, and not a friend to help you."

"And served him damned well right!" broke in Ham Sandwich. "Looky here, boys—"

From the constable: "Order! Order, gentlemen!"

A voice: "Did your uncle know what you was up to?"

"No, he didn't."

"Did he give you the matches, sure enough?"

"Yes, he did; but he didn't know what I wanted them for."

"When you was out on such a business as that, how did you venture to risk having him along—and him a *detective?* How's that?"

The boy hesitated, fumbled with his buttons in an embarrassed way, then said, shyly,

Uses for a detective

"I know about detectives, on account of having them in the family; and if you don't want them to find out about a thing, it's best to have them around when you do it."

The cyclone of laughter which greeted this naïve discharge of wisdom did not modify the poor little waif's embarrassment in any large degree.

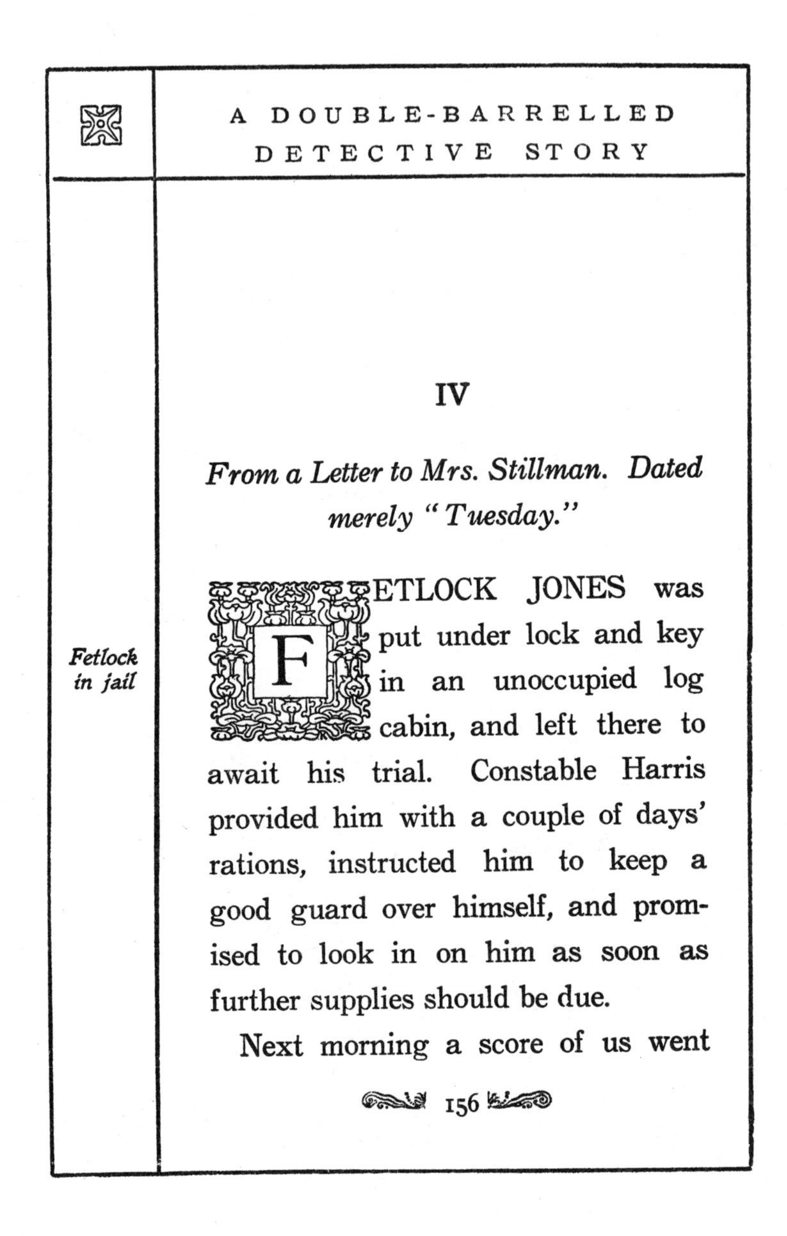

IV

From a Letter to Mrs. Stillman. Dated merely "Tuesday."

Fetlock in jail

FETLOCK JONES was put under lock and key in an unoccupied log cabin, and left there to await his trial. Constable Harris provided him with a couple of days' rations, instructed him to keep a good guard over himself, and promised to look in on him as soon as further supplies should be due.

Next morning a score of us went

with Hillyer, out of friendship, and helped him bury his late relative, the unlamented Buckner, and I acted as first assistant pall-bearer, Hillyer acting as chief. Just as we had finished our labors a ragged and melancholy stranger, carrying an old hand-bag, limped by with his head down, and I caught the scent I had chased around the globe! It was the odor of Paradise to my perishing hope!

The missing one found

In a moment I was at his side and had laid a gentle hand upon his shoulder. He slumped to the ground as if a stroke of lightning had withered him in his tracks; and as the boys came running he struggled to his knees and put up his pleading hands to me, and out of his chattering jaws he begged me to persecute him no more, and said,

"You have hunted me around the world, Sherlock Holmes, yet God is my witness I have never done any man harm!"

Driven insane

A glance at his wild eyes showed us that he was insane. That was my work, mother! The tidings of your death can some day repeat the misery I felt in that moment, but nothing else can ever do it. The boys lifted him up, and gathered about him, and were full of pity of him, and said the gentlest and touchingest things to him, and said cheer up and don't be troubled, he was among friends now, and they would take care of him, and protect him, and hang

IN A MOMENT I WAS AT HIS SIDE

any man that laid a hand on him. They are just like so many mothers, the rough mining-camp boys are, when you wake up the south side of their hearts; yes, and just like so many reckless and unreasoning children when you wake up the opposite side of that muscle. They did everything they could think of to comfort him, but nothing succeeded until Wells-Fargo Ferguson, who is a clever strategist, said,

"If it's only Sherlock Holmes that's troubling you, you needn't worry any more."

"Why?" asked the forlorn lunatic, eagerly.

Sherlock dead again

"Because he's dead again."

"Dead! Dead! Oh, don't trifle

with a poor wreck like me. *Is* he dead? On honor, now—is he telling me true, boys?"

"True as you're a-standing there!" said Ham Sandwich, and they all backed up the statement in a body.

Hanged by mistake

"They hung him in San Bernardino last week," added Ferguson, clinching the matter, "whilst he was searching around after you. Mistook him for another man. They're sorry, but they can't help it now."

"They're a-building him a monument," said Ham Sandwich, with the air of a person who had contributed to it, and knew.

"James Walker" drew a deep sigh—evidently a sigh of relief—and said nothing; but his eyes lost something of their wildness, his countenance cleared visibly, and its drawn look relaxed a little. We all went to our cabin, and the boys cooked him the best dinner the camp could furnish the materials for, and while they were about it Hillyer and I outfitted him from hat to shoe-leather with new clothes of ours, and made a comely and presentable old gentleman of him. "Old" is the right word, and a pity, too; old by the droop of him, and the frost upon his hair, and the marks which sorrow and distress have left upon his face; though he is only in his prime in the matter of years. While he ate, we smoked and chatted; and when he was finishing he found his voice at last, and of his own

Soothing the old man

accord broke out with his personal history. I cannot furnish his exact words, but I will come as near it as I can.

The "Wrong Man's" Story.

The wrong man's story

It happened like this: I was in Denver. I had been there many years; sometimes I remember how many, sometimes I don't—but it isn't any matter. All of a sudden I got a notice to leave, or I would be exposed for a horrible crime committed long before—years and years before—in the East. I knew about that crime, but I was not the criminal; it was a cousin of mine of the same name. What should I better do? My head was all disordered by fear, and I didn't know.

Under a false name

I was allowed very little time—only one day, I think it was. I would be ruined if I was published, and the people would lynch me, and not believe what I said. It is always the way with lynchings; when they find out it is a mistake they are sorry, but it is too late,—the same as it was with Mr. Holmes, you see. So I said I would sell out and get money to live on, and run away until it blew over and I could come back with my proofs. Then I escaped in the night and went a long way off in the mountains somewhere, and lived disguised and had a false name.

I got more and more troubled and worried, and my troubles made me see spirits and hear voices, and I could

Restless voices

not think straight and clear on any subject, but got confused and involved and had to give it up, because my head hurt so. It got to be worse and worse; more spirits and more voices. They were about me all the time; at first only in the night, then in the day too. They were always whispering around my bed and plotting against me, and it broke my sleep and kept me fagged out, because I got no good rest.

And then came the worst. One night the whispers said, "We'll never manage, because we can't *see* him, and so can't point him out to the people."

They sighed; then one said: "We must bring Sherlock Holmes. He can be here in twelve days."

They all agreed, and whispered and jibbered with joy. But my heart broke; for I had read about that man, and knew what it would be to have him upon my track, with his superhuman penetration and tireless energies.

Fleeing from Sherlock

The spirits went away to fetch him, and I got up at once in the middle of the night and fled away, carrying nothing but the hand-bag that had my money in it—thirty thousand dollars; two-thirds of it are in the bag there yet. It was forty days before that man caught up on my track. I just escaped. From habit he had written his real name on a tavern register, but had scratched it out and written "Dagget Barclay" in the place of it. But fear gives you a watchful eye and keen, and I read the true name through the scratches, and fled like a deer.

Hunted for years

He has hunted me all over this world for three years and a half—the Pacific States, Australasia, India—everywhere you can think of; then back to Mexico and up to California again, giving me hardly any rest; but that name on the registers always saved me, and what is left of me is alive yet. And I am *so* tired! A cruel time he has given me, yet I give you my honor I have never harmed him nor any man.

That was the end of the story, and it stirred those boys to blood-heat, be sure of it. As for me—each word burnt a hole in me where it struck.

Self-reproach

We voted that the old man should bunk with us, and be my guest and Hillyer's. I shall keep my own counsel, naturally; but as soon as he is well rested and nourished, I shall take him to Denver and rehabilitate his fortunes.

The boys gave the old fellow the bone-mashing good-fellowship handshake of the mines, and then scattered away to spread the news.

At dawn next morning Wells-Fargo Ferguson and Ham Sandwich called us softly out, and said, privately:

"That news about the way that old stranger has been treated has spread all around, and the camps

After Sherlock again

are up. They are piling in from everywhere, and are going to lynch the P'fessor. Constable Harris is in a dead funk, and has telephoned the sheriff. Come along!"

We started on a run. The others were privileged to feel as they chose, but in my heart's privacy I hoped the sheriff would arrive in time, for I had small desire that Sherlock Holmes should hang for my deeds, as you can easily believe. I had heard a good deal about the sheriff, but for reassurance' sake I asked,

"Can he stop a mob?"

"Can *he* stop a mob! Can Jack *Fairfax* stop a mob! Well, I should smile! Ex-desperado—nineteen scalps on his string. Can *he!* Oh, I *say!*"

As we tore up the gulch, distant cries and shouts and yells rose faintly on the still air, and grew steadily in strength as we raced along. Roar after roar burst out, stronger and stronger, nearer and nearer; and at last, when we closed up upon the multitude massed in the open area in front of the tavern, the crash of sound was deafening. Some brutal roughs from Daly's Gorge had Holmes in their grip, and he was the calmest man there; a contemptuous smile played about his lips, and if any fear of death was in his British heart, his iron personality was master of it, and no sign of it was allowed to appear.

A mob

"Come to a vote, men!" This from one of the Daly gang, Shadbelly

Burn him

Higgins. "Quick! is it hang, or shoot?"

"Neither!" shouted one of his comrades. "He'd be alive again in a week; burning's the only permanency for *him.*"

The gangs from all the outlying camps burst out in a thunder-crash of approval, and went struggling and surging toward the prisoner, and closed around him, shouting, "Fire! fire's the ticket!" They dragged him to the horse-post, backed him against it, chained him to it, and piled wood and pine cones around him waist-deep. Still the strong face did not blench, and still the scornful smile played about the thin lips.

"A match! fetch a match!"

Shadbelly struck it, shaded it with his hand, stooped, and held it under a pine cone. A deep silence fell upon the mob. The cone caught, a tiny flame flickered about it a moment or two. I seemed to catch the sound of distant hoofs—it grew more distinct — still more and more distinct, more and more definite, but the absorbed crowd did not appear to notice it. The match went out. The man struck another, stooped, and again the flame rose; this time it took hold and began to spread—here and there men turned away their faces. The executioner stood with the charred match in his fingers, watching his work. The hoof-beats turned a projecting crag, and now

Funeral pyre

The sheriff

they came thundering down upon us. Almost the next moment there was a shout—

"The sheriff!"

And straightway he came tearing into the midst, stood his horse almost on his hind feet, and said,

"Fall back, you gutter-snipes!"

He was obeyed. By all but their leader. He stood his ground, and his hand went to his revolver. The sheriff covered him promptly, and said:

"Drop your hand, you parlor-desperado. Kick the fire away. Now unchain the stranger."

The parlor-desperado obeyed. Then the sheriff made a speech; sitting his horse at martial ease, and not warming his words with any touch

"THE SHERIFF!"

of fire, but delivering them in a measured and deliberate way, and in a tone which harmonized with their character and made them impressively disrespectful.

"You're a nice lot—now ain't you? Just about eligible to travel with this bilk here—Shadbelly Higgins—this loud-mouthed sneak that shoots people in the back and calls himself a desperado. If there's anything I do particularly despise, it's a lynching mob; I've never seen one that had a man in it. It has to tally up a hundred against one before it can pump up pluck enough to tackle a sick tailor. It's made up of cowards, and so is the community that breeds it; and ninety-nine times out of a hundred

Cowing the mob

Names the sheriff

the sheriff's another one." He paused—apparently to turn that last idea over in his mind and taste the juice of it—then he went on: "The sheriff that lets a mob take a prisoner away from him is the lowest-down coward there is. By the statistics there was a hundred and eighty-two of them drawing sneak pay in America last year. By the way it's going, pretty soon there'll be a new disease in the doctor books—*sheriff complaint.*" That idea pleased him—any one could see it. "People will say, 'Sheriff sick again?' 'Yes; got the same old thing.' And next there'll be a new title. People won't say, 'He's running for sheriff of Rapaho County,' for instance; they'll say, 'He's running for Coward of Rapaho.' Lord, the idea of a grown-up person being afraid of a lynch mob!"

He turned an eye on the captive, and said, "Stranger, who are you, and what have you been doing?"

"My name is Sherlock Holmes, and I have not been doing anything."

A name to conjure with

It was wonderful, the impression which the sound of that name made on the sheriff, notwithstanding he must have come posted. He spoke up with feeling, and said it was a blot on the country that a man whose marvellous exploits had filled the world with their fame and their ingenuity, and whose histories of them had won every reader's heart by the brilliancy and charm of their literary setting, should be visited under the Stars and Stripes by an outrage like this. He apologized in the name of the whole nation, and made Holmes a most handsome bow, and told Constable Harris to see him to his quarters, and hold himself personally responsible if he was molested again. Then he turned to the mob and said:

Driven to their holes

"Hunt your holes, you scum!" which they did; then he said: "Follow me, Shadbelly; I'll take care of your case myself. No—keep your pop-gun; whenever I see the day that I'll be afraid to have you behind me with that thing, it 'll be time for me to join last year's hundred and eighty-two;" and he rode off in a walk, Shadbelly following.

When we were on our way back to our cabin, toward breakfast-time, we ran upon the news that Fetlock Jones had escaped from his lock-up in the night and is gone! Nobody is sorry. Let his uncle track him out if he likes; it is in his line; the camp is not interested.

Fetlock escapes

The wrong righted

V

TEN *days later.*—"James Walker" is all right in body now, and his mind shows improvement too. I start with him for Denver to-morrow morning.

Next night. Brief note, mailed at a way station.—As we were starting, this morning, Hillyer whispered to me: "Keep this news from Walker until you think it safe and not likely to disturb his mind and check his improvement: the ancient crime he

spoke of was really committed—and by his cousin, as he said. *We buried the real criminal* the other day—the unhappiest man that has lived in a century — Flint Buckner. His real name was Jacob Fuller!" There, mother, by help of me, an unwitting mourner, your husband and my father is in his grave. Let him rest.

The real criminal

THE END

AFTERWORD

Lillian S. Robinson

Mark Twain was a popular writer. The man himself might have insisted on stopping the action right here to present me with this century's Stating the Obvious or Dog Bites Man Award. Still, without an exploration of all the dimensions of this unremarkable declaration, the reader may find "The Stolen White Elephant," *Tom Sawyer, Detective*, and *A Double Barrelled Detective Story* considerably more mysterious than they need to be.

Twain's writing was popular not only in the literary sense of being rooted in the voices and experiences of the people, but also in the commercial sense of appealing to a broad contemporary audience. It was a success for which Twain consciously worked, studying and responding to public taste as reflected in his own sales and those of other authors. Twain the satirist—who, I suspect, had enough self-awareness to view these ambitions of his with the same jaundiced eye he cast on the foibles of others—might have identified the guiding precepts of commercial literature as being two in number: first of all, if what you're doing sells, keep doing it—whence popular series and recurrent characters; second, if what someone else is doing sells, do that, too—whence popular genres like the Western, the romance, and the mystery.

The three works of detective fiction collected here bear the stamp of both of these unprincipled principles. They bring back beloved characters, often placing them in situations that have worked before, and they latch onto a best-selling contemporary trend introduced by other writers. Thus, in *Tom Sawyer, Detective*, Twain casts his familiar Missouri adolescents, Tom and

Huck, in the roles of Conan Doyle's phenomenally successful detectives, Sherlock Holmes and Dr. Watson. The site of the action is the Phelps farm, where the final chapters of *Huckleberry Finn* take place, and the plot, like those of such better-known works as *Pudd'nhead Wilson* and *Those Extraordinary Twins*, relies on the device of a mix-up in identity. "The Stolen White Elephant" takes off from (and on) the popular notion of the brilliant detective whose reasoning powers far surpass those of ordinary, commonsensical mortals. And in *A Double Barrelled Detective Story*, where confusion of close relatives also figures in the narrative, the "bloodhound instinct" with which prenatal influence has (ridiculously) endowed one character is measured against the matchless deductive abilities of that quintessential human bloodhound, Sherlock Holmes himself.

But even this briefest of summaries suggests that another precept needs to be articulated fast, the rule of conduct for the satirist who is convinced that some current literary fashion is a lot of hooey. Does the quest for commercial success win out? (Precept: If it works and it's hooey, do it anyway.) If so, the quondam satirist is compelled to write hooey, too, in full awareness of the enormity of the crime. Or does the satirical vision prevail? (Precept: If it works and it's hooey, show it up.) In which case, the public's poor judgment gets thrown back in its face in the form of unmistakable parody. Among them, these three seriously flawed detective stories of Twain's combine both approaches. *Tom Sawyer, Detective* plays it (almost) straight with the Sherlock Holmes mystique, while the other two make merciless fun of self-proclaimed detective omniscience.

Tom Sawyer, Detective, a short (28,000-word) novel written in 1895 and published both as a serial and in book form (along with *Tom Sawyer Abroad* and other stories) the following year, belongs to the afterlife of Tom Sawyer and Huck Finn. Whether or not Twain ever actually articulated that first rule of the commercial writer — if it sells, keep doing it — the maxim is apparent in the reappearance of Tom and Huck in works subsequent to *Adventures of Huckleberry Finn*, most notably this novel and *Tom Sawyer Abroad*.

Of course, Tom's presence in *Huckleberry Finn* is also a "rerun," and a highly problematic one. There, however, the crux is the artistic risk Twain took in assigning so painful a role to a well-established and beloved character. Tom's insistence on implementing what he knows to be the unnecessary rescue of Jim and on modeling the rescue upon the adventure tales he has read reveals a shallowness and even a cruelty that tarnishes the image of the clever naif who is the popular hero of the novel that bears his name.

If Twain's risk-taking works in *Huckleberry Finn*, it is because Tom's failure to honor either the true story or basic human decency in his eagerness to impose a (lesser) book-derived narrative upon reality serves as a foil for Huck's sense of priorities, which places personal values above any culture he has learned. Twain makes it clear that Tom's moral failure is attributable to his readiness to rate what he reads more highly than what he sees (or might otherwise see) for himself. Huck's superiority lies in his ability to discard the accumulated teachings of civilization when they violate the truth as he experiences it. The contrast is especially striking because, throughout the rest of the novel, it is Huck who is the repository of "white" knowledge, much of it gleaned from books, albeit at second hand, while Jim disputes this knowledge with a fundamental if sometimes wrongheaded literalness.

In *Tom Sawyer Abroad*, Tom's book learning once again gives him romantic visions of heroic activity (he dreams, at the beginning, of going on a crusade to the Holy Land), but it also provides the solid, practical information about geography, mechanics, meteorology, and natural history that makes it possible for the boys, once stranded aloft, to turn their predicament into a triumph. In this context, Tom is restored to the role of genuinely rather than fictively resourceful hero, a walking compendium of the lore of literate white civilization in the nineteenth century. Huck and Jim are both (and both equally) outside of the world that the written word has opened to Tom, for good or ill, and their admiration of Tom's knowledge is often tempered by an exaggerated commonsense distrust of abstraction and theory.

Jim is absent from *Tom Sawyer, Detective*, so Huck, once more the narrator, as he was in both *Tom Sawyer Abroad* and *Huckleberry Finn*, bears the entire burden of expressing awe at Tom's extraordinary powers. Although Tom still hungers after fame and public recognition of his feats, the boys' adventure originates in something inherent in their age and condition, spring fever (Huck tellingly calls it "homesickness") and the concomitant itch to be on the move. Similarly, from the trick he uses to assure Aunt Polly's sending him on the visit to the Phelpses (a ploy reminiscent of the whitewashing scene in *Tom Sawyer*) through his arrival at the mystery's elaborate solution, Tom's abilities are represented as being as innate and natural as his youthful wanderlust. It is not that he *knows* more but that he *sees* more — notices details that others miss and understands the significance of the pattern they form.

"I never see such a head as that boy had," exclaims Huck in one of his reiterated paeans to Tom's powers of observation and deduction. "Why *I* had eyes and I could see things, but they never meant nothing to me. But Tom Sawyer was different. When Tom Sawyer seen a thing it just got up on its hind legs and *talked* to him — told him everything it knowed. *I* never see such a head" (144). Even Tom's curiosity is characterized as a natural trait. As Huck explains,

> If you'd lay out a mystery and a pie before me and him, you wouldn't have to say take your choice; it was a thing that would regulate itself. Because in my nature I have always run to pie, whilst in his nature he has always run to mystery. People are made different. (122)

The diction here is pure Huck Finn (who else would conjure up a mystery and a pie side by side on a table?), but the uncritical voice is an artifact of the commercial imperative. If Tom Sawyer's detective endeavors are to present him as a brilliant observer of the Holmes school and not just a boy who reads a lot and likes to show off, he must be cleansed (whitewashed?) not only of the serious misdeeds in *Huckleberry Finn*, but even of the mischievous acts in *Tom Sawyer*. And Huck also has to be denatured. Twain's double concession to popular taste requires him to bring back characters that worked for him before and then to dress them in the trappings that worked for other writers. So Huck still talks like Huck — intermittently, at least — but his text is that of Dr. John Watson.

The reader's outrage at this flattening of Huck into Watson may actually begin much earlier, with a more fundamental question: What is Huck doing there? After all, in the famous penultimate sentence of *Huckleberry Finn*, he says, "But I reckon I got to light out for the Territory ahead of the rest, because Aunt Sally she's going to adopt me and sivilize me and I can't stand it." Buckets of critical ink have been spilled over the implications of this passage, with commentators disagreeing on many points but always assuming that Huck did indeed light out as he claimed he'd have to. It is an article of faith with those readers interested in Twain as an artist that Huck Finn, having secured Jim's freedom, makes good his own emancipation, his escape from "sivilization."

But then, as Huck says in the final sentence of his own *Adventures*, "I been there before." By "there," he means a respectable woman's adoption and its uncomfortable consequences, but both of the popular sequels imply that Huck in fact did return "there," to the town in which he and Tom and Jim had "been before," for that is where both novels start. In our collective literary-historical imagination, Huck is off exploring the Western frontier. Meanwhile, in Twain's ongoing creative and commercial life, he not only fails to light out but becomes the faithful recorder of Tom Sawyer's feats.

Then again, just before the remark about lighting out, Huck declares that "there ain't nothing more to write about, and I am rotten glad of it, because if I'd a knowed what a trouble it was to make a book I wouldn't a tackled it and ain't agoing to no more." So the attentive reader will notice that the very existence, as well as the content, of the later potboilers violates the text of *Huckleberry Finn*. The afterlife of Tom and Huck harks back, in fact, to the very beginning of *Huckleberry Finn*, where, after reintroducing himself as a character from the book "made" by Mr. Mark Twain that is "mostly a true book, with some stretchers," Huck tells of how, not long after the narrative in Mr. Twain's book, he "lit out" from the Widow Douglas and her "sivilizing" efforts "and was free and satisfied." The denouement of this escape attempt follows immediately.

> Tom Sawyer, he hunted me up, and said he was going to start a band of robbers, and I might join if I would go back to the widow and be respectable. So I went back.

If the commercial sequels are already somehow present in the subtext of *Huckleberry Finn*'s final paragraph, the beginning of that novel shows precisely what is going to happen again: Tom Sawyer will offer a false transgression in place of Huck's contemplated true one, making of Twain's masterpiece plus its sequels a combined narrative that doubles back on itself like a kind of *Huckleberry Finnegans Wake*, and assuring Huck's availability for whatever new version of the "band of robbers" Tom thinks up next.

Not only must Huck become virtually unrecognizable in order to play Tom's rather obtuse sidekick and chronicler, but so must Tom himself. The spirit of mischief associated with the "real" Tom Sawyer flares up for a brief moment at the beginning of *Tom Sawyer, Detective*, when the boys are longing for a change of scene and the prospect of adventure it entails. But once launched upon the adventure, Tom starts taking it and himself very seriously, and Huck's attitude is shaped, willy-nilly, by his mentor's. Neither of the versions of "bad boy" once consummately represented by these two is allowed to play this new detective game. There are two turning points in the plot: one when the tale of the diamond thief escaping from his former confederates is eclipsed by the discovery of a corpse, the other when, upon Silas Phelps's arrest, the (comparative) lark of the murder story becomes a matter of life and death for the dotty old man and of justice for the community. The blunting of Tom's character and the muting of Huck's, which I have called his "denaturing," deprive these turning points of their power. His motivation may change as the narrative becomes more serious, but Tom Sawyer has — unfortunately — been "on the case" from the beginning.

As an astute reader of Sherlockian convention, Twain was aware that its success depended, in the first instance, on two literary elements: the idiosyncratic and invariably "extraordinary" character of the detective hero and the apparent opacity of the plot he is called upon to unravel through application of his powers of observation and deduction. Beyond that, runaway success such as Conan Doyle's mysteries enjoyed depended on the ability to create a *culture* of readers sharing their excitement over each plot and their eagerness for the next installment or adventure. Twain apparently set to work on fitting Tom and Huck into what a dispassionate commentator might have warned him was the Procrustean bed of the Holmes-Watson formula before he had an adequate idea of how he would develop a narrative around them, once installed. Only after a false start on a detective story called "Tom Sawyer's Mystery" did Twain hit upon a plot he believed could be adapted to the givens of the Tom-as-Sherlock situation. The novel falls flat because neither the detective character nor the plot is sufficiently compelling, separately or together. And any attempt to build a community of eager fans upon readers' familiarity with the characters of Tom and Huck fails as a result.

Twain precedes the text of *Tom Sawyer, Detective* with a note:

> Strange as the incidents of this story are, they are not inventions, but facts — even to the public confession of the accused. I take them from an old-time Swedish criminal trial, change the actors, and transfer the scene to America. I have added some details, but only a couple of them are important ones. (115)

This note may be said to bear the same relation to the famous notice at the beginning of *Huckleberry Finn* ("Persons attempting to find a motive in this narrative . . .") as the dismal sequel does to the classic text. Although it is considerably less amusing, the prefatory note in *Tom Sawyer, Detective* is in its way quite as disingenuous. For Twain's source was not the record of a "Swedish criminal trial," but a novel published in Denmark in 1829 recounting a miscarriage of justice that occurred in that country in the seventeenth century. The substitution of one Scandinavian kingdom for another is not particularly significant, but there is reason to believe the confusion was anything but accidental. It does serve to muddy the narrative's line of descent, in that the wholesale takeover of the plot of a novel published less than seventy years earlier has a rather different ring from (at least implicitly) researching and adapting the record of an obscure three-hundred-year-old trial. The obfuscation matters, because Twain, as soon as he heard the summary of *The Minister of Veilby*, the Danish novel in question, interrupted his labors on *Joan of Arc* to dash off a version featuring Tom Sawyer, apparently believing the purloined plot would assure him of a hit.

Unlike most of Twain's narratives, especially the two earlier and better-known novels featuring Tom and Huck, the plot thus acquired is very tightly constructed. Indeed, the weave is so close that the reader is able to work out the solution long before Tom Sawyer's dramatic mid-trial revelations. In arriving at our deductions, of course, we have the advantage of reading a narrative, not participating in an event, which means that we are confronted with details that we know have to be significant: the presence of twins in the Dunlap family, the fact that Jake, the "bad twin," whose existence was forgotten or unknown to his fellow townspeople, was seen by Tom, Huck, and us to be headed for his brother's place, and the continued animosity of Brace Dunlap toward Silas Phelps.

Because these details have been selected for us out of the multiplicity of data with which Tom is presented on the scene, not only are we able to put them together, but the mystery looks rather obvious. In a Sherlock Holmes narrative, Dr. Watson and we readers are given, in principle, all the information that Holmes himself has, so that we are all properly astounded together at the use to which the great detective has uniquely been able to put facts that only confuse us. Here, by contrast, Tom Sawyer knows as much as we do yet takes longer to come up with the solution. So only Huck — the new, subordinate Huck — is prepared to echo Dr. Watson's amazement at the great detective's insight. Twain has got the thing backwards, which would work in a satire, but is simply ineffective in the kind of straight adaptation he apparently intended.

Putting the guaranteed-successful Tom and Huck into a guaranteed-successful detective format, with a plot that had already worked (as both history and literature), required too many modifications to be altogether convincing. Twain's contemporaries, avid readers of both Tom and Huck and Holmes and Watson, could not bridge the gap between intention and achievement here, any more than we can a century later. And I catch myself looking for the satirist whom Twain had temporarily silenced in a misguided appeal to what he conceived to be the popular taste.

One might wish, in fact, that in *Tom Sawyer, Detective* Twain had stolen a leaf from his own book — surely a more legitimate literary theft than the one perpetrated on *The Minister of Veilby* — and adopted the attitude and tone of the admittedly slight story "The Stolen While Elephant." This tale, initially intended as a chapter in *A Tramp Abroad* but deleted from that manuscript and first published as the title piece of an 1882 collection of short fiction, is actually reprinted in the 1896 volume that was the first appearance in book form of *Tom Sawyer, Detective*. Its inclusion there, along with both novels in which Huck Finn has been reduced to Tom Sawyer's uncritical Boswell, offers a potential and highly salutary corrective to what is most distressing in Twain's "real" detective novel.

The satire in "The Stolen White Elephant" begins with yet another prefatory note in which the piece is said to have been omitted from *A Tramp Abroad*

> because it was feared that some of the particulars had been exaggerated, and that others were not true. Before these suspicions had been proven groundless, the book had gone to press. (7)

So the manifest absurdities of plot and character are presented as having been scrupulously but in the event unnecessarily held from publication pending "verification" of their nonsense. And the narrator, a "chance railway acquaintance," is similarly introduced as a genuinely good old man whose "earnest and sincere manner imprinted the unmistakable stamp of truth upon every statement which fell from his lips." What he really is, of course, is the classic unreliable narrator, not because of any premeditated falsification, but because his point of view is so uncritical.

The only place where the narrator's viewpoint could be interpreted as anything more sinister than extraordinary credulity is in his opening description of how the King of Siam came to be sending a white elephant to Queen Victoria, with its bland papering-over of assumptions too ingrained even to be called an apologia for the excesses of British imperialism. Even this may be read as an aspect of the character Twain is introducing, of whom it may be said that if he unquestioningly swallows the line that the rulers of small undeveloped countries knuckle under in disputes with colonial superpowers because they realize that the stronger country was right all along, then he will indeed swallow anything.

The brilliance — self-proclaimed and media-enhanced — of Inspector Blunt, chief of the New York City detectives, is never challenged by the narrator, even as agents in all quarters simultaneously wire back contradictory "evidence" tracing the trail of the enormous and distinctive piece of lost property. Nor is his faith in Blunt shaken by the announcement that the principal suspects on whom the detective has had his eye turn out to be dead; and the transparent explanation — "I had long suspected these facts. . . . this testimony proves the unerring accuracy of my instinct" (32) — does not appear to faze him. Similarly, when the dead elephant is discovered in the vaults beneath police headquarters, Blunt exclaims, "Our noble profession is vindicated," and the narrator, fainting, believes that it is. In the concluding paragraph, the narrator explains that Blunt's "compromise," the ransom that was to be turned over to the nonexistent thieves at a rendezvous in the basement, where instead they literally stumble over the elephant's carcass, nonetheless cost him $100,000. (The "ransom," in short, did get paid to *someone*.) Detective expenses, he tells us, covering the services — also nonexistent — of a public agency ran to another $42,000. He goes on,

> I never applied for a place again under my government; I am a ruined man and a wanderer in the earth — but my admiration for that man, whom I believe to be the greatest detective the world has ever produced, remains undimmed to this day, and will so remain unto the end. (35)

Satire has long since withdrawn from the scene, to be replaced by broad farce, with the corrupt, incompetent, and vainglorious police official as its proximate target. It should be emphasized that it is not yet detective *fiction* that is being burlesqued. Twain took his historical inspiration from the hilariously blundering efforts of the New York police to recover the corpse of one Alexander T. Stewart, which had been stolen from the family vault in November 1878. ("The Stolen White Elephant" is dated November or December of that year.) The fictional cops may have been more openly venal than those in charge of recovering Mr. Stewart's body, but both sets of officers seem about equally incompetent.

The *textual* sources that Twain lampoons in "The Stolen White

Elephant" are Allan Pinkerton's memoirs of his detective experiences. The pomposity, pretension, self-righteousness, and frequent wrongheadedness of these stories, which Twain saw as constituting a subgenre of their own, with the plot of each case following the same narrative conventions and development, would be hard to exaggerate further. Twain was able to do it only by making the stolen object in his narrative so extremely large, placing it under the cops' very noses, and focusing on the fetishization of (in this case, entirely fruitless) routine.

In 1879, possibly alluding to this short story, as well as the never completed *Simon Wheeler, Detective*, Twain wrote, "I have very extravagantly burlesqued the detective business — if it is possible to burlesque that business." For those contemporaries who saw the parallel between "The Stolen White Elephant" and Pinkerton's popular accounts of "real-life" detective exploits — and it is not clear how many did — the burlesque also makes a more serious social contribution; it provides an alternative view of those Pinkerton agents whose role in repressing the emerging labor movement was soon to become notorious. Perhaps thoughtful readers could even draw the connection between the officious attitudes Twain ridicules and the suppression of popular dissent.

The nonsense of "The Stolen White Elephant" is completely accessible to the reader who does not know or even care about the pioneering Pinkerton and his self-congratulatory memoirs. Twain's spoof can stand on its own feet — such as they are — without reference to its immediate target. By contrast, the reader who is not conversant with Sir Arthur Conan Doyle's fictions and the Sherlock Holmes mystique they engendered will miss much of the fun of *A Double Barrelled Detective Story*. The broad humor of characters with names like Fetlock Jones, Wells-Fargo Ferguson, and Ham Sandwich, not to mention the famous "solitary oesophagus" that sleeps "upon motionless wing" among the "autumn splendor" of spring flowers, is inescapable, but without reference to the literary source, the business seems rather pointless.

Fortunately, the *name* Sherlock Holmes is known to even more people than have perused the great detective's chronicled adventures. So when that name is reverently invoked in a barroom as far from the "Extraordinary

Man's" usual haunts as Hope Canyon, California, because the detective himself has shown up in that unlikely venue (only, of course, to *be* shown up), his baggage includes a clear signpost pointing us in the direction of the satire.

The composition of *A Double Barrelled Detective Story* coincides exactly with the publication of *The Hound of the Baskervilles*, which the *Strand* magazine began to serialize in August of 1901, and which represents one of the peaks of Holmes-mania on both sides of the Atlantic. If Twain's frustration peaked along with the craze, the time was clearly ripe for a satirical look at the phenomenon. One can almost hear him snorting, "'The footprints of a gigantic hound,' my eye!" and the notion of the bloodhound certainly figures prominently in Twain's effort. But *A Double Barrelled Detective Story* is more closely modeled on Doyle's first Sherlock Holmes narrative, *A Study in Scarlet*.

Because it is *A Study in Scarlet* that introduces Holmes and his methods, the text reveals Dr. Watson at his most consistently awestruck as he recounts the great detective's powers of observation and his solution — initially as baffling as the crimes — of two gruesome murders. It is this tone of breathless amazement at Holmes' detective abilities that Twain parodies. The tavern regulars in Hope Canyon, California, have been reading or at least hearing about Watson's accounts, and their version of Holmes' reputation magnifies his prowess only one further notch, from superhuman to supernatural. (Asked whether God — who, we are to understand, made the rest of us fallible creatures and created the whole world in six days — also made Holmes, Ham Sandwich says, "Not all at one time, I reckon.")

A Study in Scarlet also supplies the model for the double narrative on which Twain trains his double barrel. In Doyle's novel, after Holmes has identified, traced, and trapped the murderer, he invites his mystified companions to "put any questions that you like to me now," assuring them that "there is no danger that I will refuse to answer them." But the reader turns the page to encounter *Doyle's* apparent refusal, since the next section begins not with the promised explanation but with what seems to be an entirely unrelated narrative, a third person account of love, violence, and promised revenge in Mormon Utah; only after several more chapters, in which this grim tale works to its conclusion, are we returned to Dr. Watson's "reminiscences" and made to understand that the interpolated story is the background and provides the motives for the London murders that Holmes has just solved.

A Double Barrelled Detective Story has a similarly bifurcated structure, one part consisting of the worldwide search conducted by Archy Stillman, the young man with prenatally induced bloodhound instincts, for the father against whom he has sworn revenge, the other part being the story of the murder that Holmes attempts to solve in the Hope Canyon silver-mining camp, to which Archy's travels have also brought him. As in *A Study in Scarlet*, the murder plot is brought into relation with a dramatic (not to say melodramatic) story of love, betrayal, and vengeance.

But part of the joke resides in the differences between the two sets of narratives. In Doyle's novel, the detective plot constitutes the frame-story into which the old history containing the motive is inserted. The reader's emotional trajectory runs from fascination with Holmes' methods, combined with a (rather enjoyable) shock at the murders, to identification with the avenger and saddened relief that God's justice, rather than the less flexible human law, will prevail in this case. By contrast, Twain begins *A Double Barrelled Detective Story* with the melodrama of the star-crossed lovers, one of whom is rapidly double-crossed as her husband perpetrates the grotesque abuse that provides the motive for revenge, and that also employs the bloodhounds who endow the avenger, as yet *in utero*, with the means to track the husband down. The abuse and revenge story thus constitutes the frame into which the murder plot is inserted, and until the novel's final paragraph, Archy Stillman himself, rather than his mission, appears to be the only link connecting the two narratives.

"Emotional trajectory" is much too high-flown a term to describe the reader's reactions to Twain's narrative. Besides, there is no trajectory to speak of, since the scene-shifting itself and the events that take place in the different sites provide equal amusement. Even the reader who is drawn into the initial melodrama, despite its patent exaggerations, will find it hard to sustain any response but laughter through the first scene shift, to New England, where the child Archy learns why he is different from the other boys, to say nothing of the young man's subsequent epistolary reports to his patient mother on his progress in pursuing the brutal Jacob Fuller.

The principal narrative shift, however, paralleling Doyle's switch of voice and venue in *A Study in Scarlet*, is away from the search for Jacob Fuller to the Hope Canyon mining community. Archy's last letter to his mother in the frame-story is already datelined Hope Canyon, so the change, this time, is not one of scene. But as in Doyle, the narration shifts to the third person, conventionally "omniscient" but far from authoritative, with its wealth of confusing botanical, ornithological and anatomical detail (the "oesophagus" passage). This is funny in itself, but it has a further meaning for those familiar with Doyle's original, which also begins the interpolated narrative with a scene-setting passage, about the Alkali Plain, and which relies for its effects on descriptions of the corrupt and tyrannical social organization of the Mormon patriarchy. The point to be taken is that Doyle's narrator, like Twain's, doesn't know what he's talking about — literally or figuratively.

The real focus of the satire is the figure of Sherlock Holmes, who enters before there is a mystery to be solved, indeed before the murder is committed. Nor does his arrival on the scene serve to prevent the crime, which is never a mystery to the reader, who has been privy to the plan of the perpetrator — none other than Holmes' nephew, Fetlock Jones — from its inception. Once the news spreads that Holmes has checked in to the tavern, we hear more of Fetlock's thoughts as he pricks the balloon of detective reputation.

> "Uncle *Sherlock*! The mean luck of it! . . . But what's the use of being afraid of *him*. Anybody that knows him the way I do knows he can't detect a crime, except when he plans it all out beforehand and arranges the clews and hires some fellow to commit it according to instructions." (96)

Indeed, young Fetlock, bent on mayhem, rapidly integrates "Uncle Sherlock" into his project.

> "We'll go off for a walk, and I'll leave him in the road a minute, so that he won't see what it is I do: the best way to throw a detective off the track, any-

> way, is to have him along when you are preparing the thing. Yes, that's the safest — I'll take him with me." (97–98)

And so it transpires, with this reasoning providing a horselaugh at the trial, once Holmes' "scientific investigation," with its panoply of instruments measuring irrelevant details and its eager pursuit of red herrings, has been shown up for the fraud it is. Meanwhile, Archy Stillman, whose "bloodhound" gifts have led him to pursue and persecute the wrong man for the injury to his mother (although they've also helped him solve real problems for the Hope Canyon residents), employs common sense and *appropriate* close observation to identify the killer. Twain also takes the opportunity in this denouement — twice — to poke fun at Doyle's bringing his hero back to life, for further lucrative adventures, after supposedly killing him off at the Reichenbach Falls.

In one sense, Twain's "Sherlock Holmes" is precisely the same kind of character as Inspector Blunt of "The Stolen White Elephant." But Twain makes it clear that his intention is not merely to expose the fraudulent nature of officious detective brilliance, but to unmask something he considered even worse, meretricious literature. The sheriff who rescues Holmes from a lynch mob composed of erstwhile admirers says "with feeling" that it is

> a blot on the country that a man whose marvelous exploits had filled the world with their fame and their ingenuity and *whose histories of them had won every reader's heart by the brilliancy and charm of their literary setting*, should be visited under the Stars and Stripes by an outrage like this. (175–176, emphasis added)

In *Tom Sawyer, Detective*, Twain tries to get aboard the Sherlock Holmes bandwagon by substituting the tried and true characters of Tom and Huck for Doyle's Holmes and Watson. The attempt fails, in large measure because readers cannot share Huck's awe at Tom's detective brilliance in arriving at a solution telegraphed many pages earlier. In *A Double Barrelled Detective Story*, Twain tries to hijack that same bandwagon by holding "scientific intellection" up to ridicule and making its practitioner just as blundering, self-promoting, and ultimately corrupt as the incompetent policeman of "The Stolen White Elephant." This effort also fails, because despite the name and the deliberate literary parallels, the reader does not believe that Twain's Sherlock is anything like the "real" thing.

Twain's most interesting work with the devices and conventions of detective fiction lies elsewhere than in the three pieces that make up the present volume. One example of that work, the unfinished manuscript *Simon Wheeler, Detective*, may not be any more successful as a spoof than the two satirical stories here (though it is no less so). Unlike them, however, it also gives occasional hints of the satirical non-genre Twain novel it could have become. At the other extreme in Twain's oeuvre, *Pudd'nhead Wilson*, which was not meant as a genre novel, anticipates the police forces of the world, as well as the traditions of twentieth-century detective fiction, by introducing fingerprint evidence to resolve the plot.

It is tempting to imagine the wonderful satire of the Holmes craze that Twain might have produced had he made Tom Sawyer apply an overly literal reading of Doyle to an actual or imagined crime, and let Huck, as his uncritical Watson, chronicle the ensuing debacle. Such a strategy might even have allowed Twain to bring together the goals that eluded him when pursued separately: to capitalize on the popularity of the Sherlock Holmes stories while exposing what he saw as their literary crimes. Who knows? The result might even have been popular.

FOR FURTHER READING

Lillian S. Robinson

Because the three works of Twain's collected here are archetypes of "minor" writings by a recognized master, there is very little critical discussion of any of them. The suggestions given below should be supplemented by reading in one or more of the general critical biographies of Twain, such as Justin Kaplan, *Mr. Clemens and Mark Twain* (New York: Simon and Schuster, 1966), Hamlin Hill, *Mark Twain: God's Fool* (New York: Harper and Row, 1973), and Everett Emerson, *The Authentic Mark Twain: A Literary Biography of Samuel L. Clemens* (Philadelphia: University of Pennsylvania Press, 1984).

Not only does Mark Twain's unfinished novel *Simon Wheeler, Detective*, which he worked on intermittently between 1877 (the year he completed his play *Cap'n Simon Wheeler, the Amateur Detective*) and 1898, afford some of the same, admittedly mild literary pleasure as the two comic pieces in the present collection, but the scholarly edition of the text offers the best available introduction to the entire topic of Twain and detective fiction. In his introduction to the text (New York: New York Public Library, 1963), editor Franklin R. Rogers situates Twain's fragment in the context of the author's work, including all his recorded thoughts on detection, real and fictional.

For Twain's relationship with Arthur Conan Doyle and his works, see Howard G. Baetzhold, *Mark Twain and John Bull: The British Connection* (Bloomington: Indiana University Press, 1970), Franklin R. Rogers, *Mark Twain's Burlesque Patterns As Seen in the Novels and Narratives, 1855-1885* (Dallas: Southern Methodist University Press, 1960), and Jeanne Ritunnano, "Mark Twain vs. Arthur Conan Doyle on Detective Fiction," *Mark Twain Journal* 16, no. 1 (Winter 1971): 10–14. Each of these works also provides some general critical insight into *Tom Sawyer, Detective*, "The Stolen White Elephant," and *A Double Barrelled Detective Story*.

Anyone wishing to pursue questions raised by *Tom Sawyer, Detective* should be familiar with the other documents of what I have called the "afterlife" of Huck and Tom. These consist of the novel *Tom Sawyer Abroad*, published in 1894, and the fragments "Huck Finn and Tom Sawyer Among the Indians" (written in the 1880s and published in 1969), and "Tom Sawyer's Conspiracy" (dating from the late 1890s and also published in 1969). On the novel itself, see J. Christian Bay, "*Tom Sawyer: Detective*: The Origin of the Plot," in *Essays Offered to Herbert Putnam*, edited by William Warner Bishop and Andrew Keogh (New Haven: Yale University Press, 1929), pp. 80–88, and Daniel M. McKeithan, *Court Trials in Mark Twain and Other Essays.* (The Hague: Martinus Nijhoff, 1958).

On "The Stolen White Elephant," Howard G. Baetzhold's *Mark Twain and John Bull*, cited above, discusses the origins of the story, as does his later article "Of Detectives and Their Derring-Do: The Genesis of Mark Twain's 'The Stolen White Elephant,'" *Studies in American Humor* 2 (1976): 183–95. There are also two articles devoted to the comic opening of the Hope Canyon section of *A Double Barrelled Detective Story*. See Lane Cooper, "Mark Twain's Lilacs and Laburnums," *Modern Language Notes* 47 (1932), and W. Keith Kraus, "Mark Twain's *A Double Barrelled Detective Story:* A Source for the Solitary Oesophagus," *Mark Twain Journal* 16, no. 2 (Summer 1972): 10–12.

ILLUSTRATORS AND ILLUSTRATIONS IN MARK TWAIN'S FIRST AMERICAN EDITIONS

Beverly R. David & Ray Sapirstein

From the "gorgeous gold frog" stamped into the cover of *The Celebrated Jumping Frog of Calaveras County* in 1867 to the comet-riding captain on the frontispiece of *Extract from Captain Stormfield's Visit to Heaven* in 1909, illustrators and illustrations were an integral part of Mark Twain's first editions.

Twain marketed most of his major works by subscription, and illustration functioned as an important sales tool. Subscription books were packed with pictures of every type and size and were bound in brassy gold-stamped covers. The books were sold by agents who flipped through a prospectus filled with lively illustrations, selected text, and binding samples. Illustrations quickly conveyed a sense of the story, condensing the proverbial "thousand words" and outlining the scope and tone of the work, making an impression on the potential purchaser even before the full text had been printed. Book canvassers were rewarded with up to 50 percent of the selling price, which started at $3.50 and ranged as high as $7.00 for more ornate bindings. The books themselves were seldom produced until a substantial number of customers had placed orders. To justify the relatively high price and to reassure buyers that they were getting their money's worth, books published by subscription had to offer sensational volume and apparent substance. As Frank Bliss of the American Publishing Company observed, these consumers "would not pay for blank paper and wide margins. They wanted everything filled up with type or pictures." While authors of trade books generally tolerated lighter sales, gratified by attracting a "better class of readers," as Hamlin Hill put it, authors of subscription books sacrificed literary respectability for popular appeal and considerable profit.[1]

The humorist George Ade remembered Twain's books vividly, offering us a child's-eye view of the nineteenth-century subscription book market.

> Just when front-room literature seemed at its lowest ebb, so far as the American boy was concerned, along came Mark Twain. His books looked at a distance, just like the other distended, diluted, and altogether tasteless volumes that had been used for several decades to balance the ends of the center table . . . so thick and heavy and emblazoned with gold that [they] could keep company with the bulky and high-priced Bible. . . . The publisher knew his public, so he gave a pound of book for every fifty cents, and crowded in plenty of wood-cuts and stamped the outside with golden bouquets and put in a steel engraving of the author, with a tissue paper veil over it, and "sicked" his multitude of broken-down clergymen, maiden ladies, grass widows, and college students on the great American public.
>
> Can you see the boy, Sunday morning prisoner, approach the book with a dull sense of foreboding, expecting a dose of Tupper's *Proverbial Philosophy*? Can you see him a few minutes later when he finds himself linked arm-in-arm with Mulberry Sellers or Buck Fanshaw or the convulsing idiot who wanted to know if Christopher Columbus was sure-enough dead? No wonder he curled up on the hair-cloth sofa and hugged the thing to his bosom and lost all interest in Sunday school. *Innocents Abroad* was the most enthralling book ever printed until *Roughing It* appeared. Then along came *The Gilded Age*, *Life on the Mississippi*, and *Tom Sawyer*. . . . While waiting for a new one we read the old ones all over again.[2]

Publishers, editors, and Twain himself spent a good deal of time on design — choosing the most talented artists, directing their interpretations of text, selecting from the final prints, and at times removing material they deemed unfit for illustration.[3]

With the exception of *Following the Equator* (1897), books released in the twilight of Twain's career were not sold by subscription. Twain's later books, published for the trade market by Harper and Brothers, seldom contained more than a frontispiece and a dozen or so tasteful illustrations, rather than the hundreds of illustrations per volume that subscription publishing demanded. Illustration, however, remained a major component of Twain's later work in two important cases: *Extracts from Adam's Diary*, illustrated by Fred

Strothmann in 1904, and *Eve's Diary*, illustrated by Lester Ralph in 1906.

The stories behind the illustrators and illustrations of Mark Twain's first editions abound in back-room intrigue. The besotted or negligent lapses of some of the artists and the procrastinations of the engravers are legendary. The consequent production delays, mistimed releases, and copyright infringements all implied a lack of competent supervision that frequently infuriated Twain and ultimately encouraged him to launch his own publishing company.

In many cases, Twain took illustrations into account as he wrote and edited his text, using them as counterpoint and accompaniment to his words, often allowing them to inform his general narrative strategy and to influence the amount of detail he felt necessary to include in his written descriptions. In the most artful and carefully considered illustrated works, an analysis of the relationships between author and illustrator and between text and pictures illuminates key dimensions of Twain's writings and the responses they have elicited from readers. Examinations of even the most straightforward examples of decorative imagery yield insights into the publishing history of Twain's books and his attitudes toward the production process.

The original illustrations in Twain's works have often been replaced in the twentieth century by subsequent visual interpretations. But while Norman Rockwell's well-known nostalgic renderings of *Tom Sawyer* and *Huckleberry Finn* may tell us much about 1930s sensibilities, we would do well to reacquaint ourselves with the first American editions and the artwork they contained if we want to understand the books Twain wrote and the world they affected.

Illustrated books, like the illustrated weekly magazines that first appeared in the 1860s, were a significant source of visual images entering nineteenth-century homes. Because of their widespread popularity and the relative paucity of other sources of visual information, Twain's books helped to define America's perceptions of remote people, exotic scenes, and historic events. In addition to being an essential element of Mark Twain's body of work, illustrations are a documentary source in their own right, a window into Twain's world and our own.

NOTES

1. For background on subscription book publishing, see Hamlin Hill, *Mark Twain and Elisha Bliss* (Columbia: University of Missouri Press, 1964), chapter 1. See also R. Kent Rasmussen, "Subscription-book publishing" entry, *Mark Twain A to Z: The Essential Reference to His Life and Writings* (New York: Facts on File, 1995), p. 448.

2. George Ade, "Mark Twain and the Old-Time Subscription Book," *Review of Reviews* 61 (June 10, 1910): 703–4; reprinted in Frederick Anderson, ed., *Mark Twain: The Critical Heritage* (London: Routledge and Kegan Paul, 1971), pp. 337–39.

3. Beverly R. David, *Mark Twain and His Illustrators, Volume 1 (1869–1875)* (Troy, N.Y.: Whitston Publishing Company, 1986), discusses in detail Twain's involvement in the production of his early books.

READING THE ILLUSTRATIONS IN *THE STOLEN WHITE ELEPHANT, TOM SAWYER, DETECTIVE,* AND *A DOUBLE BARRELLED DETECTIVE STORY*

Beverly R. David & Ray Sapirstein

While *The Stolen White Elephant, Etc.* was published without illustration in the 1882 first edition, an ornate cover featured the face of a slyly malevolent elephant peering out at the reader from within a decorative frame as if sharing an inside joke. For Twain's next foray into the detective genre, his publisher commissioned the talented Arthur Burdett Frost (1851–1928), among the foremost American illustrators, to work on *Tom Sawyer, Detective* (1896).

Frost began his career as a wood engraver and lithographer, devoting his evenings to the study of drawing. In 1874 he furnished illustrations for the best-seller *Out of the Hurly-Burly*, by Max Adeler (Charles Heber Clark), and a year later he began drawing political cartoons for the *New York Daily Graphic*, the newspaper that apprenticed several of Twain's illustrators. Frost spent a year studying in England in 1877 and upon his return enrolled in the Pennsylvania Academy of Fine Arts to study painting with Thomas Eakins. He also studied with William Merritt Chase for a short time in 1891.[1] Frost's bibliography reads like a primer in English and American popular literature. He drew for innumerable periodicals and illustrated his own works and others by Dickens, Thackeray, Lewis Carroll, Albion Tourgée, Theodore Roosevelt, Frank Stockton, Joel Chandler Harris, Richard Harding Davis, Alice French, Ruth McEnery Stuart, John Kendrick Bangs, Thomas Nelson Page, Thomas Bailey Aldrich, and Owen Wister, among a myriad of authors.[2]

In addition to Frost's work on *Tom Sawyer, Detective*, Twain enlisted him as a contributor to *Following the Equator* the next year. While Frank Stockton described Frost as "essentially a humorist," an appreciative Joel Chandler Harris commented that "the pictorial art of Frost . . . rises above that of contemporary illustrators" and that "the individuality, the atmosphere are all American."[3] Famous for his landscapes and outdoor sporting scenes, Frost was known to his contemporaries as a "Mark Twain of the Illustrators."[4]

Unfortunately, by 1896 Frost's eyesight had begun to fail, and he could no longer execute the delicate pen-and-ink line drawings he had done earlier in his career. For *Tom Sawyer, Detective* he produced dark and mysterious wash drawings that were photoengraved and inserted into the book. The young detectives, Tom and Huck, are at times indistinguishable except for their costumes; Tom is identified by a vest and high cap, while Huck wears a loose shirt and flat cap.

Frost's characters are a far cry from the characters True Williams originally depicted in *The Adventures of Tom Sawyer* and the ones Edward Kemble drew for *Adventures of Huckleberry Finn*. Dan Beard, too, had illustrated Tom, Jim, and Huck, in *Tom Sawyer Abroad* in 1894. Each illustrator constructed a different vision of the characters, and each altered readers' perceptions of them. Edward Kemble indelibly forged Huck's identity, building on Twain's descriptions and etching a distinctive, convincing persona, full of character and individuality. It is worth noting that Twain selected a different illustrator to interpret each of his Tom and Huck books, and that each artist ranked among the most important and competent illustrators of the day.

Perhaps influenced most by Beard's conception in *Tom Sawyer Abroad*, or possibly because of his growing visual impairment, Frost blurred the individual identities of the characters, portraying them almost as a unit of two fairly generic and conventional young adventurers rather than as two distinct but complementary characters. Twain had continued to differentiate Tom and Huck and had steadfastly maintained his original conception of them even as he inserted them into a rather conventional potboiler without the larger themes and concerns of *Adventures of Huckleberry Finn*. Frost may not have read *Huckleberry Finn*, or like many readers, he may have perceived it as a standard boy's adventure, missing much of the drama that lay just beneath the surface. In any case, he was justified in his treatment of *Tom Sawyer, Detective* as a discrete entity, without heavy references to its predecessors.

Mark Twain did not read proof or give final approval to Frost's illustrations before publication, since he had already begun his *Following the Equator* tour, which lasted from late summer 1895 through midsummer 1896. After the publication of *Tom Sawyer, Detective*, Twain said that Frost, the "best humorous artist that I know of," had "told me three years ago that he had long had an ambition to make some illustrations for me."[5] Twain later expressed satisfaction with Frost's work, although not specifically with the drawings for *Tom Sawyer, Detective*, and solicited his services as an illustrator for a last but unrealized travel book. Unfortunately, all of Frost's illustrations were excised from subsequent American editions of *Tom Sawyer, Detective*, an omission that probably had more to do with budgetary concerns than with Twain's estimation of Frost as an illustrator.

The next tale in this volume, *A Double Barrelled Detective Story* (1902), features illustrations by Lucius Hitchcock (1868–1942), a Paris-trained painter who had illustrated *The Man That Corrupted Hadleyburg* two years earlier.[6] As in *Hadleyburg*, Hitchcock's sepia-tone illustrations were executed as finished wash drawings rather than the comparatively primitive line drawings that had decorated much of Twain's early work. From the 1880s onward, photomechanical reproduction enabled illustrators to produce paintings and delicately shaded images that could not be captured by the cruder engraving process, and that gave the books an aura of refinement.

Hitchcock's images recall the conservative pictorial sensibility of the *Saturday Evening Post* and the Brandywine school illustrators N. C. Wyeth and Howard Pyle, his contemporaries. He is equally stilted in his depiction of stalwart, archetypal characters, and his style would seem to conflict with Twain's antipathy toward moribund sentiment and morality. However, Hitchcock's conventional illustrations operate on one level to give this venture into the detective genre an authentic appearance, even as Twain gently satirizes and undermines the genre on another. Unfortunately, the book provides no image of Sherlock Holmes; the illustrator missed a golden opportunity to match Twain's comic irony.

A Double Barrelled Detective Story also features decorative layouts with illustrated capitals and marginal inscriptions that condense and index the events in the text. Several different colors of ink were used throughout the first edition, a mark of the fastidious craftsmanship with which the book was produced. Only the text was originally printed in black ink. The marginal lines and page headings appeared in red, and the illustrations were printed in chestnut brown, inset in a deep olive border (the colors have not been reproduced in this volume).

The fine bookmanship of *A Double Barrelled Detective Story* attempts to legitimize Twain's genre-bending vacillation between pulp Western and detective parody. The fine-art style of the illustrations and book design conflicts markedly with the comic-bookish content, as is evident in the awkwardly lurid "He Proceeded to Lash Her to a Tree" (facing 8) and in the conventional adolescent romanticism of such illustrations as "The Sheriff" (facing 172) and "Yes I'll Save You!" (facing 134). Not surprisingly, reviewers were not distracted by the story's elegant swaddling. A reviewer in the *Bookman* in 1904 prefaced a terse thrashing of *Extracts from Adam's Diary* by recalling, "We thought when we read 'A Double-Barrelled Detective Story' that Mark Twain could do no worse."[7]

NOTES

1. Biographical entry for Frost in Guy McElroy, ed., *Facing History: The Black Image in American Art 1710–1940* (Washington, D.C.: Bedford Arts/Corcoran Gallery, 1990), p. 105.

2. Theodore Bolton, Frost entry, *American Book Illustrators: A Bibliographic Checklist of 123 Artists* (New York: R. R. Bowker, 1938), pp. 55–62.

3. Stockton quoted in Helen L. Earle, ed., *Biographical Sketches of American Artists* (Lansing: Michigan State Library, 1924), p. 124; Joel Chandler Harris, *Atlanta Constitution*, September 26, 1892.

4. Earle, *Biographical Sketches*, p. 124.

5. Quoted in Note on the Texts, *The Adventures of Tom Sawyer*; *Tom Sawyer Abroad*; *Tom Sawyer, Detective*, ed. John C. Gerber, The Works of Mark Twain (Berkeley: University of California Press, 1982), p. 192.

6. Mantle Fielding, *Dictionary of American Painters, Sculptors, and Engravers*, enl. ed. (Greens Farms, Conn.: Modern Books and Crafts, 1974), p. 171.

7. Harry Thurston Peck, "Mark Twain at Ebb Tide," *Bookman* 19 (May 1904):

235–36; reprinted in Frederick Anderson, ed., *Mark Twain: The Critical Heritage* (London: Routledge and Kegan Paul, 1971), p.254.

A NOTE ON THE TEXT

Robert H. Hirst

The text of *The Stolen White Elephant, Etc.* is a photographic facsimile of a copy of the first American edition dated 1882 on the title page. Although books printed from the first edition plates were manufactured until at least 1894, the earliest copies of the first edition were published in June 1882. Two copies were deposited with the Copyright Office on June 12 (*BAL* 3404). §The text of *Tom Sawyer, Detective* is a photographic facsimile of that novella as printed in the first American edition of *Tom Sawyer Abroad; Tom Sawyer, Detective, and Other Stories,* dated 1896 on the title page. Although books printed from the first edition plates were manufactured until at least 1905, the earliest copies of the first edition were published in November or December 1896. Two copies were received by the Copyright Office on November 7 (*BAL* 3447). §The text of *A Double Barrelled Detective Story* is a photographic facsimile of a copy of the first American edition, all known copies of which are dated 1902 on the title page. The first edition was published in April 1902; two copies were received by the Copyright Office on April 10. The copy reproduced here has Jacob Blanck's state B endpapers, but the sequence of states A and B is not known (*BAL* 3471). §All three of the original volumes reproduced here are in the collection of the Mark Twain House in Hartford, Connecticut (810/C625st/1882/c. 6; 810/C625tos/1896/OCLC°; and 810/C625dou/1902/c. 1).

THE MARK TWAIN HOUSE

The Mark Twain House is a museum and research center dedicated to the study of Mark Twain, his works, and his times. The museum is located in the nineteen-room mansion in Hartford, Connecticut, built for and lived in by Samuel L. Clemens, his wife, and their three children, from 1874 to 1891. The Picturesque Gothic-style residence, with interior design by the firm of Louis Comfort Tiffany and Associated Artists, is one of the premier examples of domestic Victorian architecture in America. Clemens wrote *Adventures of Huckleberry Finn*, *The Adventures of Tom Sawyer*, *A Connecticut Yankee in King Arthur's Court*, *The Prince and the Pauper*, and *Life on the Mississippi* while living in Hartford.

The Mark Twain House is open year-round. In addition to tours of the house, the educational programs of the Mark Twain House include symposia, lectures, and teacher training seminars that focus on the contemporary relevance of Twain's legacy. Past programs have featured discussions of literary censorship with playwright Arthur Miller and writer William Styron; of the power of language with journalist Clarence Page, comedian Dick Gregory, and writer Gloria Naylor; and of the challenges of teaching *Adventures of Huckleberry Finn* amidst charges of racism.

CONTRIBUTORS

Beverly R. David is professor emerita of humanities and theater at Western Michigan University in Kalamazoo. She is currently working on volume 2 of *Mark Twain and His Illustrators*, and on a Mark Twain mystery entitled *Murder at the Matterhorn*. She has written a number of sections on illustration for the *Mark Twain Encyclopedia* and her *Mark Twain and His Illustrators, Volume 1 (1869–1875)* was published in 1989. Dr. David resides in Allegan, Michigan, in the summer and Green Valley, Arizona, in the winter.

Shelley Fisher Fishkin, professor of American Studies and English at the University of Texas at Austin, is the author of the award-winning books *Was Huck Black? Mark Twain and African-American Voices* (1993) and *From Fact to Fiction: Journalism and Imaginative Writing in America* (1985). Her most recent book is *Lighting Out for the Territory: Reflections on Mark Twain and American Culture* (1996). She holds a Ph.D. in American Studies from Yale University, has lectured on Mark Twain in Belgium, England, France, Israel, Italy, Mexico, the Netherlands, and Turkey, as well as throughout the United States, and is president-elect of the Mark Twain Circle of America.

Robert H. Hirst is the General Editor of the Mark Twain Project at The Bancroft Library, University of California in Berkeley. Apart from that, he has no other known eccentricities.

Walter Mosley is the author of the Easy Rawlins' mysteries *Devil in a Blue Dress* (1990), which won the Shamus award given by the Private Eye Writers of America and was named Best First Novel of 1990 by the Mystery Writers of America, *A Red Death* (1991), *White Butterfly* (1992), and *Black Betty* (1994). Other books include *RL's Dream* (1995), *A Little Yellow Dog* (1996), and *The Socrates Fortlow Stories*, forthcoming from Norton. He is president of the Mystery Writers of America. A native of Los Angeles, California, he now lives in New York City.

Lillian S. Robinson, professor of English at East Carolina University in Greenville, North Carolina, has taught graduate and undergraduate courses in detective fiction. She is the author of *Sex, Class, and Culture* (1978), *Monstrous Regiment: The Lady Knight in Sixteenth-Century Epic* (1985), and *In the Canon's Mouth* (forthcoming from Indiana University Press). Her own murder mystery, *Publish or Perish*, is forthcoming from Wildcat Publishing Company. In addition to scholarly books and articles, her (often satirical) commentaries on American cultural issues have appeared in the *Nation* and *Women's Review of Books*.

Ray Sapirstein is a doctoral student in the American Civilization Program at the University of Texas at Austin. He curated the 1993 exhibition *Another Side of Huckleberry Finn: Mark Twain and Images of African Americans* at the Harry Ransom Humanities Research Center at the University of Texas at Austin. He is currently completing a dissertation on the photographic illustrations in several volumes of Paul Laurence Dunbar's poetry.

ACKNOWLEDGMENTS

There are a number of people without whom The Oxford Mark Twain would not have happened. I am indebted to Laura Brown, senior vice president and trade publisher, Oxford University Press, for suggesting that I edit an "Oxford Mark Twain," and for being so enthusiastic when I proposed that it take the present form. Her guidance and vision have informed the entire undertaking.

Crucial as well, from the earliest to the final stages, was the help of John Boyer, executive director of the Mark Twain House, who recognized the importance of the project and gave it his wholehearted support.

My father, Milton Fisher, believed in this project from the start and helped nurture it every step of the way, as did my stepmother, Carol Plaine Fisher. Their encouragement and support made it all possible. The memory of my mother, Renée B. Fisher, sustained me throughout.

I am enormously grateful to all the contributors to The Oxford Mark Twain for the effort they put into their essays, and for having been such fine, collegial collaborators. Each came through, just as I'd hoped, with fresh insights and lively prose. It was a privilege and a pleasure to work with them, and I value the friendships that we forged in the process.

In addition to writing his fine afterword, Louis J. Budd provided invaluable advice and support, even going so far as to read each of the essays for accuracy. All of us involved in this project are greatly in his debt. Both his knowledge of Mark Twain's work and his generosity as a colleague are legendary and unsurpassed.

Elizabeth Maguire's commitment to The Oxford Mark Twain during her time as senior editor at Oxford was exemplary. When the project proved to be more ambitious and complicated than any of us had expected, Liz helped make it not only manageable, but fun. Assistant editor Elda Rotor's wonderful help in coordinating all aspects of The Oxford Mark Twain, along with literature editor T. Susan Chang's enthusiastic involvement with the project in its final stages, helped bring it all to fruition.

I am extremely grateful to Joy Johannessen for her astute and sensitive copyediting, and for having been such a pleasure to work with. And I appreciate the conscientiousness and good humor with which Kathy Kuhtz Campbell heroically supervised all aspects of the set's production. Oxford president Edward Barry, vice president and editorial director Helen McInnis, marketing director Amy Roberts, publicity director Susan Rotermund, art director David Tran, trade editorial, design and production manager Adam Bohannon, trade advertising and promotion manager Woody Gilmartin, director of manufacturing Benjamin Lee, and the entire staff at Oxford were as supportive a team as any editor could desire.

The staff of the Mark Twain House provided superb assistance as well. I would like to thank Marianne Curling, curator, Debra Petke, education director, Beverly Zell, curator of photography, Britt Gustafson, assistant director of education, Beth Ann McPherson, assistant curator, and Pam Collins, administrative assistant, for all their generous help, and for allowing us to reproduce books and photographs from the Mark Twain House collection. One could not ask for more congenial or helpful partners in publishing.

G. Thomas Tanselle, vice president of the John Simon Guggenheim Memorial Foundation, and an expert on the history of the book, offered essential advice about how to create as responsible a facsimile edition as possible. I appreciate his very knowledgeable counsel.

I am deeply indebted to Robert H. Hirst, general editor of the Mark Twain Project at The Bancroft Library in Berkeley, for bringing his outstanding knowledge of Twain editions to bear on the selection of the books photographed for the facsimiles, for giving generous assistance all along the way, and for providing his meticulous notes on the text. The set is the richer for his advice. I would also like to express my gratitude to the Mark Twain Project, not only for making texts and photographs from their collection available to us, but also for nurturing Mark Twain studies with a steady infusion of matchless, important publications.

I would like to thank Jeffrey Kaimowitz, curator of the Watkinson Library at Trinity College, Hartford (where the Mark Twain House collection is kept), along with his colleagues Peter Knapp and Alesandra M. Schmidt, for having been instrumental in Robert Hirst's search for first editions that could be safely reproduced. Victor Fischer, Harriet Elinor Smith, and especially Kenneth M. Sanderson, associate editors with the Mark Twain Project, reviewed the note on the text in each volume with cheerful vigilance. Thanks are also due to Mark Twain Project associate editor Michael Frank and administrative assistant Brenda J. Bailey for their help at various stages.

I am grateful to Helen K. Copley for granting permission to publish photographs in the Mark Twain Collection of the James S. Copley Library in La Jolla, California, and to Carol Beales and Ron Vanderhye of the Copley Library for making my research trip to their institution so productive and enjoyable.

Several contributors — David Bradley, Louis J. Budd, Beverly R. David, Robert Hirst, Fred Kaplan, James S. Leonard, Toni Morrison, Lillian S. Robinson, Jeffrey Rubin-Dorsky, Ray Sapirstein, and David L. Smith — were particularly helpful in the early stages of the project, brainstorming about the cast of writers and scholars who could make it work. Others who participated in that process were John Boyer, James Cox, Robert Crunden, Joel Dinerstein, William Goetzmann, Calvin and Maria Johnson, Jim Magnuson, Arnold Rampersad, Siva Vaidhyanathan, Steve and Louise Weinberg, and Richard Yarborough.

Kevin Bochynski, famous among Twain scholars as an "angel" who is gifted at finding methods of making their research run more smoothly, was helpful in more ways than I can count. He did an outstanding job in his official capacity as production consultant to The Oxford Mark Twain, supervising the photography of the facsimiles. I am also grateful to him for having put me in touch via e-mail with Kent Rasmussen, author of the magisterial *Mark Twain A to Z*, who was tremendously helpful as the project proceeded, sharing insights on obscure illustrators and other points, and generously being "on call" for all sorts of unforeseen contingencies.

I am indebted to Siva Vaidhyanathan of the American Studies Program of the University of Texas at Austin for having been such a superb research assistant. It would be hard to imagine The Oxford Mark Twain without the benefit of his insights and energy. A fine scholar and writer in his own right, he was crucial to making this project happen.

Georgia Barnhill, the Andrew W. Mellon Curator of Graphic Arts at the American Antiquarian Society in Worcester, Massachusetts, Tom Staley, director of the Harry Ransom Humanities Research Center at the University of Texas at Austin, and Joan Grant, director of collection services at the Elmer Holmes Bobst Library of New York University, granted us access to their collections and assisted us in the reproduction of several volumes of The Oxford Mark Twain. I would also like to thank Kenneth Craven, Sally Leach, and Richard Oram of the Harry Ransom Humanities Research Center for their help in making HRC materials available, and Jay and John Crowley, of Jay's Publishers Services in Rockland, Massachusetts, for their efforts to photograph the books carefully and attentively.

I would like to express my gratitude for the grant I was awarded by the University Research Institute of the University of Texas at Austin to defray some of the costs of researching The Oxford Mark Twain. I am also grateful to American Studies director Robert Abzug and the University of Texas for the computer that facilitated my work on this project (and to UT systems analyst Steve Alemán, who tried his best to repair the damage when it crashed). Thanks also to American Studies administrative assistant Janice Bradley and graduate coordinator Melanie Livingston for their always generous and thoughtful help.

The Oxford Mark Twain would not have happened without the unstinting, wholehearted support of my husband, Jim Fishkin, who went way beyond the proverbial call of duty more times than I'm sure he cares to remember as he shared me unselfishly with that other man in my life, Mark Twain. I am also grateful to my family—to my sons Joey and Bobby, who cheered me on all along the way, as did Fannie Fishkin, David Fishkin, Gennie Gordon, Mildred Hope Witkin, and Leonard, Gillis, and Moss

Plaine—and to honorary family member Margaret Osborne, who did the same.

My greatest debt is to the man who set all this in motion. Only a figure as rich and complicated as Mark Twain could have sustained such energy and interest on the part of so many people for so long. Never boring, never dull, Mark Twain repays our attention again and again and again. It is a privilege to be able to honor his memory with The Oxford Mark Twain.

Shelley Fisher Fishkin
Austin, Texas
April 1996